ADVANCES IN LASERS
AND APPLICATIONS

ADVANCES IN LASERS AND APPLICATIONS

Proceedings of the Fifty Second
Scottish Universities Summer School in Physics,
St. Andrews, September 1998.

Edited by

D M Finlayson — University of St. Andrews
B D Sinclair — University of St. Andrews

Series Editor

P Osborne — University of Edinburgh

Copublished by
Scottish Universities Summer School in Physics &
Institute of Physics Publishing, Bristol and Philadelphia

British Library Cataloguing-in-Publication Data:

*A catalogue record for this book is available
from the British Library*

> *ISBN 0-7503-0631-9* (hbk)
> *0-7503-0632-7* (pbk)

Library of Congress Cataloging-in-Publication Data are available.

Copublished by

SUSSP Publications
The Department of Physics, Edinburgh University,
The King's Buildings, Mayfield Road, Edinburgh EH9 3JZ, Scotland

and

Institute of Physics Publishing, wholly owned by
The Institute of Physics, London
Institute of Physics Publishing, Dirac House, Temple Back, Bristol BS1 6BE, UK
US Office: Institute of Physics Publishing, The Public Ledger Building,
Suite 1035, 150 Independence Mall West, Philadelphia, PA 19106, USA

Printed in Great Britain by J W Arrowsmith Ltd, Bristol

SUSSP Proceedings

/continued

SUSSP Proceedings (continued)

Executive Committee

Dr Bruce Sinclair	University of St Andrews	*Director and co-editor*
Dr Kishan Dholakia	University of St Andrews	*Secretary*
Dr Derryck Reid	University of St Andrews	*Treasurer*
Ms Tracy McKechnie	University of St Andrews	*Social Secretary*
Dr David Finlayson	University of St Andrews	*Editor*
Professor Alan Miller	University of St Andrews	
Professor Jim Hough	University of Glasgow	
Dr Robin Vaughan	University of Dundee	

Lecturers

John R M Barr	Pilkington Optronics
Malcolm H Dunn	University of St Andrews
Majid Ebrahimzadeh	University of St Andrews
David C Hanna	University of Southampton
Jean Pierre Hirtz	Thomson-CSF Laser Diodes
Jim Hough	University of Glasgow
Günter Huber	University of Hamburg
Ursula Keller	Swiss Federal Institute of Technology, Zurich
Wayne H Knox	Bell Laboratories, Lucent Technologies
William F Krupke	Lawrence Livermore National Laboratory
Peter Loosen	Fraunhofer Laser Institute, Aachen
Peter F Moulton	Q-Peak, Massachusetts
Lawrence E Myers	Lightwave Electronics, California
Algis Piskarskas	University of Vilnius
Sheila Rowan	Stanford University
Wilson Sibbett	University of St Andrews
Anne Tropper	University of Southampton
Rudolf M Verdaasdonk	University Hospital, Utrecht

Director's Preface

The last decade or so has seen a renaissance in the field of lasers. Diode-laser-pumped solid-state lasers have been taking a larger and larger share of research activities, and then of commercial devices. At the same time there have been major advances in the generation of ultrashort pulses, and in the materials and techniques used in optical parametric oscillators. The advances in each of these three fields have helped the others become stronger, and have allowed the development of a wide range of real-world applications.

This book is written as the proceedings of the 52nd Scottish Universities Summer School in Physics. This was a nine-day international summer school organised to run immediately before two major international conferences in this field in Glasgow. The School took as its topic "Advances in Lasers and Applications", with the emphasis being on the science and applications of solid-state lasers, optical parametric oscillators and amplifiers, and ultrashort pulse lasers. An outstanding team of international experts came to the University of St Andrews to give their presentations and to discuss their ideas with all the participants.

Although this book forms the proceedings of the School, I see it more importantly as an up-to-date review of the science, technology, and applications of modern laser systems. It will be of wide interest to scientists and engineers working with lasers and their applications. Each chapter is written by a distinguished expert as a tutorial review of their field, and covers some of the major topics that were explored at the School. As such, the chapters are suitable as introductions to the various specialist areas at a postgraduate level and beyond. I believe they will be particularly valuable in allowing experienced people in one field to learn more about relevant developments in other topics. The chapters assume a basic knowledge of lasers and nonlinear optics.

The 52nd SUSSP was held at the University of St Andrews from the 5th to the 13th of September 1998. Accommodation was in John Burnet Hall, and the formal programme was at the School of Physics and Astronomy. The 100 participants from 16 countries were able to attend the 35 lectures presented by 18 senior scientists. Most of the "young researchers" also presented their own work orally or in one of the two poster sessions. The School succeeded in bringing people together to share their knowledge and ideas in a spirit of co-operation. The following week many of the participants went on to attend the associated CLEO/Europe – EQEC conferences in Glasgow.

As well as the formal events, a busy social programme helped keep people together at the School. Putting and football competitions gave some exercise, while others explored Scott's ship Discovery. Participants were able to take an organised hike up into the Ochil hills, and many took the opportunity of walks along the St Andrews beaches. The many pubs in St Andrews encouraged toasts to be drunk to new friends. Food, drink, and music from around the world were enjoyed during the international evening, and the banquet and ceilidh on the final evening gave a Scottish flavour and brought the proceedings to a pleasant close.

The director and other organisers are grateful for the help of many organisations and individuals in the running of the School. The School was run with financial and organisational support from SUSSP. The principal financial sponsor was the European Commission through the TMR and INCO programmes, but significant financial assistance was also provided by the UK EPSRC, the Quantum Electronics Group of the UK Institute of Physics, LEOS, OSA, Elliot Scientific, and Scottish Enterprise. The Director and all participants greatly appreciated the sterling work of the local organising team, Dr Kishan Dholakia as the School secretary, Dr Derryck Reid as treasurer, Ms Tracy McKechnie as social organiser, and Professor Alan Miller for his advice and encouragement.

The concept of the School was first mooted by Professor Terry King as chairman of the Quantum Electronics Group of the IOP, and then taken further by the members of the local organising committee of CLEO/Europe - EQEC. Thanks are due to them and to the members of our International and UK Advisory Committees who helped the local team bring forward the successful proposal for the School. Professor Jim Hough and Dr Robin Vaughan provided valuable input to the bursaries committee. Staff and students of the School of Physics and Astronomy cheerfully gave their support to the Summer School, willingly assisting its running in many ways. Jackie Smith and her team at John Burnet Hall provided an excellent residential service.

We thank our team of lecturers for giving a series of great presentations, and for willingly sharing their time with others during their stay in St Andrews. We are also grateful to all the young researchers for providing the enthusiasm, energy, and team spirit that were hallmarks of the School.

Bruce Sinclair

St Andrews, March 1999

Editors' Note

The editors would like to thank all the authors for their excellent and timely contributions to this volume, and for the very considerable time and effort they expended in the production of their manuscripts. Michael Mazilu is acknowledged for his expert help in handling the many different submitted file types and converting them into the LaTex format of the final version.

David Finlayson and Bruce Sinclair

St Andrews, March 1999

Contents

A review of diode-pumped lasers

D C Hanna and W A Clarkson

University of Southampton,England

1 Introduction

The first demonstration of a diode-pumped laser was in 1964 (Keyes and Quist 1964), just two years after the first operation of a diode laser. Potential attractions of diode pumping, compared with pumping by an incandescent lamp, were evident at that time. However, it needed about 20 years before diode lasers became commercially available with long life under room-temperature operating conditions and with the power levels appropriate for laser pumping. Pioneering experiments performed using such diode lasers as pumps graphically illustrated the benefits of diode-pumping and were discussed in an early review paper (Byer 1988). These benefits include high efficiency, compactness, stable low-noise operation, reduced thermal effects in the laser medium, long-life and the prospect, if not an immediate benefit, of low cost. Decreasing costs of diode lasers and increasing diode powers have gone hand-in-hand and so fuelled this interest in diode-pumped lasers. As these trends continue, so there is an inexorable move towards diode-pumped lasers of ever greater power. The earlier prevailing notion that diode-pumped lasers are free from pump-induced thermal effects, has long gone. That was, to some extent, a consequence of the low diode powers initially available. Now, with diode lasers offering tens of Watts of pump power, thermal problems have become a key issue in the further development of diode-pumped lasers.

This question of thermal problems will form an important theme of this brief tutorial review (Section 3). First however (in Section 2), we will briefly consider the question of appropriate geometries (pumping configuration, shape of laser medium, resonator geometry) for diode-pumped solid-state lasers. In fact there is now a bewildering variety of geometries under investigation, with many being commercially available, each with its staunch champions. Clearly the optimum choice of geometry is far from being a settled question. In fact, it is interesting to note that two of the currently favoured geometries, namely the fibre laser and the face-pumped thin disc laser, represent two diametrically opposed extremes from a basic cylindrical rod geometry. With such widely divergent approaches, it is clear that there is still considerable scope for innovative ideas in this area and even relatively small changes in, for example, the available brightness or power of diode lasers, could have major consequences in terms of changing the balance between favoured geometries, or indeed, stimulating the invention of yet more contenders.

2 Choice of geometry

2.1 Longitudinal or transverse pumping

It is useful, in order to have a framework for subsequent discussion, if we first consider a laser medium having the form of a cylindrical rod, with resonator mirrors placed at each end of the rod (Figure 1). Two widely used pumping configurations are longitudinal-pumping (also known as end-pumping), and transverse-pumping (also known as side-pumping.) Heat removal is from the lateral surface of the cylinder, so the heat flow is radial (hence two-dimensional). As we shall see in more detail in Section 3 (see also Koechner 1996) this radial symmetry for the heat flow has some undesirable consequences. It leads to a radial symmetry of stress (arising from the radial temperature gradient), and hence a radial symmetry of the stress-induced birefringence. Thus, in an otherwise isotropic medium, the effect of this birefringence on the polarisation state of the laser beam passing through the rod is to produce a complex depolarisation behaviour, equivalent to the effect of retardation plates having a radially dependent value of retardation and with axes oriented in radial/tangential directions. The resulting depolarisation can cause significant loss and indeed be the barrier to further power scaling. A means of reducing the effect of this depolarisation is indicated in Section 3.

Another solution to the birefringence problems is to use a slab-shaped laser medium, and arrange for a one-dimensional heat removal (Figure 2) from the large faces of the slab. This means that the thermally-induced birefringence properties of the slab correspond to a retardation plate which is everywhere oriented in the same direction even though the magnitude of retardation varies with location along the direction of heat flow. The problems associated with the complex depolarisation behaviour of the rod are eliminated. Notice that for the slab there are two versions of transverse-pumping, either edge-pumping, that is through the small side of the slab, or face-pumping through the large face(s). That

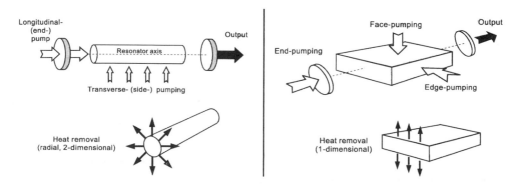

Figure 1. *Cylindrical rod geometry.*   **Figure 2.** *Planar slab geometry*

choice can depend on whether the pump absorption is adequate in the short dimension perpendicular to the face, whether the pump can be focussed into the small dimension of the edge, and whether the arrangements for heat removal, which ideally should be from the large face, are compatible with face-pumping.

2.2 Rod-lasers with longitudinal pumping

Longitudinal pumping (see Figures 1 and 2) avoids the problem of conflicting arrangements for pumping and heat removal. It has a number of other attractive features (and, as we shall see below, it also has its problems). Attractions are: (1) It can provide a long absorption path in the medium, thus helping to achieve efficient pump absorption. A longer length also reduces the heat load per unit length, thus reducing the risk of stress-induced fracture. (2) By reducing the transverse dimension of the pump beam, the gain for a given pump power can be increased and thus significantly higher gains can be achieved. For low gain laser transitions, or 3-level (and quasi-3-level) transitions where a high pump intensity is needed, longitudinal pumping is particularly valuable. (3) With a pump beam of small transverse dimensions, one can conveniently match the laser mode-size to the dimension of the pumped volume. This is helpful for selection of the lowest-order transverse mode and is also conducive to the attainment of a high conversion efficiency from pump light to laser light.

Of course there are limits to how small the transverse dimension of the pump beam can be, imposed by diffraction spread over the length of the medium (or over the pump extinction length, whichever is the shorter). Most diode lasers do not have a diffraction limited output. Their divergence is (approximately) M^2 that from a diffraction limited source, where the pump beam quality factor M^2 can be up to about 1000 (see Sasnett (1989) for a discussion of M^2). The minimum effective pump beam radius (defined as r.m.s. beam radius which results in the minimum pumped volume) is given by (Clarkson and Hanna 1998).

$$w_{p_{\min}}^2 = \frac{2\,\lambda\,M^2}{\alpha_p\,\pi\,n\sqrt{3}} \tag{1}$$

where α_p is the pump absorption coefficient, λ is the pump wavelength, and n is the refractive index of the laser medium. So, the pump beam area is proportional to M^2, and hence the threshold pump power is inversely proportional to M^2. Pump sources with a small value of M^2 are therefore desirable for end-pumping, particularly for pumping low-gain transitions. Later, we make a brief mention of shaping the highly elliptical output beam from a diode-bar, so as to achieve an essentially circular beam with a minimised M^2 value. Even so, the M^2 values achieved in this way are typically about 80. It is important to bear in mind these large M^2 values when extrapolating from the results obtained using a Ti:sapphire laser ($M^2\sim1$ typically) as a simulation of pumping by a diode-laser.

We have indicated that longitudinal pumping has a number of benefits. However, the more intense pumping also brings its own problems. For example, the thermal lensing can be very strong. If the lens was equivalent to a spherical lens then this could, in principle, be corrected by an appropriate choice of spherical mirrors or lenses in the laser resonator. It has to be recognised however, that this lens power will change with pump power. A number of papers by Magni and co-authors (see e.g. Magni 1986a,b; Magni and Zavelani-Rossi 1998) indicate how optimised resonator designs can be made. The thermal lens is incorporated in these designs which provide optimal resonator stability with respect to misalignment sensitivity and to changes of thermal lens power. The assumption made in these papers is of a spherical lensing behaviour in the laser rod. This is valid for a laser medium that is subjected to a transversely uniform pump intensity (typical of lamp pumping, and transverse pumping in general) and which would apply to end-pumping with a uniform ('top-hat') pump beam. In fact, however, pump beams typically have a Gaussian-like pump intensity distribution. This leads to a lensing behaviour with a significant non-spherical component, that is, the lens has spherical aberration. This aberration can lead to a major degradation in the M^2 value of the output beam of the diode -pumped laser.(Clarkson and Hanna 1998). In Section 3 we indicate an approach that mitigates this problem by choosing a suitable resonator design, and that has proved satisfactory (giving $M^2 \sim 1$) for end-pumped lasers of multiwatt power. For much higher powers, transverse pumping with careful attention to pump uniformity has proved very effective at power levels up about 200W (Hirano *et al.* 1998).

So far we have not given any indication of how laser material parameters enter into the design choices. This is a complex question as it involves a multi-parameter design space and we do not attempt to provide any general conclusions. We will briefly indicate some of the factors that need to be considered in practice. Thus, for example, if we go back to Equation (1), we note that a small M^2 value is beneficial in reducing the threshold power. However, even if M^2 is larger than desirable, there may be the possibility of choosing a material with a strong pump absorption (large α_p), thus still allowing a reasonably small pump spot-size to be used. The crystal $NdYVO_4$ provides a good example of the use of this strategy. It has a strong pump absorption and also has a very high gain cross-section, so that an acceptable gain can be achieved even when pumped by diodes whose output has a large M^2 (for example, after having been fibre-coupled). Absorption of pump light in a shorter length does however pose an increased challenge in terms of thermally-induced stress fracture. Other material parameters that determine how close one is to the stress fracture limit are the material tensile strength and the thermal conductivity. Thus one can see already how many material parameters begin to enter the design process. As $NdYVO_4$ is naturally birefringent, it does not suffer problems from thermally-induced stress birefringence since the overall birefringence is dominated by the material's intrinsic birefringence. $NdYVO_4$ is not, however, immune from thermally-induced lensing, which depends on thermal conductivity and the temperature coefficient of refractive index, dn/dT.

Thermal inputs depend on other factors also. For example, in Nd-doped materials there is an upconversion process involving two neighbouring Nd ions in the upper laser level which leads to one of these ions being de-excited and the energy thus lost being deposited as heat (e.g. Guy *et al.* 1998, Pollnau *et al.* 1998). This is more problematic in highly doped material, and is obviously more of a problem for end-pumped lasers because of the high inversion density. This effect is clearly seen as a degradation in performance under Q-switched conditions or in operation as an amplifier, as in both thee cases the inversion density is at a higher level than in cw operation.

So the catalogue of relevant material parameters increases, even though we have so far restricted this discussion to Nd laser transitions (We note here the valuable compilation of laser material parameters in Moulton (1987) and in Zayhowski (1997)). Other transitions, such as those of Yb offer significant differences, some beneficial, some not. Thus the Yb:YAG laser (and other Yb-doped materials) are currently exciting considerable interest. A primary initial motivation was that in Yb:YAG the energy difference between pump laser photons (at ~940nm) and the emitted laser photons (~1030nm) is about one-third of that between pump photons (807nm) and emitted photons for the most common laser transition in Nd:YAG at 1064nm. This represents a reduction of heat load by a factor of three, and so Yb:YAG presents an effective way of reducing the magnitude of the pump-induced thermal effects. Other significant differences are that the absorption and emission cross-sections are smaller than for Nd:YAG, but on the other hand, this can be offset by the fact that much higher Yb concentration can be used as the simple Yb^{3+} energy level structure does not allow the concentration-quenching mechanism that is present in Nd-doped materials. Finally it should be added that the lower laser level in Yb is significantly populated at room temperature (~5%), so that more intense pumping is needed to achieve net gain. Thus there are many factors different from Nd:YAG and there is no single universal statement that can be made along the lines that one system is better than the other. The relative merits are very much dependent on the intended operating conditions. Furthermore, the balance could be shifted significantly by other changes, such as improvements and developments in diode laser performance which may arise in a different manner for 807nm diodes for Nd pumping compared with 940nm diodes for Yb pumping.

2.3 Face-pumped thin disc lasers

The differences between Nd:YAG and Yb:YAG have been instrumental in bringing about a new design approach to high-power diode-pumped lasers. This is the face-pumped thin disc geometry, (Figure 3), which has been applied to such good effect with Yb:YAG (Giesen *et al.* 1994), in which one starts with the benefit of a smaller pump quantum defect. The basis of this approach is to avoid transverse thermal gradients by removing heat from the face of a thin disc rather than, as in the case of a cylindrical rod, from its transverse surface. By keeping the disc thin, the temperature difference between the cooled face and the pump input face is kept small, thus minimising any increase in thermal population of the lower laser level. The thin disc results in a weak pump absorption. To some extent, this can be offset by the higher concentration possible for Yb, but a multipass pump arrangement is needed to obtain reasonable pump efficiency. This completely new geometry has necessitated a series of technological developments to cope with features which are so different from the conventional geometries. The result however, is that a level of performance is now achieved (Erhard *et al.* 1999) which clearly shows the benefits of circumventing the thermal problems of Nd:YAG in a conventional rod geometry.

2.4 Fibre lasers

The thin disc laser can be viewed as a very short laser rod. The other extreme, of an extremely elongated laser rod is represented by the fibre laser. We give only a brief discussion of fibre lasers here, as they are covered in a separate chapter of this volume (Tropper 1999). The main points about fibre lasers that we wish to emphasise in this context are the dramatic move towards higher power operation (many tens of Watts

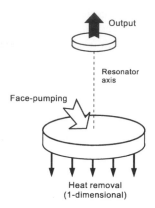

Figure 3. *Face-pumped thin disc geometry, with heat removal from the back face.*

already reported), and the added design flexibility that is conferred by virtue of the fact that very high gains are available, so that master-oscillator power-amplifier (MOPA) designs become particularly attractive.

An essential feature of the fibre laser is that the active (rare-earth) ion is doped into a guiding core (Figure 4). This is usually chosen to have a diameter of a few micrometres, so that the fibre is monomode, that is, at the laser wavelength only the lowest order mode can propagate. Problems of transverse mode control, which are characteristic of bulk lasers, are thus removed. Likewise problems of heat dissipation are removed, since heat can be dissipated over a long length of fibre, and in any case the heat has only a short distance to travel to the outer surface of the fibre. Two of the main obstacles to high-power operation with good beam quality, that are encountered with bulk lasers, are thus removed. On the other hand two new problems enter. One concerns the power limitation imposed by damage or other nonlinear limiting mechanisms, as high-powers propagate through the very small core area. The other problem is a more basic one - if a high output power is to be achieved then a high input power is required, and in this case into a monomode core of small dimensions. If end-launching into the core is used then the pump laser has to have an essentially diffraction-limited beam. However, high-power pump lasers such as diode lasers are typically a long way from being diffraction-limited.

The solution to this problem has been via the use of cladding-pumping (e.g. Po *et al.* 1993), see Figure 4, in which the pump light is launched into the cladding, either at the end, or through the side, in which case multiple pumps can be used along the fibre. Light which is launched into the cladding is guided by the interface between the cladding and its outer cladding and thus gets progressively absorbed into the core as it propagates. The result is a monomode laser output from the core, and since pump powers of many tens of Watts can be launched in this way, then tens of Watts of output is achievable, with a beam quality factor M^2 of essentially unity. Such schemes will undoubtedly have many important consequences for the future development of coherent light sources, and it will take some time for these consequences to be worked through. So far there has been relatively little published on the question of scaling limitations to fibre lasers (see however Zenteno 1993), and it is clear that much work will be needed to fully characterise these high-power fibre lasers.

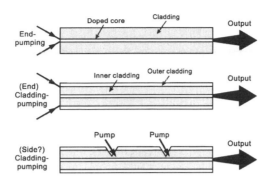

Figure 4. *Fibre laser geometries : direct end-pumping of the core and two examples of cladding-pumping.*

As progressively higher powers are sought from fibre lasers, so there will be a move to using larger core sizes to accommodate the power. This is already occurring (Offerhaus *et al.* 1998) and multimode cores are also being used but in such a way as to still achieve a monomode output (Fermann 1998). Eventually one can expect these trends to lead to fibre lasers encountering the familiar thermal problems and mode-control problems of bulk lasers. What is clear however, is that high-power fibre lasers offer many additional opportunities in the development of light sources. These include the use of stimulated Raman scattering in the fibre, which readily occurs at the power levels available, as a means of efficiently shifting the oscillation wavelength to essentially any desired value in the ∼1–2μm range (Persephonis *et al.* 1996). The availability of high brightness multiwatt sources at any specified wavelength in this range is a major novelty. Such fibre laser sources can of course be used for pumping other lasers, and one can envisage hybrid schemes in which diode-pumped fibre lasers are used to provide high brightness pumping of bulk lasers. This could, for example, allow much greater freedom in choice of pump geometry. In this way the best features of fibre lasers and bulk lasers could be combined and indeed the developments in fibre lasers can be seen as complementing and enhancing the capabilities of bulk lasers.

3 Thermal effects in bulk lasers

Having reviewed some general considerations as to the choice of pumping/resonator geometry, we now consider some specific situations as illustrative examples. These examples are chosen to show the nature and magnitude of thermal effects that can be encountered in practice and also to indicate some approaches for mitigating these effects. Some experimental results which validate these approaches are given. We confine our discussion to examples involving a cylindrical rod geometry with end-pumping. This type of geometry is the most widespread at present and certainly the most thoroughly researched. Space does not permit any detailed discussion of other geometries. It is not the intention to imply that the rod geometry is best. The examples are intended to show that with care and attention to detail, problems caused by thermal effects in laser rods can be mitigated. The same would apply to other geometries. To develop an appreciation of other geometries, the reader is directed to some references dealing with the slab geometry; some general background on slabs (Koechner 1996); the use of a slab geometry for a diode-pumped Cr:LiSAF laser (Kopf *et al.* 1997) and for a diode-pumped Nd glass laser (Aus der Au 1998); the use of a side-pumped Nd:YLF slab as an amplifier (Wall *et al.* 1999).

3.1 Transverse-mode quality of a 946nm Nd:YAG laser

As an example of a laser transition that poses a strong challenge in terms of thermal effects, we consider the 946nm transition in Nd:YAG. Interest in this transition has centred particularly on the possibility of generating blue light (473nm) by second harmonic generation (Fan and Byer 1987, Clarkson and Hanna 1996a). This transition presents two problems. First it has a small emission cross-section, being around an order of magnitude smaller than for the 1064nm transition, second, the lower laser level has a significant population at room temperature (quasi-three-level transition) resulting in an absorption of the laser radiation of about 0.8%/mm for a 1% Nd-doped YAG crystal. Intense pumping is therefore needed to produce an adequate gain for efficient operation. To achieve this intense pumping, using a high-power diode-bar, we have made use of a beam-shaping technique (Clarkson and Hanna 1996b) to transform the elongated output beam from the bar to an essentially circular beam, without significant loss of brightness. In this way, one can readily achieve the desired pumping conditions, such as a 150μm spot-size, collimated over the 5mm length of the laser rod. That such intense pumping could produce strong thermal lensing was not unexpected. Indeed, for a uniform top-hat pump beam, one can show (Koechner 1996) that the thermally-induced lens in the pumped region is of spherical form with a focal length f_t given by

$$f_t = \frac{2\pi K_c w_p^2}{P_p \gamma \eta_{abs}(dn/dT)} \tag{2}$$

where K_c is the thermal conductivity, w_p the beam radius, P_p the incident pump power, η_{abs} the fraction of pump power absorbed, and γ the fraction of this absorbed power converted to heat. One notes the w_p^2 dependence, so that for $w_p = 150\mu$m, and $P_{pabs} = 10$W, a very short focal length is produced, $f_t \approx 130$mm, but correctable in principle by the simple provision of an appropriate spherically curved element (lens or mirror) in the resonator. For a Gaussian beam it can be shown (Clarkson and Hanna 1998) that the focal length is radially dependent, that is,

$$f_t(r) = \frac{2f_t(0)}{w_p^2}\left(\frac{r^2}{1 - \exp(-2r^2/w_p^2)}\right) \tag{3}$$

where $f_t(0)$ is the focal length on axis at $r = 0$, given by

$$f_t(0) = \frac{\pi K_c w_p^2}{P_p \gamma \eta_{abs}(dn/dT)} \tag{4}$$

Thus the lens has spherical aberration, as expected. However what is less obvious is the degree of degradation in beam quality that this can produce. In fact an analysis of the effect that a spherically aberrated lens has on the quality (M^2) of a beam passing through it, has been made by Siegman (1993). Applying that analysis to the case of a Gaussian pump, one has the result that a beam of initial quality M_i^2, after passing axially through the thermal lens produced by the Gaussian pump, has an exit beam quality M_f^2 given by

$$M_f^2 = \sqrt{(M_i^2)^2 + (M_q^2)^2} \tag{5}$$

where

$$M_q^2 = \left(\frac{P_p \gamma \eta_{abs} (dn/dT)}{\lambda \, K_c \sqrt{2}} \right) \cdot \left(\frac{w_L^4}{w_p^4} \right) \tag{6}$$

Thus by making $w_L \ll w_p$ one can ensure that M_f^2 is essentially unchanged from M_i^2 (since $M_q^2 \to 0$ as $w_L/w_p \to 0$). However this is not the recipe for efficient extraction of power, for which one generally arranges that $w_L \approx w_p$. Experimental measurements (Clarkson and Hanna 1998) of this beam degradation confirm the severity of beam distortion that can occur for $w_L \simeq w_p$. Furthermore it is found that when using a conventional resonator design approach, including the presence of a thermal lens $f_t(0)$, and aimed at approximately matching the beam spot-size w_L in the laser rod to the pump spot-size, the resulting output beam is significantly distorted with $M^2 \gg 1$.

To overcome this problem we have adopted the following approach, which, while its justification has been made on an heuristic rather than analytical basis, has nevertheless proved highly effective. The basic design approach we have adopted is to choose w_L somewhat less than w_p, to reduce the effects of aberration, but also to design the resonator so that the spot-size w_L decreases as the focal length of the thermally induced lens, ff, increases, i.e. $dw_L/df < 0$. The reason for the latter condition is to discriminate against higher order transverse modes. Normally this is achieved by making w_L approximately equal in size, or even somewhat larger than the pump spot-size so that higher order modes are suppressed by the action of the lasing mode in depleting the gain in the wings of the pumped region. With w_L less than w_p some other mode-selection mechanism is needed. Since higher order modes have a larger transverse dimension, they sample regions of the aberrated lens having a longer focal length. If $dw_L/df < 0$ the spot-size of these higher order modes will be correspondingly reduced with the result that they are closer in size to the fundamental mode than would be the case for a uniform value of f . Their oscillation is thus suppressed more strongly via the gain saturation produced by the fundamental mode. Typically it is then found that an aperture is not needed to suppress the next higher order modes, thus avoiding further losses due to the truncation produced by the aperture. An aperture is, however, useful in suppressing yet higher-order modes, and generally contributes to the robustness and stability of the laser oscillation.

An example of such a resonator design is shown in Figure 5, consisting of a simple three-mirror resonator with plane end-mirrors and a concave folding mirror. This provides enough flexibility in resonator design, through varying the mirror spacings d_1 and d_2 to select both the laser spot-size w_L and a suitable negative value for dw_L/df.

Figure 6 shows the calculated variation of spot size w_L versus focal length of thermal lens, as the spacing d_2 is changed (d_1 is fixed). Suitable regions of negative dw_L/df are seen. Some sample results from such a resonator include diffraction-limited operation at 2.9W on the 946nm Nd:YAG transition and 3.2W on the 1.3μm Nd:YAG transition. The latter is another low gain transition, with an added problem of increased thermal input as the lower laser level is higher in energy than the 1064nm transition and results in a greater

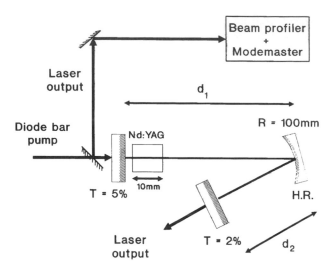

Figure 5. *Three mirror resonator suitable for selecting a desired spot-size in the laser rod and a suitable negative value of dw_L/df.*

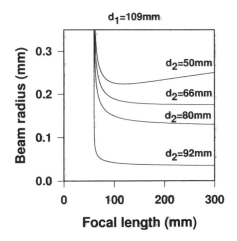

Figure 6. *Variation of spot-size w_L in the laser rod, for the resonator if fig.5, versus focal length, with d_2 as a variable parameter and d_1 fixed. Regions of negative dw_L/df are evident.*

pump quantum defect. By way of comparison, the Nd:YAG laser was also operated on the 1064nm transition, using either the above design approach or the conventional ($w_L \simeq w_p$) approach. It was found that the former gave 5.5W output with $M^2 \sim 1.1$, whereas the latter typically gave $M^2 > 2$ for the same output power, that is, a reduction in brightness by a factor of four.

Thus these results indicate on the one hand, the severity of the lensing problem that can be encountered in intense end-pumping, and in particular the severe consequences of spherical aberration, while on the other hand that techniques can be found to mitigate the aberration problem. Custom-made aspheric correction could also be considered as a means of aberration correction. It is also clear that if the spherical aberration problem is to be avoided at much higher powers in end-pumped rods, with radial heat flow, then attention must be paid to achieving a more uniform transverse pumping profile.

3.2   Thermally induced birefringence in a 946nm Nd:YAG laser

Having made the above remarks with a view to emphasising the need for a careful, critical appraisal of the thermal lensing behaviour if satisfactory laser performance is to be achieved, we now consider another problem that can pose a severe limitation on performance, namely stress-induced birefringence.

Figure 7(a) shows an end-view of the cross-section of a laser rod, assumed to be isotropic in its unpumped condition, in which as a result of pumping (whether transverse or longitudinal, uniform or not) there is a radial gradient of temperature. Consider a representative point P, as indicated in the figure, at radial distance r, with the radius making an angle ϕ with the x-axis. The thermally-induced stress at this point has a symmetry which results in a stress-induced birefringence having principal axes in the radial and tangential directions. The index difference $\Delta n(r) = n_r(r) - n_\phi(r)$ between refractive indices for light polarised in the radial ($n_r(r)$) and tangential ($n_\phi(r)$) directions will, in general, result in a change of polarisation state for light which has travelled along a

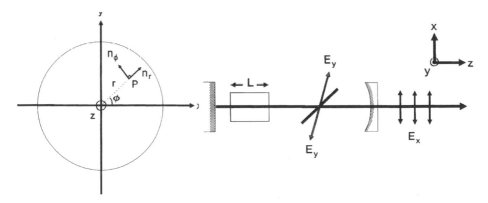

Figure 7. *(a) End view of the laser rod with the incident field E_x at point P, resolved into radial (E_r) and tangential (E_ϕ) components. (b) Schematic resonator with laser rod adjacent to end reflector and a Brewster-angle plate to select the polarisation corresponding to the electric field in the x-direction.*

path parallel to the axis and through this radial point P. In fact, if we assume the incident light is linearly polarised in the x direction, then the radial and tangential components E_r, E_ϕ into which this field can be decomposed, will, after a double pass through the rod (length L), have a difference in phase-shifts of $\Delta\theta(r) = \theta_r(r) - \theta_\phi(r) = 4\pi L\Delta n(r)/\lambda$. The polarisation state exiting the rod will therefore, in general, be changed. If the laser resonator (see Figure 7(b)) contains a polarisation-selective element such as a Brewster angle plate (which may be inserted to define a polarisation for the oscillation) then the exit light, after a double pass through the rod will suffer loss at the polariser. This depolarisation loss L_{depol} will vary with radial position of the point P, and with angle ϕ. For $\phi = 0°$, $90°$, $180°$, $270°$, the loss $L_{\text{depol}} = 0$, while for $\phi = 45°$, $135°$, $225°$, $315°$, the loss reaches its maximum value. If the total integrated loss over the whole beam is calculated, it is found that losses can rise to more than 20% (Koechner 1996). For low-gain lasers, losses of even 1% can have a serious effect on performance, so it is clearly important to find means of compensating this birefringence so as to reduce the depolarisation loss.

A well-known scheme for compensating the birefringence has been described by Scott and de Witt (1971). This involves the use of two identical laser rods, subjected to identical conditions of pumping and heat removal, with an optical rotator located between the rods, which rotates the plane of polarisation by 90°. The principle of operation is that light travelling along a path through P and parallel to the rod axis, will have its tangential field component at the exit of the rod converted by the action of the rotator to a radial field component at the entrance to the second rod, and vice versa. Thus in the passage through both rods, the same overall phase delay is experienced by both radial and tangential field components. This scheme works well, but it has the added complexity of two rods, with the requirement for well-matched conditions in each rod. Alternatives, which allow the use of only one rod include (1) the use of a Faraday rotator giving a 45° rotation per pass, with the rotator placed between the rod and an adjacent resonator end-reflector, and (2) the use of a Porro prism as the resonator end-reflector, with an appropriate wave plate located between rod and Porro prism (Lundstrom 1983). Both of these introduce significant extra loss (at the prism apex in the case of the Porro prism). A much simpler scheme (Clarkson et al. 1998) has recently been shown to be highly effective. This simply involves placing a quarter-wave plate between the rod and the resonator end-reflector. This scheme differs from the other schemes outlined above in that it cannot, even in principle, provide exact compensation. But on the other hand, it can be shown to provide extremely good compensation, with depolarisation losses of less than 0.001% measured in practice. Thus, if a moral is to be drawn from this, it may be that one should not necessarily design for perfection, but for a performance which reduces the imperfection to a level where it is no longer the dominant limitation.

An understanding of how this scheme improves the birefringence compensation, can be seen with the help of Figure 7(a). There one can see that, if the quarter-wave plate has its axes aligned to the x, y axes, then light travelling through the rod parallel to the axis and in the plane $\phi = 45°$, will have the tangential and radial components exchanged after a double-pass through the quarter-wave plate. Thus, as for the scheme of Scott and de Witt, this light will have the net birefringence cancelled. This will be true for light travelling parallel to the axis within the planes at $45°$, $135°$, $225°$, $315°$ (as well as at the usual $0°$, $90°$, $180°$, $270°$), so no depolarisation loss will be suffered by this light propagating in these planes. Thus, since without the quarter wave-plate, maximum depolarisation loss occurred at $45°$, $135°$ etc, additional compensation has now been introduced. One can in fact show that the depolarisation loss for a ray travelling axially through (r, ϕ) is $L_{\text{depol}}(r, \phi) = \sin^2 4\phi \, \sin^4(\Delta\theta/2)$. For intermediate angles, $0 < \phi < 45°$, etc., there will be depolarisation losses. However, it turns out that in practice this depolarisation loss can be extremely small. In fact, we have made a calculation of depolarisation loss for the case of uniform top-hat pumping of radius w_p, (chosen to simplify the calculation) and the result is found that the total depolarisation loss for a Gaussian laser beam, spot size

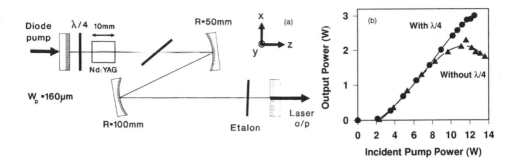

Figure 8. (a) Resonator for operation of a Nd:YAG laser at 1.32μm, showing location of quarter-wave plate. The etalon was added to prevent oscillation on another laser line nearby. (b) Laser performance at 1320nm with and without the quarter-wave plate.

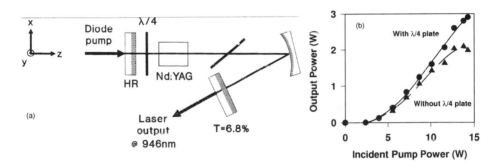

Figure 9. (a) Resonator for operation of a Nd:YAG laser at 946nm. (b) Laser performance at 946nm with and without the quarter-wave plate.

w_L, integrated over r and ϕ is given by

$$L_{\text{depol}} = \frac{3}{4A^4 + 20A^2 + 16} \tag{7}$$

where

$$A = \frac{K_c \lambda w_p^2}{P_{\text{pabs}} \gamma n^3 \alpha C_B w_L^2} \tag{8}$$

α is the expansion coefficient and C_B is the photo-elastic parameter as defined by Koechner (1996). Inserting parameters corresponding to typical operating conditions that we have used for an end-pumped 946nm Nd:YAG lasers ($P_{\text{pabs}} = 9.6$W, $\gamma = 0.14$, $A = 22$) shows that L_{depol} can be extremely small, in this case $L = 3 \times 10^{-4}$%, and in practice negligible compared to other typical losses in the laser resonator. Figure 8(a) and Figure 9(a) show examples of resonator designs used for Nd:YAG lasers operating on the low gain transitions at 1320nm and 946nm respectively. Figures 8(b) and 9(b) indicate some typical results for output power with or without the $\lambda/4$ plate, when a Brewster plate is present in the resonator. Without the quarter-wave plate a sharp roll-off in output power is evident even below the 2W output level. With the quarter-wave plate inserted this roll-off disappears. Upper limits for the depolarisation losses were determined by measuring the output power reflected off the Brewster-angle plate. For the 1320nm laser, losses of ∼0.02% were found, while for the 946nm laser the loss was a remarkably low 0.0006%. In contrast, without the quarter-wave plate the depolarisation losses for the 1320nm and 946nm lasers were 2.7% and 1.7% respectively. These results confirm the effectiveness of this simple scheme for birefringence compensation in Nd:YAG rod lasers (and by implication for other isotropic materials) which should prove helpful for further power scaling of such lasers.

3.3   Thermal lensing and avoidance of fracture in Nd:YLF lasers

As a third and final illustrative example of thermal effects and strategies for dealing with them we will consider some aspects of the operation of Nd:YLF lasers. Nd:YLF has a number of differences from Nd:YAG, some of which are beneficial, some not. It is a birefringent material, so this confers the advantage of avoiding the stress-induced birefringence effects since the natural birefringence dominates any thermally-induced component. Nd:YLF also has the advantage of a weaker thermal lensing than in Nd:YAG. This is particularly so for the 1.053μm transition (σ–polarisation). This has a lower gain than the 1.047μm transition (π–polarisation) but it is still large enough for efficient low-threshold operation. The weak lensing is due to the smallness of the temperature coefficient of refractive index, $dn/dT = -2.0\times10^{-6}\text{K}^{-1}$, and due to the fact that its negative value gives a negative lensing which is partially cancelled by the positive lensing effect from the bulging of the rod ends, caused by thermal expansion. The net effect is that the thermal lensing in Nd:YLF for the 1.053μm transition is typically more than an order of magnitude smaller than for Nd:YAG under comparable pumping conditions. The reduced lensing and the reduction in accompanying aberrations therefore make it easier to obtain good beam quality at multiwatt power levels compared with Nd:YAG. The main drawback of Nd:YLF is that it is less strong than Nd:YAG, having a fracture limit which is about five times lower than for Nd:YAG. In fact, the beneficial thermo-optical properties referred to above, compound the problem of stress-fracture since there is essentially no warning in the form of degraded optical performance as the fracture limit is approached. This has made Nd:YLF an unpopular material in some quarters where unhappy experiences of unexpected fracture have occurred.

The purpose of this last illustrative example is to reinforce the notion that a good quantitative understanding of the various processes and parameters involved in the laser's performance can allow the laser designer to operate, not only safely, but very reliably and effectively with a material that has shortcomings.

The basic strategy we have adopted is to use longer Nd:YLF crystals and to distribute the pump absorption over the full length, either by using a lower doping level, or by detuning the pump from the peak of the absorption. Of course, to distribute the pump over a longer length in a well-confined beam, requires a beam that can be collimated sufficiently well, while operating at high power. This has been achieved in our case, with diode-bar pumps using the two-mirror beam-shaper referred to earlier (Clarkson and Hanna 1996b). One should note at this point that future developments in higher brightness pump sources should allow this approach to be extended considerably, so that materials with rather poor mechanical strength should be amenable to reasonably high-power operation. Applied to Nd:YLF this approach has allowed 11.1W of TEM$_{00}$ output for 29.5W of incident pump power (from two beam-shaped 20W diode bars), (Clarkson *et al.* 1998). Under Q-switched operation 8.4W of average power was achieved at 40kHz pulse repetition rate and ~2.6mJ Q-switched pulses were obtained at 1kHz repetition rate.

Before achieving the above results a careful study had been made of the lensing behaviour. For completeness this was done for π and σ polarisation, in orthogonal planes (since Nd:YLF has anisotropic thermal properties) and under both lasing and non-lasing conditions. A striking observation from these measurements was the very large difference in lens power between lasing and non-lasing conditions, the latter having the much stronger lens. Since the non-lasing conditions, that is, conditions of high inversion density, correspond to precisely the conditions of relevance to Q-switching, or to operation as an amplifier, it is important to understand and quantify the mechanism responsible for the increased lensing. In fact it transpires that the particular mechanism responsible is energy-transfer upconversion between neighbouring excited Nd ions, thus leading to an additional heat load under non-lasing conditions. In this process one excited atom is promoted to a higher level, the other is de-excited. The promoted ion then falls back non-radiatively, depositing heat in the lattice. The net effect is that starting with two excited ions the upconversion process leaves one excited ion and the energy difference is deposited as heat. The increased lens power is also accompanied by increased stress, so that a laser which is operating perfectly satisfactorily and safely under cw conditions could be in danger of fracture if lasing were prevented while pumping continued. In fact this is a well-known experience for high-power laser systems. It is present to some degree for all Nd materials, and in fact the magnitude of the effect has been quantified for Nd:YAG (Guy *et al.* 1998). Once it is quantified, the designer can allow the necessary safety margin. By careful measurement of the lensing behaviour, we have also quantified this effect in Nd:YLF (Pollnau *et al.* 1998) and developed an analytical model that can be used in the design procedure (Hardman *et al.* 1999). The basic design strategy that is indicated by this analytical model is very straightforward. One should use a long length of laser material with the pump distributed over this length, and the pump spot-size should be large. In fact, it is also beneficial to use a low doping level rather than simply detune the pump as the means of ensuring a long absorption length. The reason for this is that with high doping levels, rapid migration of the excited state can occur from one ion to another until the necessary condition for upconversion, namely two neighbouring, excited ions is achieved. So the upconversion rate is not only dependent on excited state population density, but on total population (i.e. dopant) density.

We have used this last example to emphasise the point that a detailed quantitative understanding of a range of processes may be needed to get the best performance from a diode-pumped laser. Without this understanding the laser developer may encounter thermal effects in a rather sudden and dramatic form — laser performance seems to reach a sudden 'brick-wall'. This can be caused by the compounding effect of a number of parameters each with an unfavourable effect and some of which may have an unfavourable temperature coefficient in addition. Then an averaged treatment, assuming, for example, the pump absorption to be uniform longitudinally, can prove surprisingly over-optimistic when it comes to predicted performance. A realistic model may need to pay more attention to detail, using a numerical analysis (e.g. Pollnau et al 1998). The compounding effect referred to above was demonstrated in the numerical model. Thus with non-uniform pump absorption, the temperature at the pump input end will be higher. Since the thermal conductivity is itself temperature-dependent (unfavourably, decreasing with increasing temperature) then the temperature rise can be even greater. This can give increased lensing. If this in turn leads to a higher loss, then a greater inversion is needed. Hence more pump power, more heating, and yet more energy-transfer upconversion and so on. Conversely a small improvement in some parameter can have a seemingly disproportionate benefit. This provides a strong motivation towards an improved quantitative understanding and characterisation.

4   Closing remarks

In this 'tutorial' chapter we have had to be selective in choosing topics from the very extensive range of activities aimed at developing diode-pumped solid-state lasers. Our intention has been to indicate that this is an area where settled questions are few, where widely different approaches can co-exist, and where there is enormous scope for innovation, and radically new approaches. We have emphasised those aspects of diode-pumped lasers that relate to problems associated with power-scaling. There is no sign that power scaling is about to come to an end. The progress in development of high-power diode lasers continues at a great pace and this then drives the power-scaling of diode-pumped lasers.

Thermal problems are perhaps seen as rather mundane and unexciting. However, they will not simply go away, so solutions must be found. What we have aimed to show however, is an 'existence theorem' that is, that solutions can be found. While these will themselves fail at some, yet higher power, at least progress can be made in the meantime. We have also made the point that we need a detailed, quantitative understanding of a wide range of parameters, spectroscopic, optical, thermo-optical, mechanical, etc, if optimum designs are to be arrived at. We also hope that we have been able to convince the reader that, armed with this detailed knowledge, real and valuable progress in enhancing laser performance can be readily achieved.

References

Aus der Au J, Loesel F, Morier-Genoud F, Moser M and Keller U, 1998, *Optics Letts.* **23** 271.

Byer R L, 1988, *Science* **239** 742.

Clarkson W A, Felgate N S and Hanna D C, 1998, *CLEO-Europe* (Glasgow 1998) paper CWD1, Conference Digest, 151.

Clarkson W A and Hanna D C, 1996a, *Opt. Letts.* **21** 737.

Clarkson W A and Hanna D C, 1998, "Resonator Design Considerations for Efficient Operation of Solid-State Lasers End-pumped by High-Power Diode Bars", *Optical Resonators - Science and Engineering*NATO ASI series, eds. R. Kossowsky, M. Jelinek, J. Novak, (Kluwer Academic Publ. Dordrecht/Boston/London).

Clarkson W A and Hanna D C, 1996b, *Opt. Letts.* **21** 375.

Clarkson W A, Hardman P J and Hanna D C, 1998, *Optics Letts.* **23** 1363.

Erhard S, Contag K, Giesen A, Johannsen I, Karszewski M, Rupp T and Stewen C, 1999, *Advanced Solid State Lasers, Technical Digest* paper MC3,(Optical Society of America).

Fan T Y and Byer R L 1987, *IEEE J. of Quantum Electron.* **23** 605.

Fermann M E, 1998, *Opt. Letts.* **23** 52.

Giesen A, Hugel H, Voss A, Wittig K, Brauch U and Opower H, 1994, *Appl. Phys. B* **58** 365.

Guy S, Bonner C L, Shepherd D P, Hanna D C, Tropper A C and Ferrand F, 1998, *IEEE J. of Quant. Electron.* **34** 900.

Hardman P J, Clarkson W A, Friel G J, Pollnau M and Hanna D C, 1999, "Energy-transfer upconversion and thermal lensing in high-power cnd pumped Nd:YLF laser crystals", *IEEE J. of Quantum Electron.*to be published.

Hirano Y, Koyata Y, Yamamoto S and Kasahara K, 1998, *CLEO-US 1998* paper CPD-1

Keyes R J and Quist T M, 1964, *Appl. Phys. Lett.* **4** 50.

Koechner W, 1996, *Solid-State Laser Engineering, 4 Edn.* (Series in Optical Sciences, Springer Verlag).

Kopf D, Weingarten K J, Zhang G, Moser M, Emanuel M A, Beach R J, Skidmore J A and Keller U, 1997, *Appl. Phys. B* **65** 235.

Lundstrom E A, 1983, *US. Patent no. 4* **408** 334.

Magni V, 1986b, *Appl. Opt.* **25** 2039.

Magni V and Zavelani-Rossi M, 1998, *Optical Resonators - Science and Engineering* NATO ASI series, Eds. R. Kossowsky, M. Jelinek, J. Novak, (Kluwer Academic Publ. Dordrecht, Boston and London).

Magni V, 1986a, *Appl. Opt.* **25** 107.

Moulton P F, 1987, "Paramagnetic Ion Lasers", *Handbook of Laser Science and Technology* **1** (Solid-State Lasers, CRC Press Inc.)

Offerhaus H L, Broderick N G R, Richardson D J, Sammut R, Caplen J and Dong L, 1998, *Opt. Letts.* **23** 1683.

Persephonis P, Chernikov S V and Taylor J R, 1996. *Electronics Letts.* **32** 1486.

Po H, Cao J D, Laliberte B M, Minns R A, Robinson R F, Rockney B H, Tricca R R and Zhang Y H, 1993, *Electron. Lett.* **29** 1500.

Pollnau M, Hardman P J, Kern M A, Clarkson W A and Hanna D C, 1998, *Phys. Rev. B* **58** 16076.

Sasnett M W, 1989, *The Physics and Technology of Laser Resonators*, eds. D.R. Hall and P.E Jackson, (Adam Hilger, Bristol and New York).

Scott W C and de Witt M, 1971, *Appl. Phys. Letts.* **18** 3.

Siegman A E, 1993, *Appl. Opt.* **32** 5893.

Tropper A C, 1999, this volume.

Wall K F, Jaspan M, Dergachev A, Szpak A, Flint JH and Moulton P F, 1999, *Technical Digest, Advanced Solid State Lasers*, paper MC4,Boston 1999, (Optical Society of America).

Zayhowski J, 1997, "Miniature Solid-State Lasers", *Handbook of Photonics*, (CRC Press Inc.)

Zenteno L, 1993, *J. Lightwave Technol.* **9** 1435.

Visible cw solid-state lasers

Günter Huber

Universität Hamburg, Germany

1 Introduction

Compact and efficient cw all-solid-state lasers emitting in the visible spectral region are of interest for many applications in fields such as medicine, biology, printing technology, metrology, optical storage, and display. For large screen displays there is a need for high power cw lasers in the red, green, and blue spectral regions. For large screen displays the required laser sources would be cw (or modelocked at greater than 110MHz) systems with about 2W of red light in the 620–630nm range, 2W of green light in the 510–540nm range and 1.5W of blue light in the 430–470nm range. For a number of applications in the reprographics industry new sources with wavelengths and powers matching those of the argon ion laser are sought.

Various techniques to generate visible laser radiation have been investigated. Devices based on Second Harmonic Generation (SHG) of diode-pumped IR Nd lasers have achieved importance within a relatively short time scale, because these lasers were better developed and were, in principle, simple and compact. Systems based on Optical Parametric Oscillators (OPOs) were less attractive as these must be driven preferably in the Q-switched or modelocked regime. Up-conversion lasers are currently in the stage of fundamental research and so far need pump laser sources with excellent beam quality but have significant potential. Blue, green, and red cw semiconductor lasers with high cw output power are not expected in the near future. Frequency doubling of near infrared semiconductor lasers is a very difficult task due to the poor beam quality of such devices.

Intracavity frequency doubling of 1.06μm-Nd-lasers is an efficient method for generating green. Coherent devices in the red and blue spectral regions are much more complicated, because there are no simple infrared lasers for direct frequency doubling techniques. Frequency doubling of the Nd^{3+}-ground-state transition is a possible way to create blue coherent radiation. For the orange and red spectral region one can use sum frequency mixing of diode pumped Nd-lasers, operating at different transitions.

As an alternative, up-conversion lasers based on Er^{3+}, Tm^{3+}, and Pr^{3+} can be operated in the visible spectral region. Diode pumping of up-conversion systems with high powers seems to be difficult due to the very high power densities required for the pump radiation.

However, due to the simple design, the field of up-conversion lasers is very attractive and of high application potential—at least for the low and medium power regime.

This chapter will review work on generating visible coherent radiation through intracavity frequency doubling and mixing and by in-coherent up-conversion.

2   Intracavity frequency doubling

Frequency doubling of Nd^{3+}-lasers can produce blue and green coherent radiation. By implementing the technique of intracavity doubling high conversion efficiencies can be achieved in cw operation.

In many applications compact or even miniature microchip Nd-lasers are required (Huber 1980, Zayhowski and Mooradian 1989, Sinclair 1999). With the use of short crystals, larger cross sections and higher concentrations of the active ions are needed to improve the efficiency of these lasers. When absorption and gain coefficients are increased, it is also easier to accomplish mode matching between the pump and laser beams. It is also desirable to decrease the temperature sensitivity which is caused by the combination of narrow band absorption and temperature drift of the diode-laser pump and to increase the tolerance in the diode wavelength selection (Fan and Byer 1988, Streifer *et al.* 1988). Therefore, it is of advantage that the laser material used possesses high absorption coefficients and large absorption linewidths. For instance, in Nd^{3+}:$GdVO_4$ (GVO) and Nd^{3+}:YVO_4(YVO), both broad homogeneous absorption lines and homogeneous emission lines feature high peak cross sections (Jensen *et al.* 1994), and so may be preferable to Nd:YAG.

The spectral properties of Nd^{3+}:$LaSc_3(BO_3)_4$ (LSB) (Meyn *et al.* 1994) are also interesting for diode laser pumping. The absorption coefficient, α, of $Nd(10\%)$:$LaSc_3(BO_3)_4$ is more than three times higher than the absorption coefficient of standard $Nd(1.1\%)$:YAG because of the higher concentration of Nd^{3+}-ions in LSB.

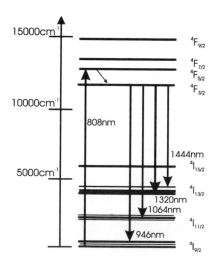

Figure 1. *Energy-level scheme of Nd:YAG.*

As well as the well-known laser transitions near 0.94μm, 1.06μm, and 1.32μm, Nd^{3+} provides additional transitions (Figure 1). There is a very weak transition at 1.44μm which has an emission cross section of 2.5 $\times$ 10^{-20}cm^2 (Kretschmann *et al.* 1997a). This transition is of interest for two reasons. It lies in the eyesafe spectral region and it can also be used for sum frequency mixing with the 1.06μm transition to obtain red light (see Section 3).

The key to efficient intracavity doubling is a low laser cavity loss. Then, even relatively small nonlinearities in the cavity can efficiently extract energy into the frequency doubled laser output.

An analysis of intracavity second-harmonic generation has been presented by Smith (1970). The steady-state round-trip saturated gain of the laser equals the sum of the linear and nonlinear losses

$$\frac{2g_0 l}{1 + I/I_s} = L + \gamma I \tag{1}$$

where g_0 is the unsaturated gain coefficient, l is the length of the laser medium, I is the circulating intensity in the laser rod, and I_s is the saturation intensity of the laser material. All linear losses at the fundamental frequency are included in the linear loss parameter L; the quantity γI is the nonlinear loss which gives the useful output at frequency 2ω. The nonlinear coupling factor γ is defined by

$$I(2\omega) = \gamma I^2(\omega). \tag{2}$$

From the theoretical treatment of intracavity doubling, it follows that a maximum value of second-harmonic power is found when

$$\gamma_{\max} = \frac{L}{I_s}. \tag{3}$$

The magnitude of the nonlinearity required for optimum second-harmonic generation is proportional to the loss, inversely proportional to the saturation intensity and independent of the gain. Thus, for a given linear loss, optimum coupling is achieved for all values of gain and hence all power levels.

It follows that if the linear loss due to the inserted second-harmonic crystal is small compared to the total internal loss, the value of γ is the same for fundamental and for second-harmonic output coupling. Also, the maximum second-harmonic power should equal the fundamental power obtainable from the same laser. However, the second-harmonic power is usually considerably below the output power which can be achieved at the fundamental wavelength. The reason for this is that the linear insertion loss of the nonlinear crystal is usually significant.

For the generation of green light, intra-cavity doubling of the 1062nm transition in Nd:LaSc$_3$(BO$_3$)$_4$ (LSB) is attractive. LSB is an extremely low loss laser material. In these lasers, the main Nd-laser transition $^4F_{3/2} \rightarrow ^4I_{11/2}$ has been frequency-doubled with high quality KTP in cw operation with a total efficiency of 30% from diode pump radiation to green laser output (Ostroumov *et al.* 1997). Due to strong absorption, materials like LSB, YVO$_4$ (YVO), and GdVO$_4$ (GVO) can also be used in microchip and sandwich geometries. In a diode pumped Nd:LSB laser sandwiched with a KTP crystal, 1.2W of green cw multi-transverse mode output was achieved (Ostroumov *et al.* 1997). Alternatively, the technique of self frequency doubling laser crystals has been investigated (Hemmati 1992, Bartschke *et al.* 1997). By the use of $Nd:YAl_3(BO_3)_4$, 225mW of green output power with a conversion efficiency of 14% have been achieved (Bartschke *et al.* 1997). The recently developed $Nd:GdCa_4O(BO_3)_3$ and $Nd:YCa_4O(BO_3)_3$ also appear promising for self frequency doubling (Mougel *et al.* 1997, Chai 1997).

Intracavity doubling of the Nd ground state transition $^4F_{3/2} \rightarrow {}^4I_{9/2}$ to obtain blue output is a more complex task. The inherent reabsorption losses in this laser system lead to high pump power thresholds in contrast to the four-level transitions around 1060nm. Thermal loading of the crystal limits the performance due to the thermal population of the lower laser level and so requires efficient heat removal. Because of their large ground state splitting and their high thermal conductivity YAG and $YAlO_3$ are promising laser hosts for this laser transition and doubling to the blue spectral region (Kellner *et al.* 1997).

$KNbO_3$ is the most commonly used nonlinear material for intracavity frequency doubling of the 946nm laser line of Nd:YAG. The high nonlinearity and the possibility of non-critical phase-matching (NCPM) as well as the small walk-off angle for critical phase-matching make this material very attractive for frequency doubling. However, $KNbO_3$ has also some disadvantages such as a very small temperature and spectral acceptance bandwidth, photorefractivity, and the possibility of domain reversal (particularly for NCPM, where heating of the crystal up to 180°C is necessary). The performance of $LiIO_3$, $\beta-BaB_2O_4$ (BBO), and LiB_3O_5 (LBO) for intracavity frequency doubling is described by Kellner *et al.* 1997.

$LiIO_3$ is the nonlinear crystal with the highest nonlinear coefficient (d_{eff}), however, it has also the smallest spectral acceptance bandwidth. BBO and $LiIO_3$ exhibit a large temperature bandwidth so there is no need for an active temperature control. The main restrictions of these crystals are the small angular acceptance bandwidth and the large walk-off angle which reduce the effective coupling constant by about an order of magnitude. LBO seems to be the most promising candidate with respect to the walk-off angle and the spectral and angular acceptance bandwidths. However, the nonlinear coefficient of LBO is less than that of BBO and $LiIO_3$, which often limits the frequency doubling efficiency.

We used a folded cavity for the intracavity frequency doubling experiments with a plane input coupler, a folding mirror with a radius of curvature of 10cm, and a curved end mirror with R = 5cm (Kellner *et al.* 1997). A scheme of the experimental setup with a composite laser rod YAG/Nd:YAG/YAG is shown in Figure 2. The mirrors were high reflection coated at 946nm (T=1.5 × 10^{-4}) and coated for high transmission at the second harmonic wavelength of 473nm. The nonlinear crystal (NLC) was placed in the focus of the laser mode between the curved mirrors. The radius of the laser mode inside the nonlinear crystal was calculated by ABCD matrix calculations to be 30μm. For this calculation a TEM$_{00}$ mode and a thermal lens of 7cm at a pump power level of 26W were taken into account. The thermal lens was determined separately in a plane parallel cavity by varying the length of the cavity. The plane parallel cavity became unstable at a length of 7cm.

The arrows in Figure 2 indicate the generated blue light output by the two counter-propagating fundamental waves inside the cavity. The output powers given in this paper are summed values of the two blue light beams. Maximum output powers of 550mW with

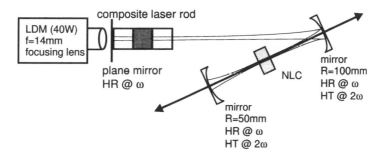

Figure 2. *Experimental setup for blue light generation by intracavity SHG. The arrows indicate the blue light output (Kellner et al. 1997). NLC: nonlinear crystal, LDM: laser diode module, HR: highly reflective, HT: highly transmissive.*

BBO and of 520mW with LiIO₃ were generated. The input pump powers were 25.8W and 24W, respectively. Using LBO the blue output power was slightly lower. An output power of 372mW at an input power of 24W was obtained. These results are summarised in Table 1 along with relevant crystal parameters. There is reasonable agreement between theory and experiment for LBO and BBO. The poorer than expected performance of LiIO₃ was due to the narrow wavelength acceptance bandwidth allowing the laser to operate on a broader bandwidth to reduce the nonlinear output coupling. We observed chaotic fluctuations of the output power in the order of 20 –30% for the three different nonlinear crystals due to longitudinal mode coupling (Baer 1986). Chaotic fluctuations could very probably be avoided by operating the laser in single frequency mode or in a broad band multi-longitudinal TEM₀₀-mode.

	LiIO₃	BBO	LBO
phase matching scheme	type I ooe	type I	type I ooe
$d_{\mathrm{eff}}(10^{-12}\mathrm{m/V})$	2.08	0.92	4.0
walk off angle	3.5°	0.7°	4.6°
n_ω	1.66	1.61	1.86
$\Delta\theta\,L(1\ \mathrm{mrad.cm})$	0.42	2.29	0.28
$\Delta\lambda\,L(1\ \mathrm{nm.cm})$	0.43	0.63	0.16
$\Delta T\,L(1\ \mathrm{K.cm})$	23	8	20
Intra-cavity power(W)	55	65	63
Expected blue output power(mW)	880	620	430
Measured blue output power(mW)	520	550	372

Table 1. *Relevant parameters and results for intra-cavity frequency-doubling the 946nm Nd:YAG laser. The nonlinear crystals were all 4mm long. Data from Kellner et al. (1997).*

3   Sum frequency mixing

Unfortunately, radiation in the spectral region between 550 and 650nm cannot be generated with second-harmonic generation because of the absence of efficient fundamental lasers. However, radiation near 620 and 590nm is for example required for display technology and medical applications, respectively. Therefore, other ways of generating coherent light at these wavelengths have to be found. One possibility is sum frequency generation, in which coherent radiation of frequencies ω_1 and ω_2 is mixed, generating radiation of frequency $\omega_3 = \omega_1 + \omega_2$.

Sum frequency mixing was first demonstrated in the 1960s (Boyd and Kleinman 1968). In the past few years, singly resonant intracavity sum frequency mixing of a Nd^{3+} laser with its pump laser diode (Baumert *et al.* 1987), with a resonantly enhanced diode (Wigley *et al.* 1995), and doubly resonant within an external resonator with its second harmonic (Kaneda and Kubota 1995) were demonstrated with output powers of as much as 186mW in the blue and the UV.

In a manner similar to that used for intracavity second harmonic generation, we can describe the efficiency of the sum frequency mixing process by the following equation:

$$I_{1+2} = \gamma' I_1 I_2 \tag{4}$$

where I_1 and I_2 are the intensities of waves at ω_1 and ω_2, respectively, and I_{1+2} is the intensity of the wave at the sum frequency. As was the case in second harmonic generation, higher efficiency is obtained with higher intensities of the fundamental beams.

3.1   Doubly resonant scheme

One way to obtain high intensities at both ω_1 and ω_2 is to produce a pair of lasers which overlap (Figure 3). Within the overlapping region of two cw lasers, intracavity mixing can be achieved using the nonlinear optical material. This doubly resonant concept combines the advantage of implementing two independent gain media with separate pump sources and partially independent resonators for 1 and 2 with the advantage of higher intracavity powers that are needed for efficient frequency conversion.

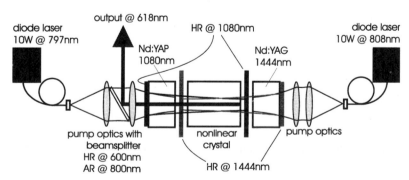

Figure 3. *Linear resonator configuration of the double resonant sum frequency mixer for 618nm (Kretschmann et al. 1997b). HR/AR highly reflective/antireflective coatings.*

Nd^{3+}:YAP (P_{nma}, k||c, E||b) at 1080nm and Nd^{3+}:YAG at 1444nm were chosen as laser materials for mixing to the red at 618nm wavelength (Kretschmann *et al.* 1997b). The YAG crystal was 1% Nd^{3+} doped ($N = 1.4 \times 10^{20}$cm^{-3}) and the YAP was 0.8% Nd^{3+} doped ($N = 1.6 \times 10^{20}$cm^{-3}). Both crystals were 7mm long, had a diameter of 3mm, and were mounted in water-cooled copper blocks. The crystals were coated to be highly reflecting for their laser wavelength on one surface and antireflecting on the other surface. In addition, the highly reflecting coating of the YAG crystal was also of low reflectance (less than 1%) near 1064nm to avoid laser oscillation at this wavelength. Both crystals were pumped with 10W fibre-coupled laser diode arrays at 797nm (YAlO$_3$) and 808nm (YAG). The pump light was focused inside each crystal with two achromatic collimating lenses. In plane-parallel resonators of 30mm length powers of 3.9W (YAlO$_3$; 2% output coupler, polarised) and 1.6W (YAG, 1% output coupler, unpolarised) were achieved at the fundamental wavelengths.

LiB$_3$O$_5$ (LBO) and KTiPO$_5$ (KTP) were investigated as nonlinear optical materials. LBO can be noncritically phase matched (NCPM) for type 1 sum frequency mixing near room temperature ($\approx 23°C$). Therefore, no walk-off is present. KTP has to be critically phase matched for type II sum frequency generation. A walk-off angle of 1.66° is present.

The mixing resonator consists of two plane-parallel lasers that are stabilised by their thermal lenses (Figure 3). Both resonators are built as short as possible to guarantee the highest intracavity powers. Both laser resonators have the same beam axis and overlap in such a way that the end mirror of one resonator is inside the other resonator. These end mirrors are coated to be highly reflecting at the laser wavelength and highly transmitting at the wavelength of the other laser. In the overlapping region of the mixer the resonator mode has a beam radius of about 210μm for both lasers and is almost collimated. In this region the nonlinear crystal is positioned. By variation of the pump power of one of the fundamental lasers, the linear dependence of the mixed radiation predicted from Equation 4 was found (Kretschmann *et al.* 1997b).

In comparison to a folded setup, the linear setup has the advantage of simple design and easy alignment. This setup is compact and could be improved to a sandwiched sum frequency mixer. A folded setup generated higher output powers (212mW) but is not as compact as the linear setup (Kretschmann *et al.* 1997b). Also, its alignment is more critical and its resonator is more sensitive to pump power or mechanical fluctuations.

It was observed that the 618nm radiation was modulated by the relaxation oscillations of the 1444nm laser at higher output powers. These oscillations are a result of the strong nonlinear coupling and the low gain of the Nd:YAG laser on this weak transition.

3.2 Singly resonant scheme

A singly resonant (at 1080nm) sum frequency mixer (Kretschmann *et al.* 1999) is shown in Figure 4. The 1080nm laser was pumped by a fibre coupled diode module (SDL, 792nm, $\varnothing$400μm, NA=0.4) producing 9W of output. With a 3% output coupler, 2.2W at 1080nm was coupled out of the cavity which fell to 1.35W when a nonlinear periodically poled LiNbO$_3$ (PPLN) crystal was placed inside the resonator. When the output coupler was replaced with a highly reflective mirror (R$\sim$99.98%), 1080nm output powers of 16mW (with intracavity beam aperture (M^2=1.1)) and 33mW (without aperture (high transversal modes)) were measured. They indicate intracavity powers of roughly 80W TEM$_{00}$ mode and 165W multi-transverse mode operation. These powers are small compared to what is expected. We assumed this might be a result of intensity dependent losses or distortion mechanisms in PPLN.

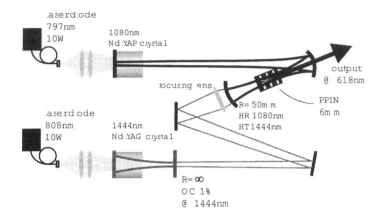

Figure 4. *Singly resonant sum frequency mixer with resonance at 1080nm (Kretschmann et al. 1999). The 1444nm Nd:YAG laser output is coupled into a 1080nm Nd:YAP resonator, where sum frequency mixing occurs in a PPLN crystal.*

The three mirror resonator of the 1080nm Nd:YAP laser generated a 60μm beam waist at the position of the nonlinear crystal . Its spectral bandwidth was measured as $\Delta\lambda_{FWHM} = 0.35$nm. A birefringent filter was then included in the laser to avoid switching of the laser wavelength induced by the additional nonlinear losses, since Nd:YAP lasers can operate at several wavelengths around 1080nm outside of the PPLN phasematching bandwidth. Up to 140mW of unwanted green radiation at 540nm were also generated inside the PPLN crystal which was not phase-matched for green generation.

The 1444nm Nd:YAG laser consisted of a 65mm plane parallel resonator and was also pumped by a fibre coupled diode module (SDL, 808nm, ∅400μm, NA=0.4). The 7mm laser crystal required a special laser coating for operation at 1444nm (compare Section 3.1 and Kretschmann *et al.* 1997a). The end coating was highly reflective for 1444nm and additionally highly transmissive at 1360nm-1320nm and 1064nm (T_{1360nm} >50%; T_{1064nm} >95%). The other coating was highly transmissive at 1444nm and 1064nm. The laser was polarised by introducing a 150μm thick glass plate at Brewster' s angle inside the resonator. From this setup up to 670mW at 1444nm with an M^2=1.1 and a spectral bandwidth of $\Delta\lambda_{FWHM}$=0.35nm were generated. The radiation from this laser was focussed through the end mirror of the 1080nm Nd:YAP laser into the PPLN crystal.

The PPLN crystal had a length of 6mm and an aperture of 0.5mm*times*3mm. Its grating period was 10.5μm. The theoretical spectral and temperature acceptance bandwidths of this crystal are 0.67nm and 5.0K, respectively. Therefore no major limitations for efficient sum frequency mixing should apply. The crystal was mounted in an oven and was held at a stabilised temperature of 136°C to achieve phase-matching and to avoid photorefractive effects. The end surfaces of the PPLN crystal were anti-reflection coated for 1080nm and 1444nm.

With a beam waist of 30μm, coherent radiation of 186mW at 618nm with M^2=1.1 and a spectral bandwidth of 0.08nm was measured. The conversion efficiency from 1444nm radiation to 618nm radiation was 28%. The noise was measured over several tens of seconds. The rms noise of the generated sum frequency output power was only 1%.

4 Up-conversion lasers

Lasers which emit at frequencies higher than that of the pump light are usually called up-conversion lasers. In these lasers, the active ion is excited by internal up-conversion of near-infrared or red light via multi-step photon excitation or co-operative energy transfer and emits anti-Stokes visible light. The advent and rapid improvement of high-power laser diodes in the red and near-infrared spectral ranges have caused new interest in the development of up-conversion lasers. The output wavelengths of laser diodes can be (temperature-) tuned to match the absorption lines of the active laser ions resulting in a substantial fraction of ions being excited into higher energy levels, thus enhancing the up-conversion process.

Visible up-conversion lasing at room temperature has been demonstrated in Tm-doped crystals (Nguyen *et al.* 1990, Thrash and Johnson 1992) and in various rare-earth-doped fluorozirconate fibres (Whitley *et al.* 1991). Er^{3+} has been identified as a very interesting ion for cw up-conversion to the green spectral region (Heine *et al.* 1994).

4.1 A perfect up-conversion laser

Figure 5 shows the energy level diagram of a simple up-conversion scheme which behaves practically perfectly. The ground state absorption (GSA) is resonant with the excited state absorption (ESA) . Fast decays to the metastable states n_2 and n_1 guarantee that no detrimental down stimulation (reverse pump processes) and corresponding reduction of the pump rates W_p occur. We also assume a fast decay of the lower laser level. With these assumptions we can write the rate equations:

$$\frac{dn_2}{dt} = \frac{W_p}{2} n_1 - \frac{n_2}{\tau_2} \tag{5}$$

$$\frac{dn_1}{dt} = \frac{W_p}{2} n_0 - \frac{W_p}{2} n_1 - \frac{n_1}{\tau_1} + \frac{\beta_{21}}{\tau_2} n_2 \tag{6}$$

$$n_0 + n_1 + n_2 = 1 \tag{7}$$

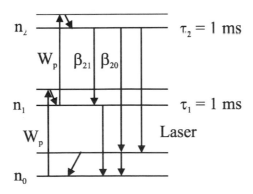

Figure 5. *Energy level diagram of a simple up-conversion laser with two resonant pump steps of GSA and ESA.*

In the steady state ($dn_i/dt = 0$) we can calculate the populations n_1, n_2, and n_3 using the parameters

pump wavelength	$\lambda_p = 970$nm,
cross section for GSA	$\sigma_{GSA} = \sigma_p = 10^{-20}$cm^2,
cross section for ESA	$\sigma_{ESA} = \sigma_p = 10^{-20}$cm^2,
branching ratios	$\beta_{21} = \beta_{20} = 0.5$,
lifetimes	$\tau_2 = \tau_1 = 1$ms,
pump beam waist	$w_0 = 10\ \mu$m.

The dependence of the populations on the pump rate and corresponding pump power are shown in Figure 6. One can see that the population of the upper laser state n_2 is already 10% at a pump power of about 50mW. This pump power and the corresponding pump intensity (or pump rate) can easily be achieved with a single mode diode laser. Thus diode laser pumping of up-conversion lasers is possible in principle.

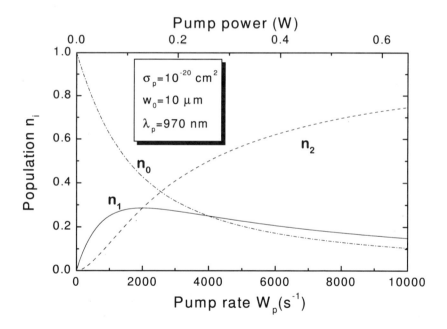

Figure 6. *Level populations n_i versus pump power of the up-conversion laser of Figure 5. For the parameter set, see text.*

4.2   Er^{3+} up-conversion scheme

Er^{3+} is an attractive laser ion with several optical transitions in the visible spectral range. Due to its complex energy level scheme with some metastable excited states, Er^{3+} gives rise to multiple co-operative energy transfer and excited state absorption processes which can be utilised for up-conversion pumping. At low Er-concentrations energy transfer processes play a negligible role.

Figure 7 shows the important processes at low Er^{3+}-concentrations. The processes of GSA and ESA are resonant. There are three metastable states from which transitions start with specific branching ratios β_{ij}.

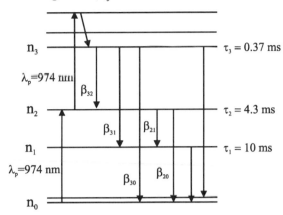

Figure 7. *Level scheme and most important processes in an Er^{3+} up-conversion laser. ESA and GSA are resonant (974nm), β_{ij} denotes the branching ratios of the involved transitions, τ_i denotes the lifetime of the corresponding metastable level.*

The rate equation system in this case is:

$$\frac{dn_3}{dt} = \frac{\sigma_{\text{ESA}}}{\sigma_{\text{GSA}} + \sigma_{\text{ESA}}} W_p n_2 - \frac{n_3}{\tau_3} \tag{8}$$

$$\frac{dn_2}{dt} = \frac{\sigma_{\text{GSA}}}{\sigma_{\text{GSA}} + \sigma_{\text{ESA}}} W_p n_0 - \frac{\sigma_{\text{ESA}}}{\sigma_{\text{GSA}} + \sigma_{\text{ESA}}} W_p n_2 - \frac{n_2}{\tau_2} + \beta_{32} \frac{n_3}{\tau_3} \tag{9}$$

$$\frac{dn_1}{dt} = \beta_{31} \frac{n_3}{\tau_3} + \beta_{21} \frac{n_2}{\tau_2} - \frac{n_1}{\tau_1} \tag{10}$$

$$n_0 + n_1 + n_2 + n_3 = 1 \tag{11}$$

As an example the spectroscopic data of Er^{3+}:YLF (see for instance Möbert *et al.* 1997) are used for the calculation:

Lifetimes	$\tau_1 = 10$ms, $\tau_2 = 4.3$ms, $\tau_3 = 370\mu$s,
branching ratios	$\beta_{32} = 0.02$, $\beta_{31} = 0.27$, $\beta_{30} = 0.67$, $\beta_{21} = 0.15$, $\beta_{20} = 0.85$,
pump wavelength	$\lambda_p = 974$nm,
GSA cross section	$\sigma_{\text{GSA}} = 0.25 \times 10^{-20}$cm^2,
ESA cross section	$\sigma_{\text{ESA}} = 1.7 \times 10^{-20}$cm^2, and
pump beam waist	$w_0 = 20\mu$m.

The resulting relative steady state populations n_0 through n_3 are shown in Figure 8. In the pump power region up to 1.4W the population of the upper laser level n_3 is less than 10%. The achievable small signal gain per round trip

$$g_0 = 2(\sigma_e n_3 - \sigma_{\text{reabs}} n_0) N_t \ell \tag{12}$$

Günter Huber

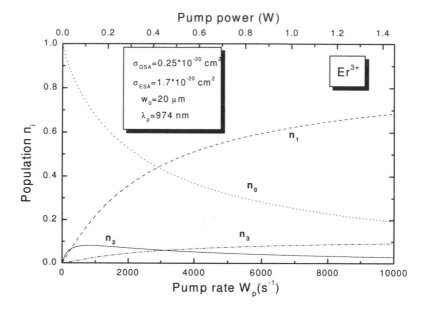

Figure 8. *Simulation of the level populations n_i of an Er^{3+} up-conversion laser . For the parameter set, see text.*

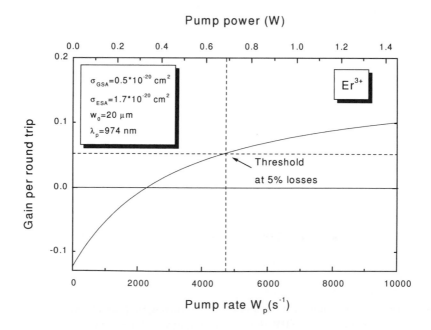

Figure 9. *Gain versus pump power in an up-conversion Er^{3+}:YLF laser.*

is given in Figure 9. The experimental parameters used for the computation of Figure 9 are:

Total Er^{3+} concentration	$N_t = 1.37 \times 10^{20} cm^{-3}$,
laser crystal length	$\ell = 2.4 mm$,
emission cross section	$\upsilon_e - 2 \times 10^{-20} cm^2$, and
effective re-absorption cross section	$\sigma_{reabs} = 2 \times 10^{-21} cm^2$.

At a round trip loss of 5%, the threshold pump power is about 700mW, which is close to the experimentally observed value of about 600mW (Möbert *et al.* 1997). In this experiment a Ti:sapphire laser was used to pump an Er:YLF laser. For 1600mW of incident pump power at 974nm, 20mW of 551nm light was generated. These results prove that the model of pure ESA-pumping at low Er concentrations is realistic.

Besides bulk crystals, waveguide structures, and fibres in particular are interesting for up-conversion lasers, because these systems create pump light confinement and allow pumping of much longer gain lengths. One example will be discussed in the next section.

4.3   ESA-pumped Yb, Pr-up-conversion laser

Recently, a cw Pr,Yb-doped up-conversion fibre laser with an output power of 1.02W at 635nm was demonstrated (Scheife *et al.* 1997, Sandrock *et al.* 1997). Pr^{3+} is an interesting laser ion because its energy levels (Figure 10) allow laser emission in the blue, green, and red spectral regions on many transitions that originate from the $^3P_0$ state (Solomon and Mueller 1963, German *et al.* 1973, Esterowitz *et al.* 1977, Kaminskii 1983, Kaminskii *et al.* 1988, Allain *et al.* 1991, Sutherland *et al.* 1996, Smart *et al.* 1991). Efficient pump sources in the blue spectral range that excite the upper laser level directly are not available. Therefore, up-conversion pumping in the near-infrared region has been investigated. As those mechanisms require high pump power densities, the use of waveguide structures such as optical fibres, which permit good confinement of both pump light and laser emission, is advantageous.

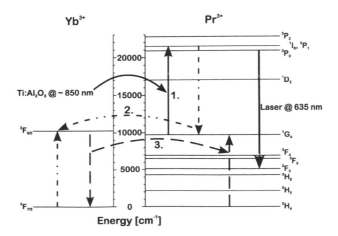

Figure 10. *Energy-level diagram of Pr,Yb:ZBLAN with the proposed photon avalanche up-conversion mechanism (Sandrock et al. 1997).*

is given in Figure 9. The experimental parameters used for the computation of Figure 9 are:

Total Er^{3+} concentration	$N_t = 1.37 \times 10^{20} cm^{-3}$,
laser crystal length	$\ell = 2.4$mm,
emission cross section	$\sigma_e = 2 \times 10^{-20} cm^2$, and
effective re-absorption cross section	$\sigma_{reabs} = 2 \times 10^{-21} cm^2$.

At a round trip loss of 5%, the threshold pump power is about 700mW, which is close to the experimentally observed value of about 600mW (Möbert *et al.* 1997). In this experiment a Ti:sapphire laser was used to pump an Er:YLF laser. For 1600mW of incident pump power at 974nm, 20mW of 551nm light was generated. These results prove that the model of pure ESA-pumping at low Er concentrations is realistic.

Besides bulk crystals, waveguide structures, and fibres in particular are interesting for up-conversion lasers, because these systems create pump light confinement and allow pumping of much longer gain lengths. One example will be discussed in the next section.

4.3   ESA-pumped Yb, Pr-up-conversion laser

Recently, a cw Pr,Yb-doped up-conversion fibre laser with an output power of 1.02W at 635nm was demonstrated (Scheife *et al.* 1997, Sandrock *et al.* 1997). Pr^{3+} is an interesting laser ion because its energy levels (Figure 10) allow laser emission in the blue, green, and red spectral regions on many transitions that originate from the $^3P_0$ state (Solomon and Mueller 1963, German *et al.* 1973, Esterowitz *et al.* 1977, Kaminskii 1983, Kaminskii *et al.* 1988, Allain *et al.* 1991, Sutherland *et al.* 1996, Smart *et al.* 1991). Efficient pump sources in the blue spectral range that excite the upper laser level directly are not available. Therefore, up-conversion pumping in the near-infrared region has been investigated. As those mechanisms require high pump power densities, the use of waveguide structures such as optical fibres, which permit good confinement of both pump light and laser emission, is advantageous.

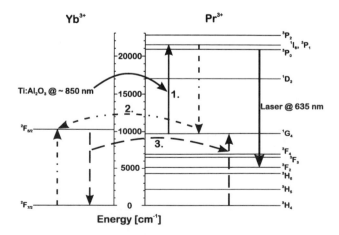

Figure 10. *Energy-level diagram of Pr,Yb:ZBLAN with the proposed photon avalanche up-conversion mechanism (Sandrock et al. 1997).*

In the experiments described by Scheife *et al.* (1997) and Sandrock *et al.* (1997), the same Pr,Yb-doped double-clad fibre manufactured by Le Verre Fluoré was used. Double-clad fibres consist of a rare-earth-doped active core with a small numerical aperture, an undoped, highly multimode inner cladding waveguide with a large numerical aperture, and a second, undoped outer cladding. Pump radiation is focused into the inner cladding instead of being launched directly into the fibre core. Therefore, the use of highly divergent high-power diode-laser arrays as pump sources is possible. The diameters of the cylindrical core, the inner cladding, and the outer cladding were 5, 125, and 200μm, respectively. The core (numerical aperture 0.2), which contained 3000ppm Pr^{3+} and 20000ppm Yb^{3+}, and the inner cladding were made of fluorozirconate glasses , whereas the outer cladding was a low-refractive-index resin. The core was decentered by 35μm and the fibre length was 51cm.

Pumping directly into the inner cladding did not yield laser action. Even the observed up-conversion fluorescence was considerably weaker than the fluorescence that occurred under direct excitation of the active core. Therefore, pump light was focused onto the fibre front face by an aspheric lens ($f = 10$mm) and launched directly into the fibre core. A plane dielectric mirror, which was highly reflective at λ_L=635nm and butted directly against this fibre end, formed the cavity, together with the cleaved fibre end face with its Fresnel reflection ($R \approx 4\%$). With this setup, laser action on the transition $^3P_0 \rightarrow^3 F_2$ could easily be demonstrated.

Fibre laser emission was achieved for pumping in the wavelength intervals 760 775nm or 795 885nm. These wavelengths correspond to the excited-state absorption transitions $^1G_4 \rightarrow {}^3P_2$ and $^1G_4 \rightarrow {}^1I_6$, respectively (Figure 10). As the latter transition is spin-allowed, its excitation yields higher output powers than pumping at shorter wavelengths.

To demonstrate the potential of the fibre laser, two Ti:sapphire lasers were used as pump sources. In order to combine the two pump beams by use of a dichroic mirror (Figure 11), the Ti:sapphire lasers were tuned to λ_{p1}=852nm and λ_{p2}=826nm, respectively. With this modified setup, an output power as high as P_{out}=1020mW was achieved (Sandrock *et al.* 1997) at an incident pump power P_{in}=5.51W, yielding a slope efficiency of η =19% with respect to the incident pump power (Figure 12).

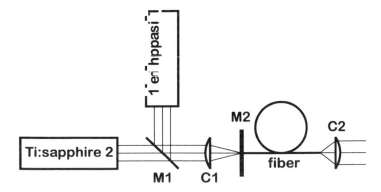

Figure 11. *Experimental setup for pumping with two Ti:sapphire lasers. M1: dichroic mirror, C1: aspheric lens, M2: plane-parallel mirror (highly reflective at 635nm, highly transmissive at 850nm), C2: collimator (Sandrock et al. 1997).*

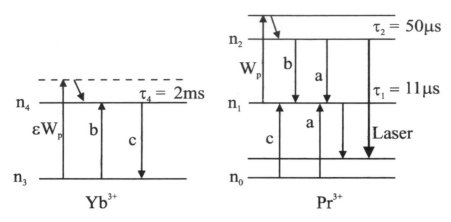

Figure 13. *Main pump steps and interactions of the sensitised avalanche up-conversion mechanism in the Yb,Pr-system as used in the rate equation model.*

rate equation system is:

$$\frac{dn_2}{dt} = W_p n_1 - \frac{n_2}{\tau_2} - a n_0 n_2 - b N_t n_3 n_2 \tag{13}$$

$$\frac{dn_1}{dt} = -W_p n_1 + 2a N_t n_0 n_2 + b N_t n_2 n_3 + c N_t n_4 n_0 - \frac{n_1}{\tau_1} \tag{14}$$

$$n_0 + n_1 + n_2 = \frac{N_{Pr}}{N_t} \tag{15}$$

$$\frac{dn_4}{dt} = \varepsilon W_p n_3 + b N_t n_2 n_3 - c N_t n_0 n_4 - \frac{n_4}{\tau_4} \tag{16}$$

$$n_3 + n_4 = \frac{N_{Yb}}{N_t} \tag{17}$$

$$N_t = N_{Yb} + N_{Pr} \text{ and } n_i = \frac{N_i}{N_t}. \tag{18}$$

As an example, we take the measured parameter set for the system Pr,Yb:YLF:

Total Yb concentration	$N_{Yb} = 7.6 \times 10^{20} \text{cm}^{-3}$,
total Pr concentration	$N_{Pr} = 0.239 \times 10^{20} \text{cm}^{-3}$,
cross relaxation	$a = N_{Pr}^{-1} \tau_{Pr,Pr}^{-1}$ with $\tau_{Pr,Pr} \sim 0$,
cross relaxation	$b = N_{Yb}^{-1} \tau_{Pr,Yb}^{-1}$ with $\tau_{Pr,Yb}^{-1} = 38000 \text{s}^{-1}$,
cross relaxation	$c = N_{Pr}^{-1} \tau_{Yb,Pr}^{-1}$ with $\tau_{Yb,Pr}^{-1} = 2850 \text{s}^{-1}$,
ESA cross section	$\sigma_{ESA} \sim (3 \times 10^{-19} \text{cm}^2$, estimated),
pump beam waist	$w_0 = 10 \mu\text{m}$, and
pump wavelength	$\lambda_p = 850 \text{nm}$.

In Figure 14 the absolute steady state population densities N_0 through N_4 are plotted versus the pump rate and pump power. In the lower pump region one can see the quadratic increase of the population density N_2 due to ESA pumping. At a pump power level of about 1W the avalanche mechanism starts and yields a strong increase of the population

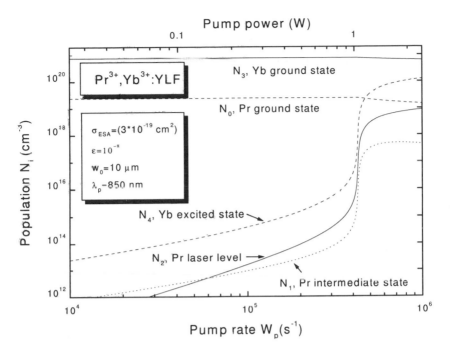

Figure 14. *Steady state population densities N_i versus pump power. The avalanche mechanism can be seen at a pump power of about 1W.*

density over several orders of magnitudes. It should be noted that this point of avalanche does not correspond to the laser threshold. The laser threshold is at somewhat higher pump powers.

In comparison to the experimental observations, the step-like avalanche process in Figure 14 seems to be too sharp. This effect could be caused by the spatially inhomogeneous pumping in longitudinal and radial directions and the corresponding inhomogeneous distribution of the avalanche working point in Figure 14.

The most characteristic features of this avalanche pump process are the existence of a pump threshold and a long fluorescence rise time which shortens at higher pump powers as observed experimentally. Up-conversion lasers such as the fibre laser described in this section are very simple devices. The 1020mW obtained at 635nm by Ti:sapphire laser pumping at 5.5W is promising. Although diode pumping of double-clad fibres has not been successful so far, high-power diode-pumping of this system seems feasible (Zellmer *et al.* 1998).

5 Conclusion

Nonlinear techniques such as intracavity second harmonic generation and sum frequency mixing as well as up-conversion pumping are efficient frequency conversion techniques. The combination with diode pumped cw solid-state lasers opens interesting perspectives for various applications in display, printing, storage, and medical techniques.

By the use of periodically poled nonlinear materials such as $LiNbO_3$, $LiTaO_3$, and KTP, and by further improvement of crystal quality and stability, the overall efficiencies of frequency doubling and sum frequency mixing schemes might be further increased.

Even though the upconversion in fibres appears comparatively simple, efficient high power diode pumping of up-conversion laser systems still has to be demonstrated. Further understanding of the fundamental processes in these complex systems will probably help to improve up-conversion lasers, which are very stable, low noise, simple, and reliable devices.

References

Allain J Y, Monerie M and Poignant H, 1991. Tunable cw lasing around 610,635, 695, 715, 885 and 910nm in praseodymium-doped fluorozirconate fibre, *Electron. Lett.* **27** 189.

Baer T, 1986. Large-amplitude fluctuations due to longitudinal mode coupling in diode-pumped intracavity-doubled Nd:YAG lasers, *J Opt Soc Am. B*, **3** 1175.

Bartschke J, Knappe R, Boller K-J and Wallenstein R, 1997. Investigation of efficient self-frequency-doubling Nd:YAB lasers, *IEEE J. Quantum Electron.* **QE-33** 2295.

Baumert J-C, Schellenberg F M, Lenth W, Risk W P and Bjorklund G C, 1987. Generation of blue cw coherent radiation by sum frequency mixing in $KTiOPO_4$, *Appl. Phys. Lett.* **51** 2192.

Boyd G-D and Kleinman D A, 1968. Parametric interaction of focused gaussian light beams, *J. Appl. Phys.* **39** 3597.

Chai B H T, 1997. Solid state laser and nonlinear optical crystals for opto-electronic applications, in *Proceedings of International Symposium on Laser and Nonlinear Optical Materials*, Singapore, 161 34.

Esterowitz L, Allen R, Kruer M, Bartoli F, Goldberg L S, Jenssen H P, Linz A and Nicolai V O, 1977. Blue light emission by a $Pr:LiYF_4$-laser operated at room temperature, *J. Appl. Phys.* **48** 650.

Fan T Y and Byer R L, 1988. Diode-laser pumped solid-state laser, *IEEE J. Quantum Electron*, **QE-24** 895.

German K R, Kiel A and Guggenheim H, 1973. Stimulated emission from $PrCl_3$, *Appl. Phys. Lett.* **22** 87.

Heine F, Heumann E, Danger T, Schweizer T, Huber G and Chai B H T, 1994. Green upconversion continuous wave $Er^{3+}:LiYF_4$ laser at room temperature, *Appl. Phys. Lett.*, **65** 383.

Hemmati H, 1992. Diode-pumped self-frequency-doubled neodymium yttrium aluminum borate (NYAB) laser, *IEEE J. Quantum Electron.* **QE28** 1169.

Huber G, 1980. Miniature Neodymium lasers, in: *Current Topics in Materials Science*, **4** 1. ed: Kaldis E, (North Holland, Amsterdam)

Huber G, Kellner T, Kretschmann H M, Sandrock T and Scheife H, 1999. Compact diode pumped cw solid-state lasers in the visible spectral region, *Optical Materials*, **11** 205

Jensen T, Ostroumov V G, Meyn J, Huber G, Zagumennyim A I and Shcherbakov I A, 1994. Spectroscopic characterisation and laser performance of diode-laser-pumped $Nd:GdVO_4$, *Appl. Phys. B* **58** 373.

Kaminskii A A, 1983. Visible lasing on five intermultiplet transitions of the ion Pr^{3+} in $LiYF_4$, *Sov. Phys. Dokl.* **28** 668.

Kaminskii A A, Kurbanov K, Ovanesyan K L and Petrosyan A G, 1988. Stimulated emission spectroscopy of Pr^{3+} ions in orthorhombic $YalO_3$ single crystals, *Phys. Status Solidi A* **105** K155.

Kaneda Y and Kubota S, 1995. Continuous-wave 355-nm laser source based on doubly resoant sum-frequency mixing in an external resonator, *Opt. Lett.* **20** 2204.

Kellner T, Heine F and Huber G, 1997. Efficient laser performance of Nd:YAG at 946nm and intracavity frequency doubling with $LiJO_3$, $-BaB_2O_4$, and LiB_3O_5, *Appl. Phys. B*, **65** 789.

Kretschmann H M, Heine F, Ostroumov V G and Huber G, 1997a. High-power diode-pumped continuous-wave Nd^{3+} lasers at wavelengths near 1.44 μm, *Opt. Lett.* **22** 466.

Kretschmann H M, Heine F, Huber G and Halldórsson T, 1997b. All-solid-state continuous-wave doubly resonant all-intracavity sum-frequency mixer, *Opt. Lett.*, **22** 1461.

Kretschmann H M, Huber G, Batchko R G, Meyn J P, Fejer M M and Byer R L, 1999. Singly resonant, continuous wave, sum frequency mixing of two Nd^{3+} lasers in periodically poled lithium Niobate, *Laser Physics* **9** 1.

Meyn J-P, Jensen T and Huber G, 1994. Spectroscopic properties and efficient diode-pumped laser operation of neodymium-doped lanthanum scandium borate, *IEEE J. Quantum Electron.* **QE-30** 913.

Möbert P E-A, Heumann E, Huber G and Chai B H T, 1997. Green Er^{3+}:$YLiF_4$ upconversion laser at 551nm with Yb^{3+} codoping: a novel pumping scheme, *Optics Letters*, **22** 1412.

Mougel F, Aka G, Kahn-Harari A, Hubert H, Benitez J M and Vivien D, 1997. Infrared laser performance and self-frequency doubling of Nd^{3+}:$Ca_4GdO(BO_3)_3$ (Nd:GdCOB), *Opt. Mat.* **8** 161.

Nguyen D C, Faulkner G E, Weber M E and Dulick M, 1990. Blue upconversion thulium laser, *SPIE Solid State Lasers*, **1223** 54.

Ostroumov V G, Heine F, Kück S, Huber G, Mikhailov V A and Shcherbakov I A, 1997. Intra cavity frequency doubled diode pumped $Nd:LaSc_3(BO_3)_4$ lasers, *Appl. Phys. B*, **64** 301.

Sandrock T, Heumann E, Huber G and Chai B H T, 1996. Continuous-wave $Pr,Yb:LiYF_4$ upconversion laser in the red spectral range at room temperature. in *Advanced Solid-State Lasers*, **1** 550 S.A. Payne and C.R. Pollock, eds., OSA Trends in Optics and Photonics (Optical Society of America, Washington, D.C.)

Sandrock T, Scheife H, Heumann E and Huber G, 1997. High-power continuous-wave upconversion fibre laser at room temperature, *Opt. Lett.*, **22** 808.

Scheife H, Sandrock T, Heumann E and Huber G, 1997. Pr, Yb-doped upconversion fibre laser exceeding 1 W of continuous-wave output in the red spectral range, in *OSA Trends in Optics and Photonics on Solid-State Lasers*, **X** 79, C.R. Pollock and W.R. Bosenberg, eds. (Optical Society of America, Washington, DC.)

Sinclair B D, 1999, Frequency doubled microchip lasers, *Optical Materials*,**11** 217.

Smart R G, Hanna D C, Tropper A C, Davey S T, Carter S F and Szebesta D, 1991. CW room temperature upconversion lasing at blue, green and red wavelengths in infrared-pumped Pr^{3+}-doped fluoride fibre, *Electron. Lett.* **27** 1307.

Smith R G, 1970. Theory of intracavity optical second-harmonic generation, *J Quantum Electron.* **QE-6** 215

Solomon R and Mueller L, 1963. Stimulated emission at 5985Åfrom Pr^{3+} in LaF_3, *Appl. Phys. Lett.* **3** 135.

Streifer W, Scifres D R, Harnagel G L, Welch D F, Berger J and Sakramoto M, 1988. Advances in diode laser pumps, *IEEE J. Quantum Electron.* **QE-24** 883.

Sutherland J M, French P M W, Taylor J R, and Chai B H T, 1996. New visible c.w. laser transitions in Pr^{3+}:YLF and femtosecond pulse generation, in *Advanced Solid-State Lasers*, **1** 277, S.A. Payne and C.R. Pollock, eds., OSA Trends in Optics and Photonics (Optical Society of America, Washington, D.C.)

Günter Huber

Thrash R J and Johnson L F, 1992. Tm^{3+} room temperature upconversion laser. in *Compact Blue-Green Lasers,* Santa Fe, New Mexico, (Technical Digest Series) 6, paper ThB3.

Tidwell S C, Seamans J F, Bowers M S, and Cousins A K, 1992. Scaling cw diode-end-pumped Nd:YAG lasers to high average power, *IEEE J.Quantum Electron.* **QE-28** 997.

Tsunekane M, Taguchi N and Inaba H, 1996. High power operation of diode-end pumped Nd:YVO$_4$ laser using composite rod with undoped end, *Electron. Lett.* **32** 40.

Weber R, Neuenschwander B, Weber H P and Albers P, 1996. High-power end-pumped composite Nd:YAG rod, in *Conference on Lasers and Electro-Optics/Europe '96* (Institute of Electrical and Electronics Engineers, New York, advance program paper CMA4.)

Whitley T J, Millar C A, Wyatt R, Brierley M C and Szebesta D, 1991. Upconversion pumped green lasing in erbium doped fluorozirconate fibre, *Electron. Lett.* **27** 1785.

Wigley P G, Zhiang Q, Miesak E and Dixon G J, 1995. High-power 467-nm passively locked signal-resonant sum-frequency laser, *Opt. Lett.* **20** 2496.

Zayhowski J J and Mooradian A, 1989. Microchip lasers, *OSA Proc. on Tunable Solid State Lasers* **5** 288 North Falmouth, 1989).

Zellmer H, Plamann K, Huber G, Scheife H and Tünnermann A, 1998. Visible double-clad upconversion fibre laser, *Electr. Lett.* **34** 565.

Fibre and waveguide lasers

Anne Tropper

University of Southampton, England

1   Introduction

The guided-wave laser is almost as old as the laser itself. The first demonstration of laser action in glass made use of a multimode waveguide; a core rod encased in a cladding of lower refractive index so that confinement of the light would counteract the effect of the poor optical quality of the available glass (Snitzer, 1961). However, although waveguiding has long been an essential feature of semiconductor diode lasers, in dielectric laser media by far the greatest research effort has gone into the development of bulk rods and slabs of high optical quality. Interest in the potential advantages of guided wave dielectric gain media was only quickened with the advent of high-quality single mode optical waveguides, especially rare earth doped silica fibres, in which the propagation losses are so low that the benefits of optical confinement are fully realised.

The most immediate advantage of the guided-wave geometry is that of reducing the cavity mode volume, and hence the pump power needed to reach threshold. Provided the guiding structure is designed so as to support only a single propagating mode at the gain wavelength, then the laser output will be spatially coherent, in a mode that is not affected, for example, by cavity misalignment. The guided-wave laser can be designed as a compact, stable, monolithic device, exploiting all the techniques of integrated optics, such as gratings, couplers and modulators. Since the active region of a guided-wave laser is typically only a few micrometres in diameter, fabrication can involve a range of deposition techniques very different from those used to grow bulk media. The resulting gain medium may have a composition or dopant concentration not available to a bulk phase. On the other hand the advantages of a guided-wave structure are cancelled if propagation losses compete too effectively with the achievable gain. A further difficulty attending these optically pumped devices is the need to couple pump light into the waveguide core. The pump sources themselves must therefore emit spatially coherent beams, and expensive micro-positioning techniques are required.

The literature on fibre and planar dielectric waveguide lasers is now so extensive that a review of this type cannot attempt to be comprehensive. My aim is rather to sketch the principles of guided-wave laser design and operation, and introduce a few selected devices of particular current interest. I shall pay particular attention to the role that guided-wave systems may play in the effort to develop compact and efficient sources emitting high power diffraction-limited beams. Diode-bar lasers emitting many tens of watts are now readily available, but it remains a challenge to convert the highly asymmetric and multimode output from such a device into a usable beam in a simple and efficient way. It is not at first sight obvious that a guided wave laser should be particularly suitable for high power operation. Waveguides are characteristically devices in which high core intensity accompanies *low* overall power, and scaling up the core area leads to multimode propagation and loss of spatial coherence. Recently, however, the technique of cladding pumping of fibre lasers has been found to be strikingly effective in overcoming these limitations.

It can be argued that *planar* waveguide structures are inherently highly compatible with high-power diode-bar pump lasers. Experimental investigation of such devices indicates that extremely compact sources, potentially able to handle 10W or more of output, can be fabricated in this way. Control of the spatial mode is a central and difficult problem, and various technical approaches will be reviewed. Equally stringent is the necessity to couple pump radiation into the guide, whether longitudinally for greater efficiency, or transversely for a less divergent output and the possibility of power scaling. We shall see that with careful positioning this can be achieved without the use of any optical components whatsoever. If, moreover, the cladding-pumping principle is employed, then the position tolerances are significantly relaxed. Alternatively it may be possible to pump through the face of the device, and we shall review some practical schemes of this type.

2   Principles of guided-wave laser devices

There exist excellent texts which give an exhaustive account of waveguide modes (e.g. Snyder and Love, 1983; Lee 1986; Nishihara *et al.* 1985). It is nevertheless worth summarising the simple underlying principles that can often be applied to yield a reasonable estimate of the performance of a device without recourse to elaborate analysis. In addition, the developer of waveguide sources may be concerned with mode properties such as beam quality, which are not discussed in the classic texts.

The simplest conceivable waveguide is the step-index symmetric planar guide, for which closed form expressions can be derived for all physical properties of interest. The discussion of guided-mode properties will therefore be illustrated with reference to this system. The generalisation of these results to asymmetric and channel guides is straightforward. Although more complex index profiles require numerical analysis, for purposes such as the estimation of lasing threshold it is often sufficient to approximate the interfaces by index steps.

2.1   Guided-wave propagation and cut-off

The polarisation eigenmodes of a waveguide structure are in general quite unlike those of free space which are purely transverse with respect to both the E- and H-fields. In the step-index planar slab, for example, it is possible for only one of these fields to have no component parallel to the propagation direction of the mode. The non-transverse component describes an ellipse in a plane which contains the propagation direction; the quarter-cycle phase shift between longitudinal and transverse components ensures that only the longitudinal component of the Poynting vector survives time-averaging, and there is no loss of power from the confined mode. Both the aspect ratio and the sense of rotation of this ellipse vary rapidly as a function of position across the mode profile.

From the viewpoint of the waveguide laser designer, who may wish, for example, to minimise the volume of a mode, it is particularly useful to bear in mind the spatial scaling properties of laser modes. The quantity λ/ϕ has the character of a scaling length, where λ is the free space wavelength of the propagating light, and ϕ is the numerical aperture (NA) of the waveguide. A waveguide with transverse dimension less than about half a scaling length will in general be monomode; all propagating modes other than the fundamental will be cut off. The NA of a waveguide can be expressed as

$$\phi = \sqrt{n_2^2 - n_1^2}$$

in terms of n_2, the maximum core index and n_1, the cladding index.

To show how the NA, a quantity which emerges from geometrical optics, comes to play such a central role in the electromagnetic theory of guided-wave propagation, we solve the Helmholtz equation for the structure sketched in Figure 1.

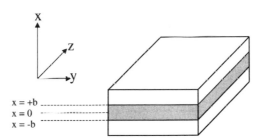

Figure 1. *Planar step-index symmetric waveguide.*

Light propagates along the z-axis guided between the core/cladding interfaces in the planes $x = \pm b$. The x-dependence of the purely transverse field amplitude of the guided solutions is shown in Table 1. It is assumed that the fields are independent of y, and have everywhere the plane-wave z-dependence $\exp(ik_z z)$.

core: n_2, $|x| < b$	cladding: n_1, $|x| > b$
even $\cos kx$	even $\cos kb \exp[-a(|x| - b)]$
odd $\sin kx$	odd $\sin kb \exp[-a(|x| - b)]$

Table 1. *Modes of a symmetric planar waveguide of depth 2b*

These are solutions of the Helmholtz equation only if k and k_z, respectively the transverse and longitudinal propagation constants of the guided mode, and a, the attenuation constant of the evanescent part of the mode are related by

$$k^2 + k_z^2 = n_2^2 k_0^2 \qquad k_z^2 - a^2 = n_1^2 k_0^2$$

where k_0 is the free space wavevector of the light. It follows that k and a satisfy

$$k^2 + a^2 = k_c^2$$

where the constant k_c, given by ϕk_0, represents an upper limit on the value of the transverse wavevector. A mode for which $k = k_c$ has a vanishingly small value of a; its evanescent wings extend indefinitely far into the cladding and it is no longer confined. A key characteristic of any propagating mode is its nearness to this cut-off point, represented by the modal parameter δ. It is convenient to define δ such that

$$k = k_c \cos(\delta\pi/2) a = k_c \sin(\delta\pi/2)$$

The entire range of modal behaviour is therefore spanned as δ takes values between 0 and 1. As δ approaches 0 the mode goes into cut-off. As δ approaches 1 the mode acquires more of a bulk character; it does not extend outside the core and its polarisation is TEM.

A guidance condition restricting the value of δ for a given guide thickness, $2b$, is imposed by the electromagnetic boundary conditions, according to which the transverse components of E and H must be continuous across the boundaries of the core. For a TE mode this is equivalent to requiring that E and $(1/\mu)dE/dx$ are continuous. At optical frequencies the magnetic permeability μ differs negligibly from its free space value, and the guidance condition may be solved to give a closed form expression for the guide thickness:

$$2b(\delta, m) = \left(\frac{\lambda}{\phi}\right) \frac{\delta + m}{2\cos(\delta\pi/2)} \tag{1}$$

where the mode index m is a positive integer or zero. The guide thickness is multiple-valued because the guide will in general support both fundamental ($m = 0$) and higher order modes; these have even or odd parity determined by whether m is even or odd. The m-th mode is just at cut-off for a guide of thickness $(\lambda/\phi)\times(m/2)$.

The TM modes do not scale rigorously with numerical aperture in this way because the continuous quantities in this polarisation are H and $(1/\epsilon)dH/dx$, and in a transparent medium the dielectric response function ϵ is the square of the refractive index. The TM guidance condition is found to take the form:

$$2b(\delta, m) = \left(\frac{\lambda}{\phi}\right) \frac{\Delta + m}{2\cos(\delta\pi/2)} \qquad \text{with} \quad \Delta = \frac{2}{\pi}\arctan\left(\frac{n_2^2(\delta\pi/2)^2}{n_1^2}\right). \tag{2}$$

Figure 2 shows the dependence of the guide thickness, in units of the scaling length, with the modal parameter δ for several TE and TM modes, calculated from Equations 1 and 2. This is much quicker than solving for δ numerically, and the curve can be used to look up values of m and δ for any guide of known thickness. The TE curves are universal in that they apply to any combination of core and cladding indices. The TM curves are specific to the values $n_1 = 1.5$ and $\phi = 0.4$; however it can be seen that the TM modal parameters are quite close to their TE values even for this extreme index difference. The TM mode is always very slightly closer to cut-off than the TE mode. For some practical purposes the difference is not significant, however it can be exploited, for example in the use of metal overlayers to make waveguide polarisers. In Section 3.2 of this chapter

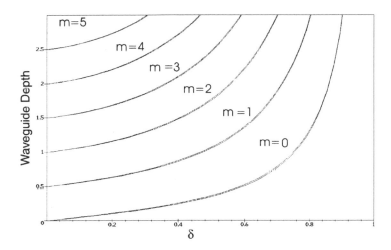

Figure 2. *Depth of a symmetric waveguide, units of (λ/NA), as a function of guidance parameter δ for TE and TM modes with mode integer m = 0,1,2,3,4,5.*

an application of this effect to spatial mode control in planar waveguide lasers will be discussed.

The eigenmodes of a circular fibre waveguide with a step-index profile can also be written in a closed form (Snyder and Love 1982), albeit a rather complex and unattractive one. The polarisation properties are particularly complicated, and both E and H will have longitudinal components unless the mode exhibits azimuthal symmetry. For very many practical devices, however, it is valid to analyse the modes in the weak-guidance approximation:

$$\Delta n = n_2 - n_1 \ll n_2$$

The radial part of the intensity distribution of each mode then resembles the form given for the step index guide in Table 1, with the sine, cosine and exponential functions replaced respectively by linear combinations of oscillatory Bessel functions of the first and second kind, and by modified Bessel functions of the first kind. In this approximation the modes can be designated LP_{qm} and their polarisation is close to TEM. The first mode integer q refers to the dependence of the intensity on the azimuthal angle χ, which may be either $\cos^2(q\chi)$ or $\sin^2(q\chi)$. The LP_{0m} modes are plane polarised. An active fibre device is ideally designed to support only the fundamental LP_{01} mode at the operating wavelength. For a step-index fibre this requires $V < 2.401$ where the dimensionless parameter V is simply the fibre radius, b, measured in units of (λ/ϕ) and multiplied by 2π. Provided the fibre NA is low, and $V > 1$, the intensity profile of the LP_{01} mode approximates closely to a Gaussian function:

$$I \approx I_0 \exp\left[-\frac{r^2}{r_0^2}\right] \quad \text{where} \quad r_0 = \frac{b}{\sqrt{2\ln V}}$$

whence one can estimate that a fraction $(1 - 1/V^2)$ of the modal power will be confined within the fibre core.

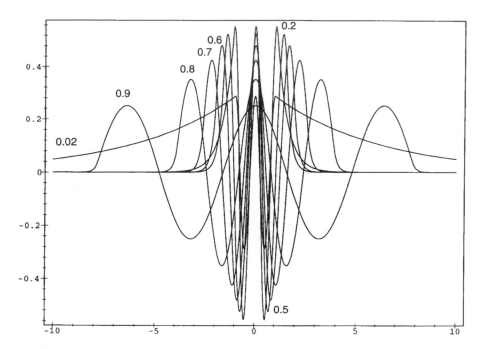

Figure 3. *E-field amplitude of a symmetric waveguide TE mode with m = 4 as a function of distance across the waveguide, x, measured in units of (λ/ϕ). The curves represent a series of waveguides of decreasing depth, supporting modes for which δ = 0.9 0.8, 0.7, 0.6, 0.5, 0.2 and 0.02.*

2.2   Beam quality of guided modes

The form of the mode profile changes strikingly as the mode approaches cut-off. This is illustrated in Figure 3, which shows the x-variation of the E-field of a TE $m = 4$ mode for different values of the guide thickness.

The amplitudes are normalised so that each curve corresponds to the same total power. As the guide gets thinner the peak amplitude initially grows with increasing confinement of the light and then drops away as the evanescent wings extend out and the mode goes into cut-off. In this regime the mode profile is strikingly unlike that of any free space propagation eigenmode, raising a general question about the beam quality of guided wave lasers.

The "quality" of a paraxial beam is usually characterised by the M^2 parameter , which quantifies by how many times the beam divergence exceeds the fundamental limit set by diffraction. For non-Gaussian beams we need a more general measure of beam radius than the $1/e^2$ intensity point. Such a measure is supplied by the square root of the variance in x, defined by

$$\langle x^2 \rangle = \frac{1}{N} \int\limits_{-\infty}^{+\infty} x^2 \left| E(x) \right|^2 dx \quad \text{where} \quad N = \int\limits_{-\infty}^{+\infty} \left| E(x) \right|^2 dx$$

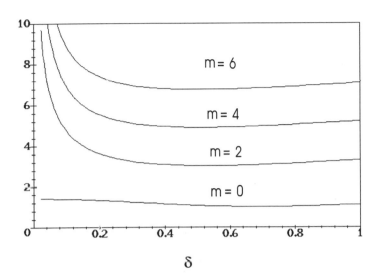

Figure 4. $M_x{}^2$ *as a function of δ for even modes of a symmetric waveguide with $m = 0$, 2, 4, 6.*

The corresponding measure of the beam divergence is the variance in the transverse wavevector, $\langle k^2 \rangle$. Let $F(k)$ be the amplitude spectrum of spatial frequencies in the beam, that is, the one-dimensional Fourier Transform of $E(x)$. Then we can define

$$\langle k^2 \rangle = \frac{1}{N} \int\limits_{-\infty}^{+\infty} k^2 \, |F(k)|^2 \, dk = \frac{1}{N} \int\limits_{-\infty}^{+\infty} \left| \frac{\partial E}{\partial x} \right|^2 dx$$

The variance in k can be related to an effective NA for the beam:

$$(NA)^2 = \frac{\langle k^2 \rangle}{k_0^2}$$

The quality of a beam is conveniently characterized by the M^2-parameter, which quantifies how many times the beam divergence exceeds the fundamental limit set by diffraction. For any paraxial beam

$$\langle k^2 \rangle \langle x_0^2 \rangle = \frac{(M^2)^2}{4}$$

where $\langle x_0^2 \rangle$ corresponds to the narrowest point of the beam at the waist. The left-hand product takes its minimum value when $M^2 = 1$ and the beam is diffraction-limited.

It is therefore straightforward to calculate the M^2-value of a beam emitted from the end of a waveguide when the distribution of E-field over the end face corresponds to a known eigenmode of the structure. In Figure 4 the variation of $M_x{}^2$ with the mode parameter δ is plotted for the first few even modes of a symmetric step-index planar guide. It can be seen that the fundamental mode is nearly diffraction-limited, with $M^2 < 1.1$, over a wide range of values of δ. A striking feature of the behaviour is that the M^2-value of each higher-order mode diverges as the mode approaches cut-off. The beam quality of a guided-wave laser, which emits mostly fundamental-mode light, may therefore be significantly degraded by the admixture of a small amount of a higher order mode close to cut-off.

2.3   Gain in active waveguides

Many dielectric laser systems are pumped longitudinally, often by diode lasers. It is interesting to consider whether more efficient conversion of the available pump power can be achieved by fabricating a waveguide in the gain medium. The aim is to reduce the threshold of the device roughly by the factor by which optical confinement reduces the volume of the lasing mode in the gain medium.

The length, l, of the gain medium in a longitudinally-pumped laser is determined by the strength of the pump absorption. Assuming a Gaussian cavity mode confocally focussed in the gain medium, the mode volume in the medium will be $\sim \lambda l^2$. If the cavity mode is confined with respect to one axis in a planar guide of numerical aperture ϕ, the mode volume is reduced relative to its bulk value by a factor of $\alpha = \phi(l/\lambda)^{1/2}$. If the mode is confined with respect to both axes, as in a fibre or channel guide, its volume is reduced by a factor of α^2. The usefulness of guided-wave designs is therefore greatest in weakly absorbing media, as strikingly exemplified by rare earth doped silica fibre lasers. Coefficients of absorption and gain per unit length in such fibres may be only $\sim 0.01 \text{cm}^{-1}$, with fibre lengths of many metres and α-values of several hundred. It is clear why these devices have no bulk counterpart.

There is an extensive literature describing the calculation of gain in guided-wave amplifiers; the reader is referred in particular to the work by Desurvire, 1994. A key result (e.g. Digonnet and Gaeta, 1985) relates the unsaturated single-pass gain exponent, γ, to an effective pump area, A_{eff}:

$$\gamma = \frac{\sigma \tau}{h\nu_p} \frac{P_{\text{abs}}}{A_{\text{eff}}} \tag{3}$$

where P_{abs} is the total power absorbed from the pump beam in the gain medium, $h\nu_p$ is the pump photon energy, and σ and τ are the emission cross-section and upper state lifetime respectively. The effective pump area is given by:

$$A_{\text{eff}}^{-1} = l \int_{\text{vol}} R(r)S(r) \, d^3r \tag{4}$$

where l is the length of the amplifier, and R, S are normalised pump-rate and signal-mode intensity distributions. In a monomode fibre or channel guide A_{eff} is of the order of the square of the scaling length $(\lambda/\phi)^2$; its precise value is determined by the degree of overlap of pump and signal modes.

In a device with a small effective area it is often possible to invert a transition strongly at quite moderate values of pump power. Equations 3 and 4 may then give a misleading overestimate of the gain, partly because saturation of the pump absorption reduces the effective overlap of pump and signal and also because parasitic effects such as upconversion may degrade the quantum efficiency of the pumping process (Guy et al. 1998).

2.4 Techniques for fabricating laser waveguides

The single most important class of active dielectric waveguide is undoubtedly that of silica fibres fabricated by chemical vapour deposition. These structures may be based on the Ge_2O_3–SiO_2, Al_2O_3–SiO_2, or P_2O_5–SiO_2 glass systems, with NA in the range 0.1–0.4, and propagation loss as low as $\sim$1dB km^{-1}. The erbium-doped fibre amplifier exemplifies all the most desirable properties of gain media of this type; smooth broad spectral features, low dopant concentration with negligible ion-ion interaction, low background loss, great mechanical robustness, and efficient amplification.

The excellent properties of silica glass fibre as a laser host medium have turned out to be actually rather specific to just a few rare earth ion laser transitions in the near-infrared region of the spectrum. The high frequency vibration modes of the silica matrix render the fibres absorbing at longer wavelengths, and also tend to quench the inversion of any shorter wavelength laser transitions. Thus most silica fibre laser development involves one of three systems: (1) the 2-μm Tm^{3+} transition, of interest for sensor applications, (2) the 0.98–1.1μm Yb^{3+} transition, well suited to high power applications both because the difference between the pump and laser photon energies is small, reducing thermal load, and because the simple level structure of Yb^{3+} confers freedom from parasitic cross-relaxation processes, and (3) the 1.55μm Er^{3+} transition, whose importance in telecommunications and as an eye-safe transition is well known. Er-doped silica fibre can include large concentrations of Yb as a co-dopant ion which will absorb 1064nm pump radiation and transfer it efficiently to the Er population, so that devices can be made short.

These three transitions have similar character; each spans the unusually large energy gap which strong spin-orbit coupling produces between the ground and first excited level manifolds in these rare earth ions with nearly filled 4f shells. Since there is no exact point symmetry at the site of a dopant ion in a glass matrix, there are also no strong selection rules, and the transition dipole moment is more or less evenly distributed between the individual transitions linking the manifolds. In the Er transition, for example, there are 56 of these, since the upper and lower manifolds contain 7 and 8 Kramers doublet levels respectively. It tends to be characteristic of the energy eigenvalues of high-angular-momentum ions in approximately octahedral fields that one level of a manifold lies well below the others, which are clustered relatively closely in energy (Yeung and Newman, 1985). This phenomenon confers the characteristic fluorescence profile, seen especially in the emission of Er and Yb, in which the lowest-to-lowest transition forms a partially resolved spike at the high energy extremity of the band, and all the other transitions combine to form a broad featureless wing extending to lower energies. The transition wing is in effect homogeneously broadened, the dense clustering of many transitions tending to obscure the inhomogeneous site-to-site variation of individual transition frequencies.

Within each manifold of levels the population relaxes to a Boltzmann distribution at the glass matrix temperature within a few picoseconds; this process is fast compared to the rate at which population relaxes between different manifolds, whether radiatively or by multiphonon emission. As a result, there is a proportional relationship between the emission and absorption cross-sections at any frequency within the transition bandwidth which has often been used, for example, to extract emission cross-section values from absorption measurements (Miniscalco and Quimby 1991). An important consequence of the fast thermalisation of population within the upper and lower manifolds of levels is that the corresponding laser transition varies smoothly from 3-level to 4-level as the transition wavelength increases. However it is often the case that the pump saturation power required to halve the total population of the ground state manifold is only a few tens of milliwatts or less in a monomode fibre, so that 3-level character is quite compatible with high efficiency.

"Soft glass" fibres, in which the vibration spectrum cuts off at energies much less than $1200 cm^{-1}$, are potentially more versatile than silica as they transmit well to longer wavelengths and allow long-lived energy levels of dopant ions. However, the materials problems which these structures present have prevented exploitation of their properties. The best-established soft fibre technology at the present time is based on a fluorozirconate glass-forming system ; ZrF_4–BaF_2–LaF_3–NaF, or "ZBLAN". There is no CVD technology for fluoride glass, so preforms of these fibres are fabricated by a rotational casting technique. Fibre losses are typically at the dB m^{-1} level, and numerical apertures up to 0.4 have been achieved. ZBLAN fibres are far more fragile than their silica counterparts, and they are potentially vulnerable to chemical attack by water. There is substantial published work on the applications of these fibres as laser sources in the 2–4μm region, and also at visible wavelengths.

One other glass fibre system has recently become commercially available after some years of development by several groups. Chalcogenide fibres with extended transparency in the infrared offer some interesting device prospects (Wang *et al.* 1997), but are even more chemically unstable and mechanically fragile than ZBLAN.

With regard to planar waveguide fabrication the situation is more complicated because no single indispensable technology has emerged comparable to that of the EDFA, although at least nine distinct techniques have been used to realise planar and channel lasers. Research in this field has been motivated by the desire to make more compact devices than glass fibre allows, to miniaturise vibronic lasers for which glass hosts do not exist, and to make fully integrated modelocked and Q-switched sources in a monolithic format.

Much published work relates to chemical surface treatments of glass or crystalline substrates, involving either ion exchange or thermal indiffusion . A planar glass technology of interest for wavelength-division-multiplexing devices in telecommunications is ion-exchange; guides of NA > 0.5 can be made by replacing Na^+ ions in the substrate with Tl ions from a bath of molten salt. Proton exchange is a classic technique of integrated optics on lithium niobate substrates, and it has also been used successfully to make laser devices. Only the extraordinary index of the uniaxial crystal is modified by the exchange, limiting the usefulness of the technique to π-polarised laser transitions; but guides of NA as high as 0.7 result.

Some work on epitaxial growth techniques can also be found in the literature. Crystalline guides of notably high quality based on $Y_3Al_5O_{12}$(YAG) have been grown by liquid phase epitaxy (LPE). In this work the active layer was doped with Ga to raise the index and Lu to achieve a lattice match. An Yb-doped structure of this kind operated as a quasi-3-level laser with near quantum-limited efficiency owing to good optical confinement and low waveguide propagation losses of $\sim$ 0.1dB cm^{-1} (Pelenc *et al.* 1995). Structures with NA up to 0.2 have been made in this way, however the Ga doping has the effect of broadening the linewidth of the laser transitions, thereby reducing the gain by a factor of about two. More recently the fabrication by molecular beam epitaxy (MBE) of crystalline fluoride waveguides has been developed (Daran *et al.* 1998), and led to the demonstration of lasing in a planar Nd-doped LaF_3 guide. This system may allow a planar technology of diode-pumped sources for the mid-infrared to be developed.

Many groups have worked to develop *physical* processes of waveguide deposition that might be applicable to a wide range of material systems. An early example was high-energy helium ion implantation which was indeed used to create waveguides in numerous crystals and glasses (Townsend 1990). Many of these structures exhibited guided-wave lasing; alas the propagation losses of the resulting guides in most cases badly compromised their performance as lasers. Subsequent work, however, suggests that these losses are not intrinsic to the implantation process, and may be substantially reduced by taking greater care over the prior preparation of the implanted surface (Peto *et al.* 1997). More recently pulsed laser deposition (PLD) has been reported as a technique by which low loss crystalline garnet guides can be made (Bonner *et al.* 1997).

The waveguide-laser community has long been interested in vibronic lasers since their inherently high threshold can in principle be markedly reduced in an optically confined structure. To be successful such a device must incorporate a high quality waveguide with low propagation loss, since the gain coefficients of these enormously broad transitions are correspondingly small. Two examples of guided-wave titanium-doped sapphire lasers have recently been reported; one fabricated by PLD on a sapphire substrate (Anderson *et al.* 1997) and the other fabricated by thermal indiffusion of titanium ions into the surface of a sapphire crystal (Hickey *et al.* 1998).

A particularly versatile and promising fabrication technique is thermal bonding which simply involves the contacting of surfaces which have been polished to an optical finish and chemically cleaned. In the low temperature version of this technique the surfaces are held together by van der Waals forces, and are not chemically modified. Chemically and structurally dissimilar materials can be bonded, with the sole proviso that the thermal expansion coefficients of substrate and superstrate must be compatible. Once the bond is formed the composite structure is strong enough to allow the superstrate to be mechanically thinned by polishing, even down to the size required for the core of a monomode waveguide. An upper cladding layer can be attached in a second bonding step. Even channel guides can in principle be fabricated. Planar laser guides made in this way have been shown to exhibit propagation losses of ~ 0.16dB cm^{-1} deduced from measurements of laser slope efficiency as a function of output coupling (Brown *et al.* 1997). Composite guides have been reported with numerical aperture values spanning a range from 0.06 (Nd-doped YAG core and undoped YAG cladding) to 0.8 (Nd:YAG core and glass cladding).

2.5 Examples of fibre and planar devices

Guided wave lasers can look very different from their free-space counterparts. Fibre lasers replace the mirrors, apertures and beam-splitters of conventional lasers with flexible fibres spliced to directional couplers and other components to form a robust system that requires no further optical alignment. An example of a mode-locked fibre laser is shown in Figure 5a (Culverhouse *et al.* 1995). The ring resonator is composed of a loop of fibre containing a gain medium formed from Er/Yb doped mono-mode fibre. Pump light is coupled in through a wavelength-dependent multiplexer, and the laser output is extracted from the cavity through the directional coupler shown at the bottom of the ring. The

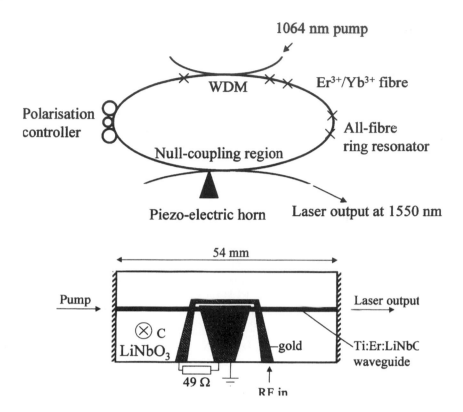

Figure 5. *Examples of waveguide lasers. (a) An all-fibre soliton laser (adapted from Culverhouse et al. 1995) and (b) a channel waveguide electro-optically mode-locked (adapted from Suche et al. 1995).*

novel loss-modulating characteristics of this coupler will be described later.

Conceptually similar resonators are constructed using planar technology, only in place of metres of fibre wrapped on drums we have a few square centimetres of substrate with lithographically patterned guides and components. For example the harmonically mode-locked $Ti:Er:LiNbO_3$ laser shown in Figure 5b is only 54mm long, yet it is reported to have generated transform-limited pulses of 3.8ps duration (Suche *et al.* 1995).

A key technology in the development of silica fibre laser sources is the fibre Bragg grating , used (i) to select specific operation wavelengths out of broad gain spectra, (ii) to enforce lasing in the far wings of a transition, (iii) as a notch filter, (iv) as a dispersion compensator and (v) in conjunction with fibre couplers to configure a variety of complex and versatile laser cavities. The grating consists of a fibre region over which the index of the core is periodically or aperiodically modulated. A grating may be fabricated by exposing a photosensitive fibre to ultraviolet light through an appropriate phase mask (Hill *et al.* 1993). An early application of the fibre grating was to the development of a distributed feedback Er-Yb fibre laser , emitting a single longitudinal mode with a frequency bandwidth of 300kHz at 1534nm (Kringlebotn *et al.* 1994).

Whereas the UV-written fibre grating has static phase properties, acoustic gratings can be used to make dynamic Bragg-coupled elements. A fibre acousto-optic frequency shifter consists of a four-port fused taper coupler made with dissimilar fibres that are so phase mismatched that coupling is zero in the absence of an acoustic wave driven into the coupling region (Birks *et al.* 1994). The coupled light is frequency-shifted, and some tuning of the acoustic frequency is possible. An example of the intriguing possibilities of this technique is provided by the sliding-frequency Er-Yb soliton laser . A ring cavity contains an Er-Yb-doped gain section and an acoustically-driven 4-port coupler of the type described above, which acts both as frequency shifter and as a tunable bandpass filter. The laser exhibits self-starting modelocking; it prefers to pulse because cw radiation at the gain peak is progressively detuned and sees greater loss than solitons which are spectrally reshaped as they circulate (Culverhouse *et al.* 1995).

The development of high-quality fluorozirconate fibres allowed the first upconversion fibre lasers to be realised in the early nineteen nineties (Allain *et al.* 1991, Smart *et al.* 1991). These are visible lasers which could be pumped in the infrared at convenient diode laser wavelengths by sequential ground and excited state absorption steps. Many upconversion fibre lasers based on ZBLAN have been reported, including transitions at violet and ultraviolet wavelengths (Funk and Eden, 1995). However many of these require pumping at a difficult wavelength, or even dual wavelength pumping. One attractively simple and versatile scheme uses the Pr^{3+} ion , which has visible laser transitions at 636nm, 615nm, 520nm and 491nm, all originating from the same metastable level. This level can be pumped by 860nm light in a fibre co-doped with Pr and Yb; the Yb ions transferring energy efficiently into the Pr system (Xie and Gosnell, 1995). A particularly efficient 480nm blue upconversion laser in Tm-doped ZBLAN fibre was first reported by Grubb (1992). A fortuitous overlap of two excited state and one ground state absorption bands allowed this laser to be pumped at a single wavelength in the 1.12–1.14μm range; Grubb reported a 32% slope efficiency with respect to absorbed pump power. Direct diode pumping has been used to achieve 102mW output power on this transition (Sanders *et al.* 1995), and with a high power bar-pumped Nd:YAG laser as the pump source, 230mW of output were observed (Paschotta *et al.* 1997).

The most attractive feature of diode-pumped upconversion fibre lasers as all-solid-state cw visible laser sources is their simplicity. No phase-matching is involved, and the broad excitation bandwidth confers a relaxed specification on the spectral characteristics of the pump diode. Although these devices are in principle spatially multimode, since the same core structure guides both signal light and pump light of much longer wavelength, in practice the emitted beam is close to diffraction-limited, presumably because the fundamental mode sees most gain. The exploitation of this technology has, however, been inhibited to date not only by the high cost of ZBLAN fibre, but also by the susceptibility of the fibre to photochromic damage induced by high pump intensities in the core. A particularly acute problem is documented for the Tm system, in which absorption at the blue transition wavelength develops in the presence of infrared pumping at a rate consistent with the formation of colour centres (Barber *et al.* 1995, Laperle *et al.* 1995, Booth *et al.* 1996, and Paschotta *et al.* 1997). The detailed behaviour of the fibre appears to be a function of its composition, suggesting that a composition might be found which is resistant to this effect.

3   Waveguides pumped by high power diodes

The appearance of commercial diode pump lasers, emitting many tens of watts continuously with great efficiency, has stimulated intense effort into scaling up the power output of diode-pumped solid state lasers, as needed for numerous applications in materials processing, medicine and information technology. The problems inherent in this process are well known. The spatial coherence of a diode bar laser in the plane of the junctions is poor, and it is difficult to use the highly divergent and asymmetric beam efficiently. Moreover, a substantial fraction of the pump power absorbed in the gain medium of the solid state laser is dissipated thermally. The resulting distributions of temperature and stress will ultimately cause stress fracture of the rod; at lower pump power inputs they give rise to thermal lensing and birefringence of the laser medium which can severely degrade the spatial quality of the output beam.

A key role for diode-pumped solid state lasers is therefore to convert the power supplied by a bar into a diffraction-limited beam which can be focused, launched into a monomode fibre or used to pump processes of nonlinear frequency conversion to wavelengths of interest. This function is sometimes characterised as brightness conversion, where the term "brightness" has its photometric sense, and may be defined for a paraxial beam in terms of the M^2 beam quality parameter introduced earlier as $P/(\lambda^2 M_x^2 M_y^2)$, where P is the power in the beam. Thus whereas the brightness of a gas laser source can easily be many tens of $\text{Wsr}^{-1}\mu\text{m}^{-2}$, the brightness of a diode bar is typically in the range 10–$20\text{mWsr}^{-1}\mu\text{m}^{-2}$.

3.1   Brightness conversion and cladding pumping

Consider a conventional fibre laser with a single cladding region. If it is to be pumped with a multimode beam of quality M_p^2, which must be focussed to a spot equal roughly in size to the fibre radius b, then the fibre must be designed with a numerical aperture at least equal to the pump divergence;

$$\frac{M_p^2 \lambda_p}{\pi b} \approx \phi$$

It follows that the fibre V-values at pump and laser wavelengths are given by:

$$V_{\text{pump}} \approx 2M_p^2 \qquad \text{and} \qquad V_{\text{laser}} \approx 2\frac{\lambda_p}{\lambda_l}M_p^2$$

The laser emission may be usefully improved in spatial coherence relative to the pump beam if its wavelength is significantly longer, and the fundamental mode of the fibre may see more gain and extract the stored energy more effectively. However in general such a device is not a particularly effective brightness converter.

This limitation is dramatically overcome if the fibre is fabricated for cladding pumping, with multimode pump radiation guided in a large-area inner cladding region of high V-number. Nested within the inner cladding is a monomode core structure doped with the laser ion. The effective absorption length for the pump radiation is thus increased by approximately the ratio of the cross-section areas of the inner cladding and core regions. An output of 35W in a diffraction-limited beam has been reported for a cladding-pumped

Yb-doped fibre laser operating at the long wavelength end of the gain bandwidth, where ground state reabsorption is negligible. This device is pumped by low-brightness diodes with a slope efficiency of 65% (Muendel *et al.* 1997). A recent review outlines the exciting prospects for compact sources offering high power and high pulse energy using cladding-pumped fibre technology (Richardson *et al.* 1997).

A cladding-pumped fibre laser achieves brightness conversion by greatly extending the length of the gain medium. This leads to a satisfactorily low thermal loading per unit length, but may impair the efficiency of the laser unless the fibre can be fabricated with sufficiently low propagation losses in the core and inner cladding. In particular there is a trade-off between pump launch efficiency and pumping rate, so that the cladding pumping technique is only likely to perform well for 4-level laser transitions. For those laser transitions in which intense pumping is essential, and stress fracture sets a limit to the thermal loading which can be applied, the slab geometry has often been preferred to the rod. The reason for this is apparent if one compares the temperature differences, ΔT_{rod} and ΔT_{slab} which develop across the rod and the slab respectively, when each experiences a power P/unit length deposited uniformly through its volume:

$$\Delta T_{\text{rod}} \approx \frac{1}{4\pi\kappa} \times P \qquad \Delta T_{\text{slab}} \approx \frac{1}{2\kappa} \times P \times \frac{t}{W}$$

where κ is the thermal conductivity of the gain medium and t and W are the thickness and width respectively of the slab, assumed cooled from one side only. It is thus possible to control the temperature excursion using the aspect ratio of the slab. The limiting case of a wide, thin slab is clearly a planar waveguide.

3.2 Diode-pumped planar waveguide sources

These ideas motivate an investigation of the potential of planar waveguide lasers as brightness converters for high-power diode bar lasers . Coupling of the diode bar radiation into a planar waveguide structure is a relatively straightforward task: there is no need to symmetrise the diode output, and cylindrical coupling lenses of short focal length and high numerical aperture are readily available. On the other hand, the gain region is correspondingly asymmetric, and it becomes difficult to extract the power in a good quality beam. One of the first experiments of this type therefore used a high-brightness broad-stripe diode to pump a Nd:YAG planar waveguide amplifier . The diffraction-limited signal ensured an output of good beam quality, even though the gain region was ~220μm wide in the unguided plane, but had only the 3.8μm depth of the near single-mode waveguide. Up to 550mW of the output from the 1.2W diode was absorbed in the 3.8μm deep waveguide which had been fabricated by LPE to have a numerical aperture of 0.23. The dependence of output power on signal input power is shown in Figure 6 (Shepherd *et al.* 1997). With an input of 90mW it was possible to extract 290mW, or 36% of the absorbed pump power.

At small signal input power levels the highest measured value of amplifier gain was 28.8dB. It was shown that this value was significantly reduced by Auger upconversion and cross-relaxation; processes which transfer energy between ions and quench the upper level population (Guy *et al.* 1998). Thus although it is possible to amplify a diffraction-limited beam efficiently in a planar waveguide, parasitic effects tend to dissipate the high

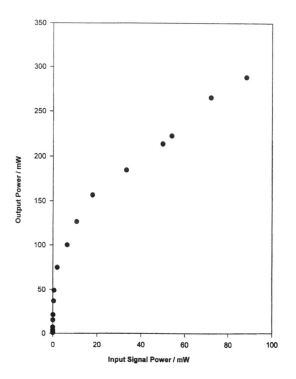

Figure 6. *Output power at 1.064μm as a function of input signal power for an LPE-grown planar Nd:YAG waveguide pumped with 550mW of absorbed power.*

inversion densities that are theoretically attainable in a waveguide. In practice it may be preferable to configure the planar guide as a laser oscillator, with the inversion clamped at a moderate value, and use other techniques to control the spatial mode of the output.

A further problem which arises in the attempt to scale planar amplifiers up in power to the multi-watt level is associated with amplified spontaneous emission (ASE). As a rough guide, ASE will saturate the gain of an amplifier above the limit

$$\frac{G\Omega}{4\sqrt{\ln G}} \approx 1$$

where Ω is the solid angle over which the full amplifier gain appears. Consider an amplifier of length l, pumped by a multimode beam which has beam quality parameter value of M_y^2 in the unguided plane. If the pump is confocally focused with spot size W_p at the waist in a guide of index n and numerical aperture ϕ, the values of Ω and l are given by

$$\Omega \approx \pi \times \frac{\phi}{n} \times \frac{M_y^2 \lambda_p}{n\pi \, W_p} \qquad l \approx \frac{2\pi n W_p^2}{M_y^2 \lambda_p}$$

It follows that the greatest value of gain achievable by such an amplifier, $G_{\max}$, is set by the lateral beam quality parameter of the diode pump source:

$$\frac{G_{\max}}{(\ln G_{\max})^{1/2}} \approx \frac{2n}{\phi\pi} \sqrt{\frac{2\pi n l}{\lambda_p M_y^2}}$$

bar & fibre lens

LPE waveguide

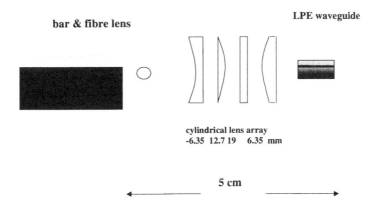

cylindrical lens array
-6.35 12.7 19   6.35 mm

5 cm

Figure 7. *Diode bar coupled to planar Nd:YAG waveguide by cylindrical lens assembly*

The value of M_y^2 for the broad stripe diode used to pump the amplifier described above is ~ 34. The corresponding value of G_{max} is 28dB, showing that this device operates very close to the limit imposed by ASE. One may estimate the value of G_{max} achieved by a similar planar amplifier, again with $l \sim 5$ mm and numerical aperture 0.23, only now pumped by a more powerful diode bar with a lateral beam quality parameter $M_y^2 \sim 2300$. In this case G_{max} is only 15dB, showing that with pump sources of such low brightness is not feasible to make high gain planar amplifiers by confining the pump light tightly in the direction in which it is spatially coherent.

A first demonstration of a high power planar waveguide source pumped by a diode bar therefore involved a laser oscillator rather than an amplifier (Bonner *et al.* 1998). A simple cylindrical lens assembly was used to couple radiation from a diode bar source into an LPE-grown planar Nd:YAG waveguide with coated endfaces. A diagram of the device is shown in Figure 7.

The output from the diode bar was collimated by a fibre lens into a sheet of radiation, $\sim 320\mu$m deep and up to 18W in power, with beam quality parameter values $M_x^2 \sim 2.3$ and $M_y^2 \sim 2300$. The waveguide was more highly multimode than the pump characteristic required; with a depth of 80μm it supported TE and TM laser modes up to $m = 8$, despite a low NA of 0.06. The laser output power is shown as a function of diode operating current in Figure 8. The greatest observed output power was 6.2W, corresponding to an overall optical-to-optical conversion efficiency of 31%.

The output from this laser was in a multimode beam, with characteristics that were sensitive, for example, to the effectiveness of the contact between substrate and heatsink. M^2-values were estimated for the beam based on intensity profile measurements. These ranged from 3–5 in the vertical dimension and from 140–160 in the lateral dimension. The best observed values of beam quality parameter therefore corresponded to a brightness enhancement of

$$\left(\frac{6.2}{18}\right)\left(\frac{2.3 \times 2300}{3 \times 140}\right)\left(\frac{808}{1064}\right)^2 = 2.5$$

relative to the fibre-lensed diode pump beam.

Anne Tropper

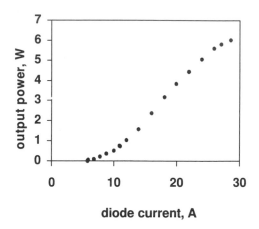

Figure 8. *Planar waveguide laser output power at 1.064μm as a function of diode bar operating current.*

It was clear that the benefits of the guided wave geometry would only be fully exploited in these experiments if thin waveguides, of depth < 10μm, were used. Such guides would need to have large NA to capture the highly diverging diode output efficiently. The first experiment of this type accordingly used a thermally bonded planar waveguide with a Nd:YAG core bounded by sapphire cladding regions. This combination of materials gives guides with NA ∼0.46. Guides with depth as low as 4μm can be fabricated by thermal bonding, with propagation losses measured to be less that 0.2dB cm^{-1} by investigating the fibre-laser performance when pumped by a Ti:sapphire laser. A particularly simple assembly of coupling optics was used to launch light from a 20W diode bar into these guides. A single graded-index rod lens was used to launch the collimated sheet of radiation from the fibre-lensed bar into the guide, while a single 19mm focal length cylindrical lens was used to focus the light in the unguided lateral direction to a region about 4mm wide. The distance from the output facet of the diode to the input face of the waveguide was 3cm. With this arrangement it was possible to couple 73% of the diode output into an 8μm deep guide, and 60% into a 4μm guide (Shepherd *et al.* 1999). In a preliminary laser experiment an output power of 3.7W was observed from the 8μm guide, although waveguide length and output coupling were not optimised in this device.

Most recently it has been shown that diode bars can be coupled efficiently into planar waveguides without any optical components at all - by proximity coupling (Bonner *et al.* 1999). Precision manipulators were used to position the emitting facet of a 10W diode bar within ∼20μm of the end face of an 8μm deep planar guide. A contact-bonded Nd:YAG-on-sapphire waveguide was used for this experiment, for the sake of the high NA of ∼0.45. By measuring the total pump power transmitted through the 2mm long structure, both guided and in radiation modes, and comparing it to that emitted by the diode, it was calculated that 45% of the incident power was absorbed in the waveguide. The efficiency of coupling into the waveguide was thereby estimated to be high, probably ∼90%. The guide was observed to lase transversely, emitting about 0.5W for 6W of pump power, in a resonator that was far from being optimised. These results suggest that proximity coupling may form the basis for a family of highly compact and efficient sources, especially bearing in mind that position tolerances for a cladding-pumped device might be significantly relaxed.

3.3 Mode control in planar laser resonators

Since these sources are extremely compact, and potentially simple and efficient, it is important to investigate the extent to which their spatial mode quality can be improved. The guided and unguided directions present separate problems.

Mode-control in the guided direction

Although it is often stated that for the fast-diverging plane of a diode bar $M^2 = 1$, in practice the initial divergence is so great that after the fibre lens the M^2-value is typically 2–3. It may therefore not be possible to couple the pump light into a guide which is monomode at the laser wavelength. An interesting possibility for such devices is to control the quality of the spatial mode in the guided direction using the evanescent field. An experimental study has shown that gold overlays can be used to create excess propagation losses for higher order TE modes, with considerable enhancement of beam quality (Brown *et al.* 1998). A 50nm thick gold coating was applied to an LPE-grown guide consisting of an 8.3μm thick YAG layer doped with Nd, Ga and Lu grown on an undoped YAG substrate. There was no superstrate, and the gold was applied directly to the upper surface of the core. The NA of the guide was 0.226, so that the guide depth was equivalent to 1.8 scaling lengths, and the guide supported TE and TM modes up to $m = 3$. An experimental comparison was made between gold-coated and uncoated regions of the guide, measuring the beam quality of the lasing output in the guided direction. It was possible to adjust the focusing of the titanium sapphire pump laser into the guide so that the higher order guided modes at the laser wavelength saw at least as much gain as the fundamental. Under these conditions whereas the output of the bare guide exhibited M^2-values in the range 1.52–2.5, the value from the gold-coated region was 1.17. The gold coating had the further effect of suppressing lasing on all TM modes, so that the output became linearly polarised, with the ratio of maximum to minimum values of power transmitted through a rotating linear polariser of 27dB, compared with 7dB in the uncoated region. An analysis of propagation in this system showed that the gold layer introduced excess losses of only 0.04dB/cm for the fundamental TE mode, but 0.5dB/cm for the highest order TE mode, and 1dB/cm for the fundamental TM mode.

An alternative approach to the problem of mode control in the guided direction is to use the technique of cladding-pumping, to which planar thermally-bonded structures are particularly well adapted. Since the M^2-value of an unsymmetrised bar pump beam is quite low in the vertical direction, the inner cladding depth need only be 2–3 times that of the core, so that the overall device length can be kept down to a few millimetres. Thermal bonding is particularly well-suited to assembling structures in which the NA of the core/inner cladding interface is small, whereas that of the inner cladding/outer cladding interface is large. The doped YAG/undoped YAG/sapphire material combination, with its low loss, mechanical strength and high thermal conductivity is at the time of writing under study as a particularly interesting system for this purpose.

Mode-control in the unguided plane

The planar laser devices which have been constructed to date typically have M^2-values of 100 or more in the unguided plane; this is clearly the characteristic which is in most urgent need of improvement. The plane/plane Fabry-Perot resonator used hitherto does little for the lateral spatial coherence; however it is desirable to keep the compact and rugged monolithic character of these devices rather than incorporate an external resonator.

It is evident that a significant reduction in M^2-value can be achieved by restricting the aperture for laser emission. Planar waveguide lasers with lateral M^2 of ~30 have been made by orienting the cavity at right angles to the direction of the incident pump beam. In this side-pumped geometry the width of the gain medium can be substantially reduced, though at the expense of a very much smaller overlap between the pump and laser modes. In an alternative approach, a sliding output mirror was used to construct a primitive form of unstable resonator. The thin, highly reflecting mirror, held against the output facet of a longitudinally-pumped planar guide by the surface tension of a drop of index-matching fluid, could be slid laterally to uncover an emission aperture of variable width. With this type of cavity output power was cut down from 3.7W to 2.4W, but with considerable improvement in M^2, from 85 to 22 in the unguided direction. A future solution to the problem of lateral beam quality in planar waveguide lasers is therefore likely to be the monolithic planar unstable resonator. Such a device might incorporate cladding pumping to give an output of several Watts, in a beam with M^2-values of 2–3 in the lateral direction and <1.2 in the guided direction.

3.4   Face-pumped lasers

Although side-pumping, with the pump laser light propagating across the laser resonator, suffers in efficiency compared to longitudinal pumping because the overlap of pump and signal is unavoidably weak, it has the attractive property of *scalability*. The power of a side-pumped device can in principle be increased up to the limit set by optical damage simply by multiplying up the length of the gain medium and the pump units coupled to it; the thermal load along the extended gain medium can everywhere be the same. Pumping through the *face* of a planar structure is equally scalable, and has the further highly attractive feature that the pump light is not launched into a waveguide, and positioning with micrometre accuracy is not required.

A fundamental difficulty attending face-pumped lasers , whether waveguide or slab, is the weak absorption of pump light by typical dielectric gain media. Any attempt to exploit face pumping in these systems therefore involves multiple passes of the pump through the gain medium. We shall examine two such schemes. In the first, due to Brauch *et al.* (1994), a thin Yb:YAG disc plays the role of an amplifying mirror in an otherwise conventional laser cavity. Diode bar pump light is delivered by a fibre bundle, and an array of mirrors re-image the reflected pump back on to the active area of the device. Heat can be removed with great efficiency from the thin slab, and thermal gradients are, to first order, directed along the axis of the laser cavity, greatly reducing their adverse effect on the quality of the output beam. Many tens of Watts can be extracted from these devices in beams of excellent quality; the most difficult part is the alignment of the multiple pump mirrors. These lasers are, of course, not guided wave devices.

In the second scheme the laser light propagates in the plane of the pumped face, and a slotted mirror achieves the necessary multipass pumping (Faulstich *et al.* 1996). The pump source is a stack of diode bar lasers, their emitting regions aligned immediately behind slots in a highly reflecting mirror. The gain medium, a slab of Nd-doped glass, is proximity-coupled to the pump, the rapid divergence of the uncollimated diode beams ensuring a uniform distribution of pump energy. A second reflecting surface below the slab sets up the multiple pump reflections, and spaces either side allow the passage of coolant. In this way a transfer efficiency of the pump light of 84% has been achieved into a slab only 0.45mm thick with 32% single-pass absorption of the pump. Pulses of 65mJ have been extracted from this system. The slotted mirror technique has also been used to pump a multimode thermally-bonded Nd:YAG on YAG waveguide, which emitted 9.3W average power at 8% duty cycle.

Finally, it is worth noting that the limitation of weak pump absorption which motivates these elaborate multipass schemes is absent in semiconductor gain media, in which absorption coefficients are measured in μm^{-1} rather than cm^{-1}. It is interesting that the active mirror concept embodied in the Yb:YAG disc laser described above has also been employed in an optically-pumped semiconductor quantum well laser, from which $\sim$0.5W of output power was extracted in a diffraction-limited beam (Kuznetsov *et al.* 1997). The semiconductor wafer in this device incorporated 14 InGaAs quantum wells pumped by optical absorption in the barriers; a Bragg grating structure provided one mirror of the resonator, which was completed by an external spherical mirror. The laser design shares many features with its Yb:YAG counterpart, however the gain medium absorbs $\sim$85% of the incident pump light in a single pass.

It is clear that planar geometries, whether or not guided-wave, are particularly well adapted to high power pumping by low-brightness diode bar sources. The most attractive designs will be simple and compact and will not involve excessively tight alignment tolerances. Proximity coupling, in various guises, scores highly by these criteria, and this may be a prominent theme of future work. Moreover the optical pumping of semiconductors - long derided as the last resort of electrically defective structures - may turn out to be significant, bringing a new lease of life to designs developed for high power dielectric lasers through the great versatility of quantum well gain media.

References

Allain J Y, Monerie M and Poignant H, 1991, *Electron. Lett.* **27** 1156.

Anderson A A, Eason R W, Hickey L M B, Jelinek M, Grivas C and Gill D S, 1997, *Opt. Lett.* **22** 1556.

Barber P R, Paschotta R, Tropper A C and Hanna D C, 1995, *Opt. Lett.* **20** 2195.

Birks T A, Farwell S G, Russell P St J and Pannell C N, 1994, *Opt. Lett.* **19** 1964.

Bonner C L, Anderson A A, Eason R W, Shepherd D P, Gill D S, Grivas C and Vainos N, 1997, *Opt. Lett.* **22** 988.

Bonner C L, Brown C T A, Shepherd D P, Clarkson W A, Tropper A C, Hanna D C and Ferrand B, 1998, *Opt. Lett.* **23** 942.

Bonner C L, Bhutta T, Shepherd D P, Tropper A C, Hanna D C and Meissner H E, 1999, *Proceedings CLEO-US*.

Booth I J, Archambault J -L and Ventrudo B F, 1996, *Opt. Lett.* **21** 348.

Brauch U , Giesen A, Karszewinski M, Stewen C and Voss A, 1995, *Opt. Lett.* **20** 713.

Brown C T A, Bonner C L, Warburton T J, Shepherd D P, Tropper A C, Hanna D C and
    Meissner H E, 1997, *Appl. Phys. Lett.* **71** 1139.
Brown C T A, Harris R D, Shepherd D P, Tropper A C, Wilkinson J S and Ferrand B, 1998,
    Phot. Tech. Lett. **10** 1392.
Culverhouse D O, Richardson D J, Birks T A and Russell P St J, 1995, *Opt. Lett.* **20** 2381.
Daran E, Shepherd D P and Schweizer T, 1998, *Proceedings of CLEO-Europe* paper CThF7.
Desurvire E, 1994, *Erbium-doped fiber amplifiers: principles and applications* (John Wiley and
    sons).
Digonnet M F J and Gaeta C J, 1985, *Applied Optics* **24** 333.
Faulstich A, Baker H J and Hall D R, 1996, *Opt. Lett.* **21** 594.
Funk S and Eden J G, 1995, *IEEE J. Sel. Topics Quantum Electron.* **1** 784.
Grubb S G, Bennet K W, Cannon R S and Humer W F, 1992, *Electron. Lett.* **28** 1243.
Guy S, Bonner C L, Shepherd D P, Hanna D C, Tropper A C and Ferrand B, 1998, *IEEE J.
    Quantum Electron.* **34** 900.
Hickey L M B, Quigley G R, Wilkinson J S, Moya E G, Moya F and Grattepain C, 1998,
    Proceedings CLEO-Europe paper CThF3.
Hill K O, Malo B, Bilodeau F, Johnson D C and Albert J, 1993, *Appl. Phys. Lett.* **62** 1035.
Kringlebotn J T, Archambault J-L, Reekie L and Payne D N, 1994, *Opt. Lett.* **19** 2101.
Kuznetsov M, Hakimi F, Sprague R and Mooradian A, 1997, *IEEE Phot. Tech. Lett.* **9** 1063.
Laperle P , Chandonnet A and Vallée R, 1995, *Opt. Lett.* **20** 2484.
Miniscalco W J and Quimby R S, 1991, *Opt. Lett.* **16** 258.
Muendel M, Engstrom B, Kea D, Laliberte B, Minns R, Robinson R, Rockney B, Zhang Y,
    Collins R, Gavrilovic P and Rowley A, 1997, *CLEO-US* paper CPD30-1.
Nishihara H , Haruna M and Suhara T, 1985, *Optical Integrated Circuits* (McGraw-Hill.)
Paschotta R, Moore N, Clarkson W A, Tropper A C, Hanna D C and Mazé G, 1997, *IEEE J.
    Sel. Topics Quantum Electron.* **3** 1100.
Pelenc D, Chambaz B, Chartier I, Ferrand B, Wyon C, Shepherd D P, Hanna D C, Large A C
    and Tropper A C, 1995, *Opt. Comm.* **115** 491.
Peto A, Townsend P D, Hole D E and Harmer S, 1997, *J. Mod. Optics* **44** 1217.
Richardson D, Minelly J and Hanna D C, 1997, *Laser Focus World* Sep. 87.
Sanders S, Waarts R G, Mehuys D G and Welch D F, 1995, *Appl. Phys. Lett.* **67** 1815.
Shepherd D P, Brown C T A, Warburton T J, Hanna D C and Tropper A C, 1997, *Appl. Phys.
    Lett.* **71** 876.
Shepherd D P, Bonner C L, Brown C T A, Clarkson W A, Tropper A C, Hanna D C and
    Meissner H E, 1999, *Opt. Comm.* **160** 47.
Smart R G, Hanna D C, Tropper A C, Davey S T, Carter S F and Szebesta D, 1991, *Electron.
    Lett.* **27** 1309.
Snitzer E, 1961, *Phys. Rev. Lett.* **7** 444.
Snyder A W and Love J D, 1983.*Optical Waveguide Theory* (Chapman and Hall.)
Suche H, Wessel R, Westenhöfer S, Sohler W, Bosso S, Carmannini C and Corsini R, 1995, *Opt.
    Lett.* **20** 596.
Townsend P D, Chandler P J and Zhang L, 1990, *Optical Effects of Ion Implantation* (Cambridge
    University Press.)
Wang J, Hector J R, Brady D, Hewak D, Brocklesby B, Kluth M, Moore R and Payne D N,
    1997, *Appl. Phys. Lett.* **71** 1753.
Xie P and Gosnell T R, 1995, *Opt. Lett.* **20** 1014.
Yeung Y Y and Newman D J, 1985, *J. Chem. Phys.* **82** 3747.

Optical parametric oscillators

Malcolm H Dunn and Majid Ebrahimzadeh

University of St Andrews, Scotland

1 Introduction; basic principles

The basic principles behind the operation of optical parametric oscillators (OPOs) have been known since the early days of nonlinear optics, and many different configurations were explored by the early investigators. However, it is only within the course of the last decade or so that OPOs have been developed as significant practical sources of coherent radiation. This has come about largely through substantial improvements in optically nonlinear materials, combined with an appreciation of the opportunities offered by the different cavity configurations that can be employed. It is now the case that OPOs are capable of generating coherent radiation across all timescales from femtosecond pulses to true continuous-wave (cw). In addition, the spectral coverage of these devices is exceptional, extending all the way from the ultraviolet (200nm) to the mid–infrared ($>5\mu$m), in many cases with a single device providing more than a decade of spectral coverage (see Figure 1).

A basic, if somewhat naïve, idea as to how an OPO operates can be obtained from Figure 2. Here, an incoming photon of light from the pump laser (pump photon, frequency ν_p) enters an optically nonlinear crystal ($\chi^{(2)}$ nonlinearity). The nonlinearity causes the photon to "crack" into two photons that together in frequency (and hence in energy) add up to the frequency (energy) of the pump photon (energy conservation). The resulting photon of higher frequency (ν_s) is called the signal photon, while that of lower frequency (ν_i) is called the idler photon. Hence we have

$$\nu_p = \nu_s + \nu_i \tag{1}$$

Where the crack occurs is governed by another constraint on the process, that of momentum conservation (also called phase matching). The sum of the momenta of the signal and idler photons inside the crystal must add up to the crystal momentum of the original pump photon. This constraint may be expressed in terms of the in-vacuo wavenumbers (k) and refractive indices (n) associated with the different frequencies involved, namely

$$k_p n_p = k_s n_s + k_i n_i \tag{2}$$

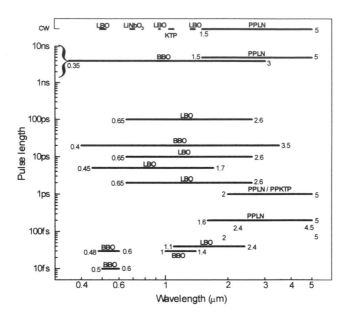

Figure 1. *The spectral and temporal coverage now available from OPOs. Optically nonlinear materials: LBO: lithium triborate; BBO; beta barium borate; PPLN: periodically poled lithium niobate; KTA: potassium titanyl arsenate; KNbO₃;potassium niobate; PPKTP: periodically poled potassium titanyl phosphate; KTP: potassium titanyl phosphate.*

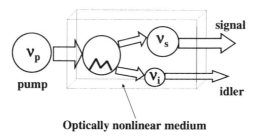

Figure 2. *Simple principles of operation of an optical parametric oscillator.*

It is the possibility of changing these refractive indices, either through changes in propagation direction through the crystal (angle tuning) or temperature (temperature tuning), that leads to the wide tuning ranges generally associated with OPOs. Essentially in these approaches, birefringence is exploited to offset the crystal dispersion. Figure 3 shows an example of the extensive tuning ranges possible for a single device, in this case through using angle tuning of a crystal of lithium triborate pumped at 355nm. By changing the propagation direction in the nonlinear crystal over a range of some fifteen degrees, it is possible to generate continuously-tunable radiation in either the signal or the idler wave from the deep blue around 455nm to the near infrared around 2μm.

An entirely adequate description of the process of parametric down conversion avoids the idea of photons altogether, and is based on a wave picture. The nonlinear crystal when

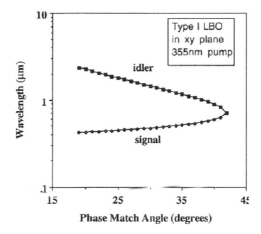

Figure 3. *Tuning range of an optical parametric oscillator based on lithium triborate and pumped at 355nm (frequency-tripled Nd laser). A vertical line drawn from a particular phase match angle intersects the lower branch of the curve at the signal wavelength and the upper branch of the curve at the corresponding idler wavelength.*

immersed in the pump wave exhibits optical gain to waves at the signal and idler wave frequencies. Hence when the crystal is incorporated into an optical cavity with appropriate optical feedback, oscillation can occur when the parametric gain exceeds the cavity loss (Figures 4 and 9). Cavities can be designed to resonate either the signal or idler waves (singly-resonant device, SRO), or both waves together (doubly-resonant device, DRO). By changing the angle of the crystal with respect to the cavity axis, tunable output ensues.

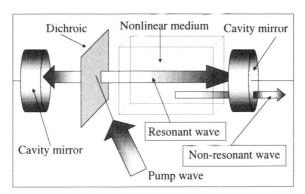

Figure 4. *A singly-resonant optical parametric oscillator. The pump wave is coupled into the nonlinear medium by the dichroic mirror and then undergoes a single pass of the cavity. Only one of the down converted waves is resonant.*

In the last few years an alternative means of phase matching in nonlinear crystals has had a significant impact on the development of OPOs. This is the technique of quasi-phase-matching, which is discussed in the context of devices later on. However, see Myers (this volume) for a comprehensive review of this important approach.

In this chapter we will pay most attention to the development of continuous-wave OPOs (cw OPOs). Before doing so we briefly discuss recent progress with pulsed OPOs where pulse durations are typically in the range from 1 to 10ns. In general these devices are pumped by Q-switched pump lasers.

2   Nanosecond optical parametric oscillators.

This class of OPO has recently been developed to the extent that many commercial versions are now available, including single-frequency versions with linewidths close to the transform limit. These devices are routinely used in many spectroscopic applications where broad tuning ranges are required. Moulton (this volume) reviews their operation in molecule-specific Lidar schemes for example. Substantial progress took place in the development of this class of OPO following the introduction of BBO (beta barium borate) and LBO (lithium triborate), and through the wider availability of KTP (potassium titanyl phosphate), as optically nonlinear materials. There has been wide investigation of pulsed devices, pumped by Q-switched pulses generated by Nd lasers (latterly also including diode-laser-pumped versions of these lasers) either directly or following frequency-doubling or frequency-tripling. Pump energies to reach oscillation threshold in these devices can be less than 1mJ and devices generating output pulses at the 100mJ to 1J energy level have been demonstrated. Generally the oscillation threshold for these devices is determined by the time required for the generated pulse to build-up from noise, and so short cavities, which ensure a rapid return of the circulating field to the nonlinear crystal, are important.

At the point in the tuning curve where the frequency of the signal and idler become equal, called the degeneracy point, the linewidth of the simple OPO generally becomes very large. For this reason much attention has been paid to techniques for narrowing the linewidth of the OPO. Two general approaches have been employed so as to control the resonant frequency of the radiation within the cavity. In injection seeding techniques, narrow linewidth radiation is injected into the OPO cavity from another source so as to be close to resonance with one of the cavity modes. In dispersive cavity techniques, frequency selective elements (gratings, etalons) are incorporated into the OPO cavity. Both these techniques are capable of providing single frequency radiation (i.e. radiation where only one cavity mode is oscillating), with linewidths close to the transform limit determined by the pulse duration.

A particularly successful example of a dispersive cavity due to Bosenberg and Guyer (1993) and now commercially available, is shown in Figure 5. This device, in which a nonlinear crystal of KTP is pumped with pulsed radiation at 532nm, from a Q-switched, frequency-doubled Nd laser, is tuned by means of a grazing incidence grating/tiltable tuning mirror combination. By appropriate control of crystal angle, cavity length and angle of incidence of radiation on to the grating, pulsed single frequency radiation may be generated with a linewidth of less than 200MHz. The output may be tuned over the ranges 700–900nm (signal wave) and 1.4–2.2nm (idler wave). Continuous tuning over a range in excess of 100cm^{-1} is possible without any frequency hops.

An example of the injection seeding technique is shown in Figure 6, and is due to Boon-Engering *et al.* (1995). The OPO itself is based on a singly-resonant ring cavity,

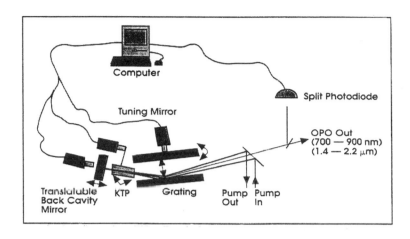

Figure 5. *Pulsed OPO which is line-narrowed and tuned by a grating/mirror combination due to Bosenberg and Guyer, 1993. (Reproduced with permission)*

with the nonlinear crystal BBO pumped at 355nm by a frequency-tripled, Q-switched Nd:YAG laser. Injection seeding of the ring cavity is by a tunable, single-frequency diode laser operating at around 830nm. The ring cavity, which resonates the idler wave, is held on to resonance at this wavelength by a suitable servo-locking loop with feedback to the piezo-mounted cavity mirror. Bandwidths of the order of 350MHz averaged over multiple pulses have been demonstrated, and the device widely used for high-resolution spectroscopy

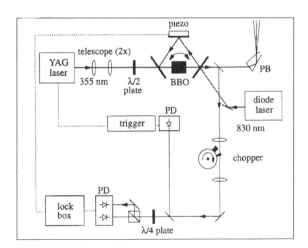

Figure 6. *Pulsed optical parametric oscillator which is frequency-controlled by injection seeding from a diode laser due to Boon-Engering et al. 1995. (Reproduced with permission)*

3   Introduction to CW optical parametric oscillators.

Continuous wave OPOs are not yet commercially available. However, over recent years considerable progress has taken place with regard to these devices, and truly practical sources of coherent radiation by this means are now close to realisation. In this section we explore recent progress with regard to the impact both of new nonlinear materials, in particular those based on quasi-phase-matching, as well as innovative cavity geometries.

Two considerations are particularly important in relation to the operation of cw OPOs. One is the pumping power required to reach oscillation threshold associated with the different cavity geometries, and the other is the constraints that the particular cavity geometry adopted places on the smooth (or otherwise) tuning of the OPO. In the previous section it was pointed out that devices can be operated with either one or other of the down-converted waves resonant in the cavity (SRO), or with both waves resonant at the same time (DRO). In the latter case, the nonlinear crystal experiences two intense waves (the resonated signal and idler waves), whereas in the former case, the crystal experiences only the one intense wave (whichever of the two is resonated). This has the consequence that, other things being equal, the pump power to reach oscillation threshold in the case of the DRO is substantially less than that in the case of the SRO. The following table summarises the expressions for oscillation threshold for different types of cw OPO.

Configuration	Threshold	Values
SRO	$K\pi/F_{i/s}$	Watts
DRO	$K\pi^2/[F_s F_i]$	10's mW
SRO + pump enhancement	$K\pi/[2F_i F_p]$	100's mWs

In this table F_s and F_i are the finesses of the signal idler cavities respectively, E_p is the pump enhancement factor and the factor K is given by

$$K = \frac{n_p^2 \varepsilon_0 c^4}{2\pi^2 L d_{\text{eff}}^2 \nu_p^3}$$

where L is the length of the nonlinear xstal, d_{eff} is its effective nonlinear coefficient, and n_p is its refractive index at the pump frequency ν_p. (Confocal focusing close to degenerate operation assumed). Single pass pump in both the SRO DRO configurations assumed.

It may be seen from this table that the oscillation threshold for the DRO is reduced by a factor of (π/F), where F is the finesse exhibited by the cavity at the additional resonance, compared to the SRO. In a well-designed cavity F may be of the order of several hundred, so the reduction in threshold is a substantial one. This is all the more important when the absolute values of oscillation threshold are calculated. For typical birefringent nonlinear materials, effective nonlinear coefficients, d_{eff}, are of the order of 1pm/V, which implies oscillation thresholds of the order of 1–10W of pump power in the case of the singly-resonant geometry. This can be reduced to the order of tens of milliwatts by employing a doubly-resonant geometry which is a substantial advantage.

An alternative approach is to simultaneously resonate the pump wave in the OPO cavity along with only one of the down converted waves. This is referred to as a pump-enhanced OPO (PE OPO). Threshold reduction in this case is given by (1/2E), where E is the pump enhancement factor. Typically this may be of the order of 20 to 30, so leading to significant threshold reduction, but generally somewhat less that that achievable with the doubly-resonant device. The DRO and the PE OPO show quite different tuning constraints, as will be discussed shortly.

Yet another approach is to obtain a high intensity pump wave by placing the OPO within the cavity of the pump laser itself (where the circulating pump field is of course much higher than that coupled out of the cavity). This class of device is referred to as in intracavity OPO (IC OPO). Since the operation of the OPO itself affects the circulating pump field within the pump laser, the output characteristics of this device are rather different from the cases above where the OPO is totally separated from the pump laser, as will be discussed below.

The different cavity geometries associated with cw OPOs are summarised in Figure 7.

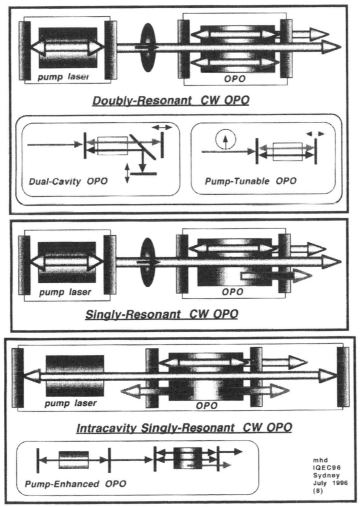

Figure 7. *Summary of the different cavity geometries employed with continuous-wave optical parametric oscillators.*

4   Singly-resonant optical parametric oscillators.

The advent of quasi-phase-matched nonlinear materials has had a substantial impact on
the development of the singly-resonant optical parametric oscillator. This is particularly
the case for OPOs based on periodically-poled lithium niobate (PPLN) and operating in
the mid infrared. Two examples of innovative devices due to Bosenberg *et al.* (1996) are
illustrated in Figure 8, one of which employs a ring cavity, the other a standing wave
cavity. When pumped by a diode-laser-pumped Nd:YAG laser and incorporating a 50mm
length of PPLN, the ring cavity based OPO exhibits a threshold of < 4W with only the
signal wave resonant. At pump powers approaching 10W, pump depletions of the order of
90% are obtained, with close to 3W of power being coupled out in the non-resonant idler
wave. (The overall efficiency for generation of useful output at the signal wavelength is
much less since the signal wave is resonant in a cavity with low output coupling in order
to keep the threshold down, and hence is subject to high parasitic losses). By varying the
period of the poling in the PPLN crystal, from 28 to 29.5μm, an idler tuning range from
3.2 to 4.1μm has been reported.

In general, single longitudinal mode operation on the resonant signal wave is obtained
without any tuning elements being required in the cavity. Under these conditions the
linewidth of the idler wave reflects the linewidth of the pump laser, in this particular case
2GHz. By the incorporation of an intracavity etalon, it proved possible to tune the signal
wave over 150 GHz without mode hops. The use of a single frequency pump laser would
result in single frequency idler output, which would then reflect the above tuning range
associated with the signal. Much progress continues to be made in the development of the
singly resonant OPO based on PPLN, for example pumping at 532nm using a frequency-
doubled, diode-laser-pumped Nd:YAG has been reported (Batchko *et al.* 1998).

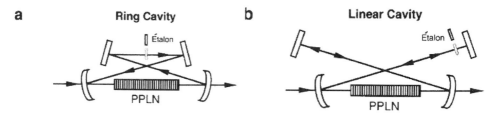

Figure 8. *Singly-resonant cw optical parametric oscillators based on periodically-poled
lithium niobate due to Bosenberg et al. 1996. (Reproduced with permission)*

With further progress in cavity engineering, in order to achieve reliable widely-tunable,
single-frequency operation, the singly-resonant PPLN OPO looks promising as a valuable
spectroscopic source for the mid infrared in the near future. However, it still exhibits
a comparatively high oscillation threshold, which is demanding on the pump laser with
regard to the power required. Furthermore the pump laser must operate on a single
frequency, if single frequency output on the non-resonant wave is required.

5 Doubly-resonant optical parametric oscillators.

The doubly-resonant oscillator is much less demanding with regard to the pump power required to reach oscillation threshold since the nonlinear conversion processes are enhanced by the presence of two intense fields within the nonlinear medium. For example, if the cavity loss associated with the second resonated wave is around 2% per round trip, then the threshold is reduced by a factor of the order of 100. In the case of the singly-resonant PPLN OPO discussed above, this would correspond to a reduction in the threshold from 4W to 40 mW.

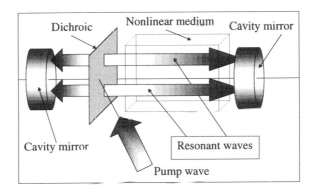

Figure 9. *A doubly-resonant optical parametric oscillator. Both of the down-converted waves are resonated within the cavity.*

If the two cavities (i.e that for the signal wave and that for the idler wave) are defined by common mirrors (see Figure 9), then the device is over constrained, and oscillation is not always possible for any arbitrary mirror separation. That this is the case may be seen as follows. Both signal and idler waves must be simultaneously resonant. In the case of cavities defined by the same pair of mirrors this means that there must be an integral number of half wavelengths between these two mirrors for both the signal and idler waves simultaneously. In addition the signal and idler frequencies must add up to equal the pump frequency. Suppose that these three constraints are fulfilled for a particular mirror spacing and that the device is oscillating. If it is now required to tune the device smoothly by moving one of the cavity mirrors, then if, for example, the cavity length is increased, this will require both the signal and idler wavelengths to increase in order to maintain oscillation on the same signal/idler mode pair. However, since the sum of the frequencies of signal and idler must also stay equal to the pump frequency, it is apparent that the two conditions cannot be simultaneously satisfied. The consequence is that either oscillation ceases altogether, or the device hops to another signal/idler mode pair. What in fact happens depends on a more detailed consideration of the circumstances than that entered into here. In any event the smooth tuning behaviour required is not attained.

One option is to overcome a constraint by splitting the single cavity into twin cavities so that the signal wave and idler wave have their mode frequencies defined by separate, and hence independently translatable, mirror surfaces. The other is to lift the constraint imposed by the pump frequency by making this frequency itself tunable.

By such means smooth frequency tuning at the single frequency level becomes possible, when a particular advantage of the doubly-resonant approach becomes apparent. This is that additional frequency selective elements may not be required within the resonant cavities in order to maintain single frequency operation, the double resonance condition itself being sufficient. This may be most readily seen by reference to Figure 10. In this figure the allowed mode frequencies of the signal and idler waves, determined by the

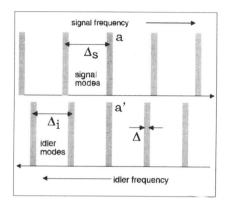

Figure 10. *Mode selection in a doubly-resonant optical parametric oscillator. Because the signal and idler modes experience different free-spectral ranges (Δ_s and Δ_i respectively), only the mode pair marked a-a' can oscillate. For strong single-frequency selection, the cavity linewidth Δ must not exceed the difference in the free spectral ranges.*

appropriate cavity resonance conditions, are plotted as a function of frequency. Since increasing frequency for the signal wave is plotted from left to right in the diagram and for the idler wave from right to left, at all points along the frequency axis, the signal and idler frequencies add up to the pump frequency (i.e. all frequencies along the axis are accessible for oscillation). In general, and importantly, the free spectral ranges of the signal and idler cavities are different. This is despite the two cavities having common mirrors and hence a common geometrical length, and arises from the different refractive indices experienced by the two waves in the crystal. Since oscillation only occurs when signal and idler waves are simultaneously in resonance, a pair of modes must line-up in the diagram (a-a' in Figure 10). The resulting displacements between the other mode pairs, due to a difference in free spectral range between signal and idler, can then ensure that only single frequency oscillation ensues. For this vernier-type effect to provide strong discrimination it is apparent that the difference in free spectral ranges must exceed the linewidths associated with the passive cavities. Further, it is important that another alignment does not occur within the phase-match bandwidth of the nonlinear medium, where competitive oscillation might take place (what are called cluster hopping effects thus ensuing). It is also apparent from the diagram that stringent requirements are placed on cavity stability in order to maintain oscillation on the selected mode pair, particularly when tuning is also involved. These issues have been discussed by Padgett *et al.* (1994), and will not be considered further here. We now review practical examples of the two approaches introduced above for the smooth tuning of the doubly-resonant device.

Figure 11 shows a dual-cavity OPO in which signal and idler fields are resonated in cavities which have a common section containing a common mirror and the nonlinear

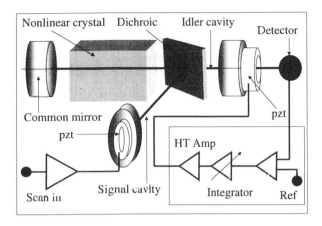

Figure 11. *Mode selection in a doubly-resonant optical parametric oscillator. Because the signal and idler modes experience different free-spectral ranges (Δ_s and Δ_i respectively), only the mode pair marked a-a' can oscillate. For strong single-frequency selection, the cavity linewidth Δ must not exceed the difference in the free spectral ranges.*

crystal and which have separate mirrors at the other end. Independent tuning of signal and idler waves is now possible so that the OPO is no longer over-constrained. By using a servo-loop applied to one of the cavity lengths to maintain the output, the OPO can now be smoothly and continuously tuned by varying the length of the other cavity. In this way we have been able to demonstrate smooth tuning over some 400 MHz (about one quarter of the free spectral range of the separate cavities)(Colville *et al.* 1994). The particular device in which we demonstrated this was based on the nonlinear material LBO pumped by a single-frequency argon laser operating at 363.8nm. The signal output was tunable over the range 502–494nm (by temperature tuning of the LBO crystal), while the idler tuned over the range 1.32–1.38μm. The pump power to reach oscillation threshold was around 100mW (LBO exhibits an effective nonlinear coefficient of 1pm/V which is low in comparison with the new quasi-phase matched nonlinear materials). When pumped at around 1W, in excess of 100mW of down-converted power was generated. A critical element in the design of such a device is the beamsplitter, which must have a low insertion loss at both the signal and idler wavelengths in order to maintain low threshold. Since type II phase matching was involved in the present arrangement (with the signal and idler waves orthogonally, linearly polarised), an enhanced Brewster plate polariser which gave greater that 99.7% efficiency at both wavelengths was used.

The alternative approach for obtaining smooth tuning of the doubly-resonant oscillator, namely the use of a tunable pump laser, has been particularly successful. In giving flexibility to the constraint that the sum of the signal and idler frequencies must be equal to the pump frequency, continuous tuning of a single frequency output can be obtained without the need for a dual-cavity geometry. The tuning range of the pump laser does not have to be particularly extensive in order to access large tuning ranges with the OPO. A requirement for the pump laser is that its tuning range is sufficient to tune the OPO over one free spectral range (typically 5–10GHz). Then by repeatedly mode hopping the

OPO to an adjacent signal/idler mode pair and re-setting the pump laser to the start of its tuning range, it is possible to end-on a series of spectral scans to attain overall an extensive tuning range (Gibson *et al.* 1998). Most recently we have been able to demonstrate the potential for continuous tuning of a single-frequency output around a pre-selected wavelength using an OPO based on the new quasi-phase-matched nonlinear material, periodically-poled KTP (PPKTP) (Gibson *et al.* 1999).

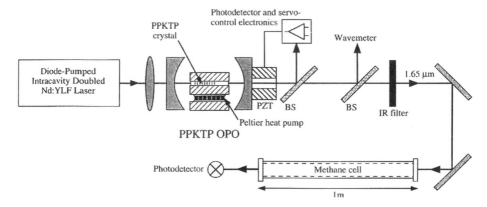

Figure 12. *A doubly-resonant OPO with a tunable pump laser. The servo-loop maintains the double-resonance condition as the pump laser frequency is tuned.*

The device is illustrated in Figure 12. The OPO was based on a 9mm long, 1mm aperture PPKTP crystal located in a 15mm cavity. The grating period of 9.55μm in the PPKTP was specifically chosen so that the idler wave of the OPO, when pumped at 523nm using an all-solid-state, diode-pumped Nd:YLF laser with an intracavity frequency-doubler, was at a wavelength close to 1.65μm, so as to access a strong absorption line in methane. The mirrors were coated for high reflectivity at the signal and idler wavelengths, save for the output coupler which had 1% transmission at the idler wavelength, since this was the required output wave. Fine temperature tuning of the PPKTP, in order to reach the exact wavelength required, was effected using a Peltier heat pump (periodically-poled KTP, unlike conventional KTP, shows temperature tuning). Considerable care was taken in the design of the cavity to ensure that the device was strongly mode selective, that is, that the device only resonated on a single signal/idler mode pair with strong discrimination against mode or cluster hopping. As has been discussed above, this requires that the cavity length change leading to a mode hop be a number of times greater than that required to keep the OPO on double resonance (determined by the linewidths of the cavity resonances). If this is the case the OPO exhibits discrete peaks in the output power as the cavity length is changed, and these provide clear features on which to lock the output of the OPO so ensuring single-frequency operation throughout the tuning cycles. A fuller consideration of these aspects is given in Padgett *et al.* (1994). The resulting device demonstrated a low pump power threshold of 25mW, and provided some 10mW of single-frequency output tunable around 1.65μm for a pump power of 200mW. The OPO cavity was locked-up using a servo control loop in order to maintain the double resonance condition as the pump laser was tuned. Smooth tuning of the idler output over 3GHz, which is the free spectral range of the OPO idler cavity, was accomplished by tuning the pump laser over about 10GHz. An additional useful feature of devices such as this is that it is possible to systematically mode-hop the cavity across the phase match bandwidth of the nonlinear medium (Gibson *et al.* 1998). The OPO can be hopped from adjacent mode to adjacent mode many times in succession. By the sequence of locking up the cavity at each mode pair in turn, scanning the OPO over just over a free spectral range, and then hopping to the next mode pair, it is possible to join up systematically a sequence of spectral scans. In this fashion, a total tuning range of some 0.3THz has been covered.

6   Pump-enhanced optical parametric oscillators.

In the pump enhanced optical parametric oscillator the cavity is made resonant at the pump wavelength as well as at either the signal or idler wavelength. (Triply resonant oscillators where all three waves are simultaneously resonant have been reported, but will not be discussed further here). In this way it is possible to greatly enhance the circulating pump intensity within the nonlinear crystal. Importantly the alignment procedures required to bring the pump field into resonance can be carried out with the OPO below threshold, and, further, since the pump field is always present, being generated by a separate source, the cavity can be readily locked-up and hence stabilised. With common mirrors defining the pump cavity and the resonant down-converted wave cavity, the OPO is then also aligned for the latter as well as the former. On increasing the pump power to above the threshold value, the OPO oscillates regardless of the mirror spacing since the device is not over constrained as is the case in the doubly-resonant oscillator. Further, stabilising the cavity length on the pump field transfers the stability of the pump laser frequency to the resonant down-converted wave which shares a common cavity. Similarly, if the pump frequency is tuned so scanning the cavity length, the resonant wave frequency tunes in step, at least until a mode hop occurs.

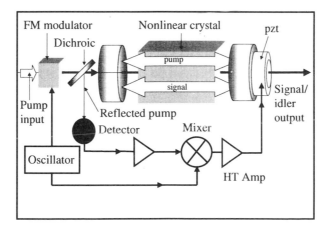

Figure 13.   *The basic principles of the pump-enhanced optical parametric oscillator. The Pound-Drever servo-loop maintains the cavity on-resonance with the single-frequency pump laser. The cavity also resonates one of the down-converted waves (in this case the signal wave)*

A generic arrangement for a pump-enhanced OPO is shown in Figure 13. The non-linear crystal is incorporated in a two mirror cavity where the left-hand-side mirror, of reflectivity r_1 and transmissivity t_1 serves as the input mirror for the pump wave. If the effective transmission of the crystal due to linear loss is t (round-trip) and due to nonlinear loss (due to parametric down-conversion) is t_{opo}, and if the reflectivity of the back mirror is r, then defining

$$r_m = t\, t_{opo} r,\tag{3}$$

the circulating power, P_c, at the pump wavelength in the resonant cavity may be written as

$$P_c = \frac{t_1 P_I}{\left(1 - \sqrt{r_1 r_m}\right)^2},\tag{4}$$

where P_I is the incident pump power. (NB: we have made the simplifying assumption that the circulating power is constant throughout the cavity on the basis that the fractional round-trip loss is small). Now at and above the OPO threshold, P_c is clamped at its value at threshold, P_{cc}, determined by the condition that the OPO gain equals OPO loss, namely

$$\kappa P_{cc} = \beta_{opo},\tag{5}$$

where β_{opo} is the round-trip linear loss of the resonated down-converted wave in the OPO, and κ is a constant depending on the effective nonlinear coefficient, the focussing arrangement of the pump, etc.

Once the incident pump power is fixed, all quantities in Equation 4 are fixed except for t_{opo}, which is hence determined by this equation. However, the down-converted power is given by

$$P_{DC} = (1 - t_{opo}) P_{cc}.\tag{6}$$

Hence we obtain

$$P_{DC} = P_{CC}\left\{1 - \frac{\left[1 - \sqrt{t_1 P_I/P_{cc}}\right]^2}{r_1 r\, t}\right\}.\tag{7}$$

If we restrict attention to the case of a perfect input mirror (where $r_1 + t_1 = 1$), then on differentiating the above equation with respect to t_1, the optimum transmission required for the pump-input mirror and the optimum down-converted power thereby generated under such conditions can be determined as

$$t_1^{opt} = P_I/P_{cc},\tag{8}$$

and

$$P_{DC}^{opt} = P_{cc}\left\{1 - \frac{[1 - (P_I/P_{cc})]}{r\, t}\right\}.\tag{9}$$

From the above it is apparent that if

$$r\, t = 1,\tag{10}$$

so that there is no parasitic loss experienced by the pump either due to leakage through the back mirror or through linear absorption within the cavity, then the down-converted power under optimum pump coupling conditions becomes equal to the input pump power. *Hence the pump-enhanced OPO can exhibit 100% efficiency for parametric down conversion (internal) if there is no parasitic linear loss of the pump wave from the cavity.*

As an example of the effects of such parasitic loss in a real system, suppose that a pump source is available that provides 300mW, and that the OPO requires 6W in the nonlinear medium in order to reach threshold. Further suppose that round-trip losses (linear) due to all causes apart from the transmission of the input mirror are 2%. Then in this case $rt = 0.98$. From Equation 8, it can be seen that the optimum transmission of the input mirror, $t_1{}^{opt}$,(assumed to be lossless) should be chosen to be 0.05 (i.e. 5% transmitting), resulting in a 20-fold enhancement of the intracavity field on-resonance and *with the OPO oscillating.* Interestingly, prior to the pump wave being of sufficient power for the OPO to reach oscillation threshold, the field enhancement is greater since there is then no nonlinear loss from the cavity. From Equation 9, it then follows that 180mW of down-converted power is generated, corresponding to an efficiency of around 60%. The remaining pump power of 120mW is lost parasitically, as may be seen by multiplying the intracavity power of 6W by the parasitic loss of 2%. Note that if this calculation is done more precisely, the sum of the down-converted pump power and the pump power lost parasitically exceeds the input pump power. The discrepancy lies in our earlier assumption of a constant intracavity pump power throughout the round trip, which is clearly only an approximation in a cavity with localised losses.

It should be noted that under the conditions discussed above the cavity is impedance matched with no back reflection of the incident pump wave.

The above expressions may readily be extended to the case where the input coupling mirror may exhibit pump-wave absorption.

Figure 13 illustrates a cw OPO with pump enhancement and based on the nonlinear crystal LBO operating in a Type-I noncritical-phase-matching configuration (Robertson *et al.* 1994). The cavity length is controlled by a servo-locking arrangement to maintain pump resonance when the device is pumped by a single-frequency argon ion laser at 514.5nm. The cavity was configured to be resonant at the signal wavelength (i.e. singly-resonant with regard to the down-converted waves), as well as providing the pump enhancement. The device reached threshold for an input pump power of 1W, and generated in excess of 1W of down-converted power in signal (946nm) and idler (1130nm) when pumped at 3W. Temperature tuning of the phase matching over 20^0C from room temperature gave a signal tuning range of 930–950nm and a corresponding idler tuning range of 1150–950nm; these ranges are proportionally extendable by increasing the temperature range. The signal wave was observed to be single frequency most of the time without the need to introduce frequency selective elements into the cavity.

More recently, a pump-enhanced cw OPO based on PPLN has been described (Schneider *et al.* 1997a), including its application to spectroscopy. Pumped by a single-frequency diode-laser-pumped Nd:YAG laser at 1064nm, and singly-resonant at the signal wavelength, an external pump power of 250mW was required to reach oscillation threshold. When pumped with 800mW, some 140mW of idler-wave power was generated. By a combination of temperature tuning and changing the period of the grating, tuning ranges of 1450–1990nm for the signal and 2290–2960nm for the idler were attained. When the cavity was servo-locked to the pump frequency, a frequency-stability in the single-frequency idler and signal waves of <10MHz/min was obtained, demonstrating the particular advantage of the pump-enhanced OPO in this respect, as discussed above. It proved possible, through tuning of the pump frequency, to tune the single-frequency idler output through in excess of 2 GHz of continuous tuning, so allowing a number of mid-infrared spectroscopy experiments to be carried out. A similar pump-enhanced OPO based on MgO:LiNbO$_3$ and pumped by a frequency-doubled Nd:YAG laser at 532.nm has also been reported (Schneider *et al.* 1997b). This device exhibited a low pump power threshold of 200mW, a phase-match tuning range of 1100–1135nm, and single-frequency output continuously tunable over 2GHz with a linewidth of < 160kHz.

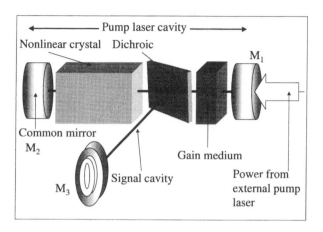

Figure 14. *The basic principles of the intra-cavity optical parametric oscillator. The dichroic beamsplitter allows the pump wave to resonate between mirrors M_2 and M_1 and the signal wave to resonate between mirrors M_2 and M_3.*

7 Intracavity CW optical parametric oscillators.

A natural extension from the pump-enhanced OPO concept is to directly access a high pump field by placing the OPO within the cavity of the pump laser itself. This results in the configuration referred to as the intracavity OPO (IC OPO). With regard to the down-converted signal and idler waves, there are the options of making the device either singly or doubly resonant (the latter, of course, showing yet further reduction in threshold). However, it was demonstrated theoretically by Oshman and Harris (1968), that the doubly-resonant intracavity OPO suffers from a number of stability problems, including both an inefficient operating regime and self-pulsing even at moderate pumping levels. In a recent analysis of the singly-resonant OPO (Turnbull *et al.* 1998), we have been able to show theoretically, and subsequently confirm experimentally, that this device in contrast operates without such instabilities or inefficiencies at all pump power levels above threshold. The generic arrangement of the intracavity OPO is shown in Figure 14. The nonlinear crystal is located within the cavity of the pump laser formed by mirrors M_1 and M_2. The beam-splitter is highly transmitting at the pump wavelength, but highly reflecting at the wavelength of the down-converted wave that is to be resonated; in the case illustrated, the signal wave. The signal wave cavity is then completed by mirror M_3, sharing a common mirror M_2 with the pump cavity.

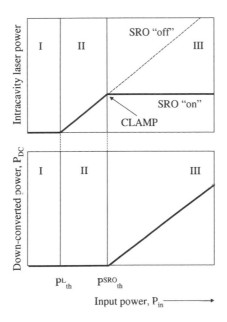

Figure 15. *Operating characteristics of the intracavity optical parametric oscillator.*

Three operating regimes of the device can be identified, as illustrated in Figure 15. In regime I, the external pump power (i.e.the pump power from another unspecified laser that is being employed to power the pump laser itself) is below that value, P_{th}^{L}, required for the pump laser itself to reach oscillation threshold . As a result there is neither an intracavity field present nor any down-converted power generated. In regime II, the external pump power is now in excess of P_{th}^{L}, and is hence sufficient to bring the pump laser into oscillation. As a result the intracavity field of the pump laser increases linearly with the external pump power. However, the strength of this intracavity field is insufficient to bring the OPO itself above threshold, so no down-converted power is generated. In regime III, the external pump power now exceeds the critical value, P_{th}^{SRO}, such that the associated intracavity pump field generated is now sufficient to bring the OPO above threshold. At this point down-converted power begins to be generated, thereafter increasing linearly with pump power, and the intracavity field is clamped at its value corresponding to the OPO threshold. This occurs because under steady-state operating conditions the OPO gain saturates to just equal the linear loss of the OPO cavity (i.e. the linear loss of the cavity associated with the resonated down-converted wave). The mechanism of saturation is the clamping of the pump field to its value at OPO threshold. This clamping is brought about by an increasing nonlinear loss associated with the OPO within the cavity of the pump laser as the external pump power is increased and the down-converted power increases. By a fuller consideration of these ideas, it may be shown that the linear increase in the total down-converted power, P_{DC}, (i.e. signal power plus idler power) with the external pump power, P_{in} , is given by

$$P_{DC} = \sigma_{max} \left(P_{in} - P_{th}^{SRO} \right) \left\{ 1 - (P_{th}^{L}/P_{th}^{SRO}) \right\}, \tag{11}$$

where σ_{max} is the output slope efficiency of the pump laser with optimum output coupling for the given parasitic loss (Colville *et al.* 1997, Turnbull *et al.* 1998).

The total down-converted power may be optimised by suitable choice of the parameters in the above equation. It may readily be shown that this optimisation occurs when

$$P_{\text{th}}^{\text{SRO}} = \sqrt{P_{\text{th}}^{\text{L}} P_{\text{in}}}. \tag{12}$$

In other words optimisation occurs when the external pump power required for the intracavity OPO to reach threshold is equal to the geometric mean of the external pump power for the pump laser itself to reach threshold and the available external pump power at which it is required to operate the OPO. Under such conditions it may readily be shown that the total down-converted power thereby generated is equal to that power that could have been coupled out of the pump laser itself under conditions of optimal (linear) output coupling. *Hence the intracavity OPO has the potential for being 100% efficient with regard to the down-converted power that it generates.*

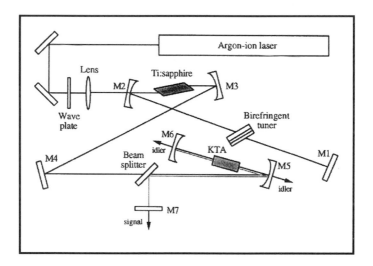

Figure 16. *Intracavity optical parametric oscillator based on a Ti:sapphire laser, with KTA as the nonlinear medium.*

We have been able to demonstrate almost this ideal efficiency in a practical device in which KTA was used as the nonlinear crystal within the cavity of a Ti:sapphire laser (Edwards *et al.* 1998a). An argon ion laser was used to pump the Ti:sapphire laser (external pump power). The arrangement is shown in Figure 16. The KTA crystal used was of length 11.5mm and was cut for Type-II noncritical-phase-matching for propagation along the optical x-axis, with anti-reflection coated end faces for both the pump and resonated signal wave. Figure 17 shows the operating characteristics of the device. The Ti:sapphire laser itself reaches threshold at an external pump power from the argon ion laser of about 0.8W. The intracavity power within the Ti:sapphire laser, which is the pump power for the OPO itself, then increases linearly with external pump power until the OPO reaches oscillation threshold at an intracavity pump power of around 15W, corresponding to an external pump power of about 3W from the argon ion laser. As the external pump power is further increased, the intracavity pump power becomes clamped to the threshold

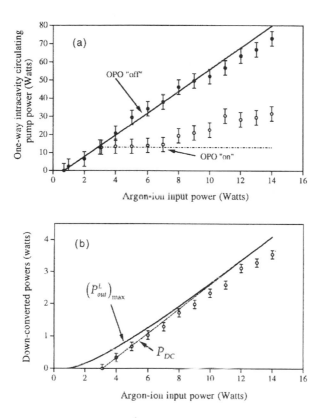

Figure 17. *Operating characteristics of a continuous-wave intracavity optical parametric oscillator based on KTA as the nonlinear medium within the cavity of a Ti:sapphire laser*

value, in contrast to the linear increase it continues to show with increasing external pump power if the OPO is maintained in the 'off' condition. At the same time the down-converted power rises linearly with the external pump power. In the lower figure this experimentally-observed down-converted power is plotted as a function of the external pump power (circles), along with the output power expected from the Ti:sapphire laser at the pump wavelength if it were operated under conditions of optimal output coupling, $(P^L_{out})_{max}$ (solid-line), and the down-converted power expected to be generated by the OPO according to Equation (10) above. As predicted by Equation (11), these latter two quantities are the same at an external pump power of 12W. It may be seen from the figure that the observed down-converted power closely approaches the 100% efficiency value expected. This behaviour is more clearly illustrated in Figure 18, where the down-conversion efficiency, defined as $P_{DC}/(P^L_{out})_{max}$, is plotted as a function of the external pump power. Down-conversion efficiencies of the order of 90% are approached.

Figure 19 illustrates the tuning behaviour of the intracavity OPO as the Ti:sapphire laser is tuned across part of its gain bandwidth (760nm to 860nm) and where the external pump power from the argon ion laser is 14W. Both signal and idler powers simultaneously remain in excess of 0.5W for tuning ranges of 1120–1200nm and 2450–2850nm respectively.

In this particular device the pump wave (intracavity field of the Ti:sapphire laser) was

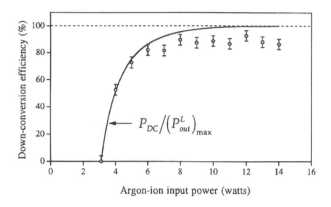

Figure 18. *Down-conversion efficiency of the KTA intracavity OPO.*

multi-longitudinal-mode (20GHz), while the resonated signal wave was generally single-frequency, with the idler wave hence acquiring the linewidth of the pump wave. A ring version (travelling-wave) of both the Ti:sapphire and OPO cavities is currently under investigation, so that through attaining single-frequency oscillation of the pump wave, the idler wave also becomes single-frequency.

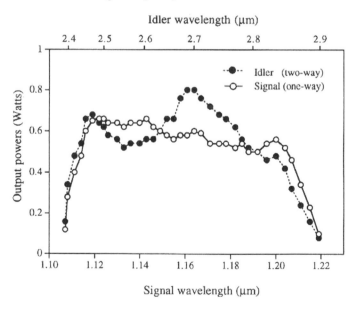

Figure 19. *Signal and idler tuning ranges of the KTA intracavity OPO.*

The use of a traditional birefringent material in the intracavity OPO highlights an important advantage of this approach. The threshold of the singly-resonant OPO based on this nonlinear material when operated outside of the pump laser cavity is prohibitively high, being of the order of 15W (see above). On the other hand this threshold can be exceeded when the device is operated intracavity with only 3W of pump power from the primary pump laser (the argon ion laser).

Intracavity OPOs within Ti:sapphire lasers based on KTP (Colville *et al.* 1997), PPLN (Turnbull *et al.* 1997), PPRTA (Edwards *et al.* 1998b), and PPKTP (Ebrahimzadeh *et al.* 1999) in addition to the KTA described above have now been demonstrated. Performance characteristics with regard to output powers and both demonstrated and anticipated tuning ranges are summarised in Table 1. In addition we have recently demonstrated devices in which the OPO is placed within the cavity of cw Nd lasers pumped by laser diodes, and in which either KTA (Colville *et al.* 1998) or PPLN (Stothard *et al.* 1998) is used as the nonlinear medium. In the former case output powers of the order of 1W in the idler wave at 3.5μm have been demonstrated when pumping the Nd:vanadate gain medium with 8W of external pump power from a diode laser. In the latter case, a 1W diode-laser also pumping Nd:vanadate as the gain medium sufficed to generate in excess of 70mW in an idler wave that was tunable over 3 to 4μm (temperature/grating-period tuning), in a compact device. A review of the current progress with regard to the development of intracavity OPOs is to be found in Ebrahimzadeh *et al.* (1999).

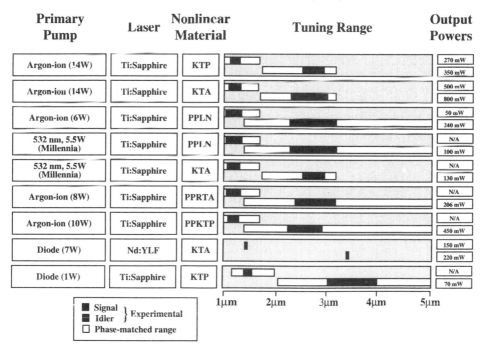

Table 1. *Characteristics of a number of different intracavity parametric oscillators*

Acknowledgements

Much of the work described in this chapter has been carried out within the Nonlinear Optics Group in the University of St Andrews, and we would like to acknowledge the contributions of the other members of the group, both past and present, including: F G Colville, T Edwards, G Gibson, A J Henderson, I Lindsay, M J Padgett, G Robertson, D Stothard, G Turnbull and J Zhang,

References

Batchko R G, Weise D R, Plettner T, Miller G D, Fejer M M and Byer R L, 1998, "Continuous-wave 532nm pumped singly resonant optical parametric oscillator based on periodically poled lithium niobate" *Opt. Lett.* **23** 168.

Boon-Engering J M, van der Veer W E, Gerritsen J W and Hogervorst W, 1995, "Bandwidth studies of an injection-seeded beta-barium borate OPO" *Opt. Lett.* **20** 380.

Bosenberg W R and Guyer D R, 1993, "Broadly tunable single frequency optical parametric frequency conversion system" *J. Opt. Soc. Am. B* **10** 1716.

Bosenberg W R, Drobshoff A, Alexander J I, Myers L E and Byer R L, 1996, "93% pump depletion, 3.5W continuous-wave, singly resonant optical parametric oscillator based on periodically poled lithium niobate" *Opt. Lett.* **21** 1336.

Colville F G, Padgett M J and Dunn M H, 1994, "Continuous-wave, dual-cavity, doubly-resonant. optical parametric oscillator" *Appl. Phys. Lett.* **64** 1490.

Colville F G, Dunn M H and Ebrahimzadeh M, 1997, "Continuous-wave, singly-resonant, intra-cavity parametric oscillator" *Opt. Lett.* **22** 75.

Colville F G, McGuckin B T, Dennis R, Ebrahimzadeh M and Dunn M H, 1998, *CLEO/Europe Technical Digest* p138 (Paper JTuA5).

Ebrahimzadeh M, Turnbull E A, Edwards T J, StothardD J M, Lindsay I D and Dunn M H, 1999, "Intracavity continuous-wave singly resonant optical parametric oscillators" *J. Opt. Soc. Am.* **3** Special Issue on Optical Parametric Devices, to be published.

Edwards T J, Turnbull G A, Dunn M H and Ebrahimzadeh M, 1998a, "High-power,continuous-wave, singly resonant, intracavity optical parametric oscillator" *Appl. Phys. Lett.* **72** 1527.

Edwards T J, Turnbull G A, Dunn M H, Ebrahimzadeh M, Karlsson H, Arvidsson G and Laurell F, 1998b, " Continuous-wave singly-resonant optical parametric oscillator based on periodically poled RbTiOAsO$_4$" *Opt. Lett.* **23** 837.

Gibson G M, Dunn M H and Padgett M J, 1998, "Application of a continuously tunable, cw optical parametric oscillator for high-resolution spectroscopy" *Opt. Lett.* **23** 40.

Gibson G M, Ebrahimzadeh M, Padgett M J and Dunn M H, 1999, "Continuous-wave optical parametric oscillator based on periodically poled KTiOPO$_4$ and its application to spectroscopy" *Opt. Lett.* **24** 397

Oshman M K and Harris S E, 1968, "Theory of optical parametric oscillation internal to the laser cavity" *IEEE J. Quantum Electron.* **4** 491.

Padgett M J, Colville F G and Dunn M H, 1994, "Mode selection in doubly-resonant optical parametric oscillators" *IEEE J. Quantum Electron.* **30** 2979.

Robertson G, Padgett M J and Dunn M H, 1994, "Continuous-wave singly resonant pump-enhanced Type II LiB$_3$O$_5$optical parametric oscillator" *Opt. Lett.* **19** 1735.

Schneider K, Kramper P, Schiller S and Mylynek J, 1997, "Towards an optical sythesiser;A single frequency parametric oscillator using periodically poled LiNbO$_3$" *Opt. Lett.* **22** 1293.

Schneider K and Schiller S, 1997, "Narrow line-width pump enhanced singly resonant parametric oscillator pumped at 532nm" *Appl. Phys. B* **65** 775.

Stothard D J M, Ebrahimzadeh M and Dunn M H, 1998, " Low pump threshold continuous-wave singly resonant optical parametric oscillator" *Opt. Lett.* **23** 1895.

Turnbull G A, Edwards T J, Dunn M H and Ebrahimzadeh M, 1997, "Continuous-wave singly-resonant intra-cavity optical parametric oscillator based on periodically poled LiNbO$_3$" *Electron. Lett.* **33** 1817.

Turnbull G A, Dunn M H and Ebrahimzadeh M, 1998, Continuous-wave intracavity optical parametric oscillator; an analysis of power characteristics" *Appl. Phys. B* **66** 701.

Ultrashort pulse generation

Ursula Keller

Swiss Federal Institute of Technologies

1 Introduction

Since 1990 we have observed tremendous progress in short and ultrashort pulse generation using solid-state lasers (Figure 1). Until the end of the 1980s, ultrashort pulse generation was dominated by dye lasers (Shank 1988) which produced pulses as short as 27fs with a typical average output power and repetition rate of about 10mW and 100MHz respectively. Shorter pulse durations, down to 6fs, were achieved only through additional amplification and fibre-grating pulse compression at much lower repetition rates. Flashlamp-pumped solid-state lasers typically produced pulse durations of ≈100ps with Nd:YAG and ≈30ps with Nd:YLF using active modelocking (Figure 2).

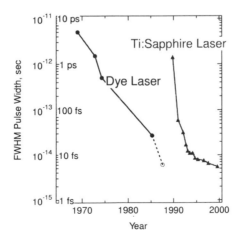

Figure 1. *History of ultrashort pulse generation*

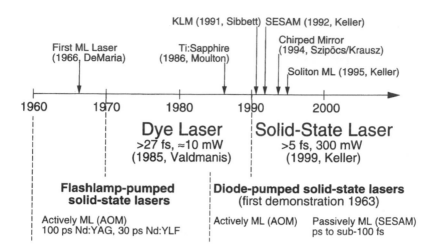

Figure 2. *Historical overview of modelocking.*

The development of higher average-power diode lasers in the 1980s stimulated a strong interest in diode-pumped solid-state lasers. Diode laser pumping provides dramatic improvements in efficiency, lifetime, size and other important laser characteristics. For example, actively modelocked diode-pumped Nd:YLF lasers generated sub-10ps pulses for the first time (Weingarten *et al.* 1990). Before 1990 however, all attempts to passively modelock solid-state lasers with long upper state lifetimes (*i.e.* μs to ms) resulted in Q-switching instabilities which at best produced stable modelocked pulses within longer Q-switched macropulses (*i.e.* Q-switched modelocking). In Q-switched modelocking, the modelocked pico- or femtosecond pulses are inside much longer Q-switched pulse envelopes (typically in the μs-regime) at much lower repetition rates (typically in the kHz-regime). The general understanding in the scientific community was that the theory of Q-switching and modelocking was well developed and mature. No big surprises were expected.

This situation changed drastically with the commercialisation of the Ti:sapphire laser, which was the first solid-state laser that was able to support ultrashort pulses without cryogenic cooling. However, the existing modelocking techniques were inadequate because of the much longer upper state lifetime of Ti:sapphire (*i.e.* in the μs regime) compared to dyes (*i.e.* in the ns regime). Therefore, passive pulse generation techniques had to be re-evaluated with new laser material properties in mind. The strong interest in an all-solid-state ultrafast laser technology was the driving force and formed the basis for many new inventions and discoveries. This has resulted in ultrashort pulses in the two-cycle regime and in practical compact all-solid-state ultrafast lasers. Today, passively modelocked Ti:sapphire lasers can produce pulses of about 5fs, shorter and more intense than ever previously generated directly from a laser (Sutter *et al.* 1999a and b, Morgner *et al.* 1999). As with dye lasers, shorter pulses down to about 4.5 to 5fs have been achieved only with additional amplification and pulse compression at kHz (Nisoli *et al.* 1997) and MHz (Baltuska *et al.* 1997) pulse repetition rates or with optical parametric amplifiers (Shirakawa *et al.* 1999). Furthermore, reliable and compact all-solid-state pulsed lasers are available now with pulse durations ranging from picoseconds to sub-100fs (Table 1). In addition, very compact passively Q-switched diode-pumped solid-state lasers can generate pulses in

Laser	Pulse	ML	Reference
diode pumped lasers			
Nd:YAG	6.8ps*	—	Appl. Phys. B 58, 347,1993
@1.06 μm	8.7ps	—	Opt.Lett.18, 640,1993
			Appl.Phys.B 58,347,1993
Nd:YLF	3.3ps*	—	Opt. Lett. 17, 505, 1992
@1.06μm	5.1ps	—	Opt.Lett.18, 640,1993
			Appl.Phys.B 58,347,1993
Nd:YLF	5.7ps	—	Opt. Lett. 21, 1378, 1996
Nd:YVO$_4$@1.3μm	4.6ps	—	Opt. Lett. 21, 1378, 1996
Nd:LSB@1.06μm	2.8ps	—	Opt. Lett. 21, 1378, 1996
ultra-short pulsed lasers			
Nd:glass	60fs, 80mW	SESAM	Opt. Lett. 22, 307, 1997
	175fs, 1W	SESAM	Opt. Lett. 23, 271, 1998
Yb:YAG	540fs, 150mW*	SESAM	Opt. Lett. 20, 2402, 1995
	340fs*	SESAM	IEEE JSTQE 2, 435, 1996
Yb:glass	160fs, 250mW*	KLM	Opt. Lett. 22, 408, 1997
	58fs, 65mW	SESAM	Opt. Lett. 23,126, 1998
Cr:LiSAF	20fs, 1mW	KLM	Appl. Phys. B 65, 227,1997
	45fs, 60mW	SESAM	Opt. Lett. 22, 621, 1997
	110fs, 500mW	SESAM	Appl. Phys. B 65, 235,1997
Cr:LiSGAF	100fs, 40mW	SESAM	IEEE JSTQE 2, 465, 1996
	14fs *	KLM	CLEO Pacific Rim 97

Table 1. *Summary of diode-pumped ps and solid-state fs lasers. ML: modelocking mechanism; KLM: Kerr lens modelocking; SESAM: soliton modelocking. (* denotes a diffraction-limited pump source such as Ti:sapphire or Krypton ion laser.)*

the nanosecond regime to as short as 37ps (Spűhler *et al.* 1999) This new all-solid-state ultrafast laser technology is based on diode-pumped solid-state lasers and solid-state saturable absorbers (Keller *et al.* 1992, 1996, 1999). Stable self-starting passive modelocked solid-state lasers using intracavity saturable absorbers have been demonstrated.

A typical set-up of a femtosecond diode-pumped solid-state laser is shown in Figure 3 (Aus der Au *et al.* 1997). The gain material (e.g. Nd:glass) is inserted at Brewster's angle and is cw pumped by two high-power high-brightness diode laser arrays. An intracavity saturable absorber, in this case a semiconductor saturable absorber mirror (SESAM—see Keller *et al.* 1992, 1996, 1999 and Section 3), was used for passive modelocking and an intracavity prism pair for dispersion compensation (Fork *et al.* 1984) (see Section 4). For picosecond pulse generation no additional dispersion compensation is required.

The goal of this Chapter is to provide a short overview of the physics of these new all-solid-state ultrafast lasers. For detailed derivations we refer the interested reader to the references provided. We restrict the following sections to pulse generation directly from a laser. Further pulse compression, frequency conversion, high-harmonic generation, optical parametric oscillators and amplifiers will not be discussed here.

We note in passing that numerous applications require ultrashort pulses with higher energy than can be obtained from an oscillator. In such cases amplification is required.

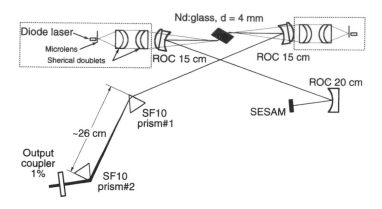

Figure 3. *Typical cavity set-up of a diode-pumped femtosecond solid-state laser (in this case a Nd:glass laser). The prism pairs are used for dispersion compensation in femtosecond lasers.*

A technique that has proved extremely successful for this purpose is called chirped-pulse amplification (CPA) (Strickland and Mourou 1985). The concept is that the ultrashort pulses from the oscillator are temporarily stretched, usually using a pair of suitably oriented diffraction gratings. This can typically expand the pulse duration by about four orders of magnitude, (e.g. from 100fs to 1ns). These stretched pulses are then amplified at an intensity level that is not hazardous to the amplifier medium (by avoiding self-focusing, for example). They can then be compressed temporally (using a grating pair) back to near their original duration. The pulses are therefore frequency-chirped then amplified and subsequently De-chirped. This explains the terminology for this chirped-pulse amplification scheme.

This amplification process can be carried out regeneratively at kilohertz repetition rates if the pulse energies are in the microjoule regime (*i.e.* peak optical powers of gigawatts). If Joule-level pulses with terawatt peak powers are required the amplification is normally carried out at a lower repetition rate of a few Hertz. Exceptionally, femtosecond pulses with peak powers as high as the petawatt regime have become available using the larger laser facilities (Perry *et al.* 1996). The peak intensity of such pulses means that they can be exploited in a wide range of applications in nuclear physics, astrophysics and nonlinear quantum electrodynamics (Umstadter *et al.* 1998).

2 Physics of pulse generation

2.1 Haus's master equation

For ultrashort pulse generation we rely on cw modelocking, where many axial laser modes are locked together in a well defined phase relationship to form a short pulse. A homogeneously broadened laser normally tends to oscillate in only one axial mode when far above the laser threshold. Additional energy has to be transferred to adjacent modes by either (a) active loss or phase modulators or (b) passive self-amplitude modulation (SAM). For reliable passive cw modelocking, the pulse generation starts from normal laser noise within less than about 1ms. The required SAM is typically obtained with intracavity saturable absorbers.

The physics of laser pulse generation is very well explained with Haus's master equation formalism (Haus 1984). This is based on linearised differential operators that describe the temporal evolution of a pulse envelope inside the laser cavity. At steady-state we then obtain the differential equation:

$$T_R \frac{\partial A(T,t)}{\partial T} = \sum_i \Delta A_i = 0 \tag{1}$$

where A is the pulse envelope, T_R is the cavity round-trip time, T is the time that develops on a time scale of the order of T_R, t is the fast time of the order of the pulse duration, and ΔA_i are the changes of the pulse envelope due to various elements in the cavity (such as gain, loss modulator or saturable absorber, dispersion etc.) (Table 2). Equation 1

Gain	$\Delta A \approx \left[g + D_g \frac{\partial^2}{\partial t^2} \right] A, \quad D_g \equiv \frac{g}{\Omega_g^2}$		Eq. 10
Constant loss	$\Delta A \approx -lA$		
Loss modulator	$\Delta A \approx -M_s t^2 A$	$M_s \equiv \frac{M\omega_m^2}{2}$	Eqs. 11, 12
Fast saturable absorber	$\Delta A \approx \gamma \|A\|^2 A$	$\gamma \equiv \frac{q_0}{I_{sat,A}}$	Eqs. 13, 14
2nd order dispersion	$\Delta A \approx iD \frac{\partial^2}{\partial t^2} A$	$D \equiv \frac{1}{2} k_n'' z$	Eqs. 18, 19
SPM	$\Delta A \approx -i\delta \|A\|^2 A$	$\delta \equiv k n_2 z$	Eqs. 22, 25

Table 2. *Summary of linearised operators that model the change in the pulse envelope for each element in the laser cavity.*

basically means that the changes in the pulse envelope after one laser round-trip time have to equal the sum of all the (small) changes due to the various elements in the cavity. Each element is modelled as a linearised operator; this will be discussed in more detail below. At steady state all these changes of the pulse envelope have to add up to zero after one cavity round-trip.

The pulse envelope is normalised such that $|A(t)|^2$ is the pulse intensity $I(t)$:

$$E(z,t) = \frac{\sqrt{2}}{\sqrt{c_n \varepsilon \varepsilon_0}} A(z,t) e^{i[\omega_0 t - k_n(\omega_0)z]} \quad \Rightarrow \quad |A(z,t)|^2 = I(z,t) \tag{2}$$

where $E(z,t)$ is the electric field, ε_0 is the electric permittivity of free space, ε determines the refractive index $n = \sqrt{\varepsilon}$, $c_n = c/n$ with c the vacuum light velocity, ω_0 is the centre frequency of the pulse spectrum and $k_n = nk$ with $k = 2\pi/\lambda$ the wave vector and λ the vacuum wavelength. Before we discuss the different modelocking models we briefly discuss the linearised operators for the differential equations.

2.1.1   Gain

A homogeneously broadened gain medium is described by a Lorentzian lineshape for which the frequency dependent gain coefficient $g\left(\omega\right)$ is given by

$$g\left(\omega\right) = \frac{g}{1 + \left(\dfrac{\omega - \omega_0}{\Omega_g}\right)^2} \approx g\left(1 - \frac{\Delta\omega^2}{\Omega_g^2}\right), \quad \text{for} \quad \left(\omega - \omega_0/\Omega_g\right)^2 \ll 1 \qquad (3)$$

where $\Delta\omega = \omega - \omega_0$, g is the saturated gain coefficient for a cavity round trip and Ω_g is the HWHM (half width half maximum) of the gain bandwidth in radians. In the frequency domain the pulse envelope after the gain medium is given by

$$\tilde{A}_{\text{out}}\left(\omega\right) = e^{g\left(\omega\right)}\tilde{A}_{\text{in}}\left(\omega\right) \approx \left[1 + g\left(\omega\right)\right]\tilde{A}_{\text{in}}\left(\omega\right), \quad \text{for} \quad g \ll 1 \qquad (4)$$

where $\tilde{A}\left(\omega\right)$ is the Fourier transform of $A\left(t\right)$. Equations 3 and 4 then give

$$\tilde{A}_{\text{out}}\left(\omega\right) = \left[1 + g - \frac{g}{\Omega_g^2}\Delta\omega^2\right]\tilde{A}_{\text{in}}\left(\omega\right) \quad \Rightarrow \quad A_{\text{out}}\left(t\right) = \left[1 + g + \frac{g}{\Omega_g^2}\frac{\partial^2}{\partial t^2}\right]A_{\text{in}}\left(t\right) \qquad (5)$$

where we used the knowledge that a factor of $\Delta\omega$ in the frequency domain produces a time derivative in the time domain. For example for the electric field we obtain:

$$\frac{\partial}{\partial t}E\left(t\right) = \frac{\partial}{\partial t}\left\{\frac{1}{2\pi}\int\tilde{E}\left(\omega\right)e^{i\omega t}d\omega\right\} = \frac{1}{2\pi}\int\tilde{E}\left(\omega\right)i\omega e^{i\omega t}d\omega \qquad (6)$$

and

$$\frac{\partial^2}{\partial t^2}E\left(t\right) = \frac{\partial^2}{\partial t^2}\left\{\frac{1}{2\pi}\int\tilde{E}\left(\omega\right)e^{i\omega t}d\omega\right\} = \frac{1}{2\pi}\int\tilde{E}\left(\omega\right)\left[-\omega^2\right]e^{i\omega t}d\omega \qquad (7)$$

and similarly for the pulse envelope:

$$\frac{\partial}{\partial t}A\left(t\right) = \frac{\partial}{\partial t}\left\{\frac{1}{2\pi}\int\tilde{A}\left(\omega\right)e^{i\Delta\omega t}d\omega\right\} = \frac{1}{2\pi}\int\tilde{A}\left(\omega\right)i\Delta\omega e^{i\Delta\omega t}d\omega \qquad (8)$$

and

$$\frac{\partial^2}{\partial t^2}A\left(t\right) = \frac{\partial^2}{\partial t^2}\left\{\frac{1}{2\pi}\int\tilde{A}\left(\omega\right)e^{i\Delta\omega t}d\omega\right\} = \frac{1}{2\pi}\int\tilde{A}\left(\omega\right)\left[-(\Delta\omega)^2\right]e^{i\Delta\omega t}d\omega. \qquad (9)$$

For the change in the pulse envelope $\Delta A = A_{\text{out}} - A_{\text{in}}$ after the gain medium we then obtain:

$$\Delta A \approx \left[g + D_g\frac{\partial^2}{\partial t^2}\right]A, \qquad D_g \equiv \frac{g}{\Omega_g^2} \qquad (10)$$

where D_g is the gain dispersion.

2.1.2   Loss modulator

A loss modulator inside a modelocked laser cavity is often an acousto-optic modulator and produces a sinusoidal loss modulation given by a time dependent loss coefficient:

$$l\left(t\right) = M\left(1 - \cos\omega_m t\right) \approx M_s t^2, \qquad M_s \equiv \frac{M\omega_m^2}{2} \qquad (11)$$

where M_s is the curvature of the loss modulation, $2M$ is the peak modulation depth and ω_m is the modulation frequency which, in fundamental modelocking will correspond to the axial mode spacing. In fundamental modelocking we have only one pulse per cavity round trip. The change in the pulse envelope is then given by

$$A_{\text{out}}(t) = e^{-l(t)} A_{\text{in}}(t) \approx [1 - l(t)] A_{\text{in}}(t) \quad \Rightarrow \quad \Delta A \approx -M_s t^2 A. \tag{12}$$

2.1.3 Fast saturable absorber

In the case of an ideal fast saturable absorber we assume that the loss q recovers instantaneously and therefore shows the same time dependence as the pulse envelope. We assume:

$$q(t) = \frac{q_0}{1 + I(t)/I_{\text{sat,A}}} \approx q_0 - \gamma I(t), \qquad \gamma \equiv \frac{q_0}{I_{\text{sat,A}}}. \tag{13}$$

The change in the pulse envelope is then given by

$$A_{\text{out}}(t) = e^{-q(t)} A_{\text{in}}(t) \approx [1 - q(t)] A_{\text{in}}(t) \quad \Rightarrow \quad \Delta A \approx \gamma |A|^2 A. \tag{14}$$

2.1.4 Group velocity dispersion (GVD)

The frequency dependent wavevector $k_n(\omega)$ in a dispersive material depends on the frequency and can be approximately written as:

$$k_n(\omega) \approx k_n(\omega_0) + k'_n \Delta\omega + \frac{1}{2} k''_n \Delta\omega^2 + \cdots \tag{15}$$

where

$$\Delta\omega = \omega - \omega_0, \quad k'_n = \left.\frac{\partial k_n}{\partial \omega}\right|_{\omega=\omega_0} \quad \text{and} \quad k''_n = \left.\frac{\partial^2 k_n}{\partial \omega^2}\right|_{\omega=\omega_0}.$$

In the frequency domain the pulse envelope in a dispersive medium after a propagation distance of z is given by:

$$\tilde{A}(z,\omega) = e^{-i[k_n(\omega) - k_n(\omega_0)]z} \tilde{A}(0,\omega) \approx \left\{1 - i[k_n(\omega) - k_n(\omega_0)]z\right\} \tilde{A}(0,\omega). \tag{16}$$

Taking into account only the first order and second order dispersion terms in Equation 15 we obtain:

$$\tilde{A}(z,\omega) \approx \left[1 - ik'_n \Delta\omega z - i\frac{1}{2} k''_n \Delta\omega^2 z\right] \tilde{A}(0,\omega). \tag{17}$$

The linear term in $\Delta\omega$ determines the propagation velocity of the pulse envelope (*i.e.* the group velocity v_g) and the quadratic term in $\Delta\omega$ determines how the pulse envelope gets deformed due to second order dispersion. The influence of higher order dispersion can be considered with more terms in the expansion of $k_n(\omega)$ (Equation 15). However, higher order dispersion become important only for ultrashort pulse generation with pulse durations below approximately 20fs.

Normally we are only interested in the changes of the pulse envelope and therefore it is useful to restrict our observation to a reference system that is moving with the pulse envelope. In this reference system we need only consider second and higher order dispersion. In the time domain we then obtain for second order dispersion:

$$A\left(z,t\right) \approx \left[1 + iD\frac{\partial^2}{\partial t^2}\right] A\left(0,t\right), \qquad D \equiv \frac{1}{2}k_n'' z \tag{18}$$

where D is the dispersion parameter. Therefore we obtain for the change in the pulse envelope:

$$\Delta A \approx iD\frac{\partial^2}{\partial t^2} A. \tag{19}$$

2.1.5 Self-phase modulation

The Kerr effect introduces a space and time dependent refractive index:

$$n\left(r,t\right) = n + n_2 I\left(r,t\right) \tag{20}$$

where n is the linear refractive index, n_2 is the nonlinear refractive index coefficient and $I\left(r,t\right)$ is the intensity of the laser beam which typically has a Gaussian beam profile. For laser host materials, n_2 is typically of the order of $10^{-16}\text{cm}^2/W$ and does not change very much for different materials. For example, for sapphire $n_2 = 3 \times 10^{-16}\text{cm}^2/W$, fused quartz $n_2 = 2.46 \times 10^{-16}\text{cm}^2/W$, Schott glass LG-760 $n_2 = 2.9 \times 10^{-16}\text{cm}^2/W$, YAG $n_2 = 6.2 \times 10^{-16}\text{cm}^2/W$, and YLF $n_2 = 1.72 \times 10^{-16}\text{cm}^2/W$. The nonlinear refractive index produces a nonlinear phase shift during pulse propagation:

$$\phi\left(z,r,t\right) = -k\,n\left(r,t\right) z = -k\left[n + n_2 I\left(r,t\right)\right] z = -k\,nz - \delta\left|A\left(r,t\right)\right|^2 \tag{21}$$

where δ is the self-phase modulation (SPM) coefficient:

$$\delta \equiv k\,n_2 z. \tag{22}$$

Then the electric field during propagation is changing due to SPM:

$$E\left(z,t\right) = e^{i\phi} E\left(0,t\right) = e^{i\phi} A\left(0,t\right) e^{i\omega_0 t} = e^{-i\delta\left|A(t)\right|^2} A\left(0,t\right) e^{i\omega_0 t - ik_n(\omega_0)z}. \tag{23}$$

For $\delta\left|A\right|^2 \ll 1$ we obtain:

$$A\left(z,t\right) = e^{-i\delta\left|A(t)\right|^2} A\left(0,t\right) e^{-ik_n(\omega_0)z} \approx \left(1 - i\delta\left|A\left(t\right)\right|^2\right) A\left(0,t\right) e^{-ik_n(\omega_0)z}. \tag{24}$$

After one cavity round-trip we then obtain

$$\Delta A \approx -i\delta\left|A\right|^2 A. \tag{25}$$

2.2 Active modelocking

Short pulses from a laser can be generated with a loss or phase modulator inside the resonator. For example, the laser beam is amplitude-modulated when it passes through an acousto-optic modulator. Such a modulator can modulate the loss of the resonator at a period equal to the round-trip time T_R of the resonator (i.e. fundamental modelocking). The pulse evolution in an actively modelocked laser without self-phase modulation

(SPM) and group-velocity dispersion (GVD) can be described by Haus's master equation (Haus 1975a). Taking into account gain dispersion and loss modulation we obtain from Equations 10 and 12 the following differential equation:

$$\sum_i \Delta A_i = \left[g \left(1 + \frac{1}{\Omega_g^2} \frac{\partial^2}{\partial t^2} \right) - l - M \frac{\omega_m t^2}{2} \right] A(T,t) = 0. \tag{26}$$

Typically we obtain pulses which are much shorter than the round-trip time in the cavity and which are placed in time at the position where the modulator introduces the least amount of loss. Therefore, we were able to approximate the cosine modulation by a parabola (Equation 11). The only stable solution to this differential Equation is a Gaussian pulse shape with a pulse duration:

$$\tau_p = 1.66 \times \sqrt[4]{D_g/M_s} \tag{27}$$

where D_g is the gain dispersion (Equation 10) and M_s is the curvature of the loss modulation (Equation 11). Therefore, as an actively modelocked laser starts and moves towards steady state conditions the pulse duration becomes shorter until the pulse broadening of the gain filter is balanced by the pulse shortening of the loss modulator. Basically, the curvature of the gain is given by the gain dispersion $D_g \equiv g/\Omega_g^2$ and the curvature of the loss modulation is given by M_s. Thus, the pulse duration scales with only the fourth root of the saturated gain (*i.e.* $\tau_p \propto \sqrt[4]{g}$) and the modulation depth (*i.e.* $\tau_p \propto \sqrt[4]{1/M}$) and with the square root of the modulation frequency (*i.e.* $\tau_p \propto \sqrt{1/\omega_m}$) and the gain bandwidth (*i.e.* $\tau_p \propto \sqrt{1/\Delta\omega_g}$). A higher modulation frequency or a higher modulation depth increases the curvature of the loss modulation and a larger gain bandwidth decreases gain dispersion. Therefore, we obtain shorter pulse durations in all cases. At steady-state the saturated gain is equal to the total cavity losses. Therefore, a larger output coupling will result in longer pulses. Thus, higher average output power is generally obtained at the expense of longer pulses.

The results that have been obtained in actively modelocked flashlamp-pumped Nd:YAG and Nd:YLF lasers can be very well explained by this result. For example, with Nd:YAG at a lasing wavelength of 1.064μm we have a gain bandwidth $\Delta\lambda_g = 0.45$nm . With a modulation frequency of 100MHz (*i.e.* $\omega_m = 2\pi \cdot 100$MHz), a 10% output coupler (*i.e.* $2g \approx T_{out} = 0.1$) and a modulation depth, we obtain a FWHM pulse duration of 93ps (Equation 27). For example, with Nd:YLF at a lasing wavelength of 1.047μm and a gain bandwidth $\Delta\lambda_g = 1.3$nm we obtain with the same modelocking parameters a pulse duration of 53ps (Equation 27). These pulse-widths agree with experimental results.

It has been well known for quite some time that the addition of a medium with a nonlinear refractive index to a passively or actively mode-locked laser system can lead to shorter pulses. The bandwidth limitation that results from gain dispersion can be partially overcome by the spectral broadening caused by the nonlinearity. We can extend the differential Equation 26 with the additional terms for SPM (Equation 24):

$$\sum_i \Delta A_i = \left[g \left(1 + \frac{1}{\Omega_g^2} \frac{\partial^2}{\partial t^2} \right) - l - M \frac{\omega_m t^2}{2} - i\delta |A|^2 + i\psi \right] A(T,t) = 0. \tag{28}$$

In this case, however, we have to include an additional phase shift ψ to obtain a self-consistent solution. So far, we have assumed $\psi = 0$. This phase shift is an additional degree of freedom because the boundary condition for intracavity pulses only requires that the pulse envelope is unchanged after one cavity round trip. The electric field, however, can have an arbitrary phase shift ψ after one round trip. The solution of Equation 28 is a chirped Gaussian pulse

$$A(t) = A_0 \exp\left[-\frac{1}{2}\frac{t^2}{\tau^2}(1 - ix)\right] \tag{29}$$

assuming a parabolic approximation for $|A|^2$ in Equation 28,

$$|A|^2 = |A_0|^2 e^{-t^2/\tau^2} \approx I_0\left(1 - \frac{t^2}{\tau^2}\right) \tag{30}$$

with the chirp parameter

$$x = \frac{\tau^2 \phi_0}{2D_g} \tag{31}$$

and the FWHM pulse duration

$$\tau_p = 1.66\tau = 1.66 \times \left(\frac{D_g}{M_s + \phi_0^2/4D_g}\right)^{1/4} \tag{32}$$

where ϕ_0 is the nonlinear phase shift per cavity round-trip:

$$\phi_0 = 2kL_g n_2 I_{0,g}. \tag{33}$$

$2L_g$ is the length of the SPM medium in the laser resonator after one round trip and $I_{0,g}$ is the peak intensity inside the SPM medium (this is often the laser gain medium).

This analytical result explains our experiments with an actively modelocked diode-pumped Nd:YLF laser (Braun *et al.* 1995). For this example, the lasing wavelength is 1.047μm, the gain bandwidth is $\Delta\lambda_g = 1.3$nm , the pulse repetition rate is 250MHz, the output coupler has a transmission of 2.5%. The measured average output power is 620mW, the mode radius inside the 5mm long Nd:YLF crystal is 127μm x 87μm and the loss modulation of the acousto-optic modelocker is about 20%. The theoretical analysis above indicates that we should expect to obtain a FWHM pulse duration of 17.8ps (Eq.32).This agrees well with the experimentally observed pulse duration of 17ps. Without SPM we would predict a pulse duration of 33ps (Equation 27).

Equation 32 predicts that more SPM will further reduce the pulse duration. However, too much SPM will ultimately drive the laser unstable. This has been shown by the numerical simulations of Haus and Silberberg (1986) which predict that pulse shortening in an actively modelocked system is limited to a factor of approximately two in the case of SPM only. They also showed that the addition of negative GVD can undo the chirp introduced by SPM, and therefore both effects together may lead to stable pulse shortening by a factor of 2.5.

However, experimental results with fibre lasers and solid-state lasers indicate that soliton shaping in the negative GVD regime may lead to pulse stabilisation and considerable more pulse shortening. We have extended the analysis of Haus and Silberberg by investigating the possible reduction in pulse width of an actively mode-locked laser as a result of soliton-like pulse formation, *i.e.* , the presence of SPM and an excessive amount of negative GVD (Kärtner *et al.* 1995). We show, by means of soliton perturbation theory, that beyond a critical amount of negative GVD a soliton-like pulse is formed and kept stable by an active modelocker. If the bandwidth of the gain is large enough, the width of this solitary pulse can be much less than the width of a Gaussian pulse generated by the active modelocker and gain dispersion alone. We established analytically that the pulse shortening possible by addition of SPM and GVD does not have a firm limit of 2.5. These analytical results are confirmed by numerical simulations and experiments with a regeneratively actively mode-locked Nd:glass laser (Kopt *et al.* 1994). The pulse-width reduction achievable depends on the amount of negative GVD available. For an actively mode-locked Nd:glass laser a pulse shortening up to a factor of 6 may result, at which stage instabilities arise.

2.3 Passive modelocking

2.3.1 Overview

In passive modelocking the loss modulation is obtained by self-amplitude modulation (SAM), where the pulse saturates an absorber, for example. In the best case, the SAM follows the intensity profile of the pulse. This is the case of a fast saturable absorber. In this case, SAM produces a much larger curvature of loss modulation than in the sinusoidal loss modulation of active modelocking, because the modelocked pulse duration is much shorter than the cavity round-trip time. Therefore, we would expect from the previous discussion of active modelocking that we would obtain much shorter pulses with passive modelocking. This is indeed what is observed.

In passive modelocking, modelocking starts from normal noise fluctuations in a laser. One noise spike is strong enough to start to saturate the absorber which results in lower loss and therefore more gain in the round trip. Thus, this noise spike begins to grow, and becomes shorter until a stable pulse duration is obtained. However, the parameters of the saturable absorber have to be chosen to ensure that the modelocking is self-starting (*i.e.* able to start from the normal intensity noise of the laser) and stable, that is, no Q-switching instabilities occur. For example, if the loss modulation becomes too large it can drive the laser unstable. The mechanism is as follows. (see also Section 3, Equation 47). The loss saturation increases the intensity inside the laser cavity. The gain then needs to saturate more strongly to compensate for the reduced loss and to keep the intensity inside the laser cavity constant. If the gain cannot respond fast enough, the intensity continues to increase as the absorber is bleached which leads to self-Q- switching instabilities or in the best case to stable Q-switched modelocking. In the latter case, the modelocked pulse train is strongly modulated at close to the relaxation oscillation frequency of the laser (typically in the kHz range).

Passive modelocking mechanisms are well-explained by three fundamental models: slow saturable absorber modelocking with dynamic gain saturation (New 19984) (Figure 4a), fast saturable absorber modelocking (Haus 1975b) (Figure 4b) and soliton modelocking (Kärtner and Keller 1995) (Figure 4c). In the first two cases, a short net-gain window forms and stabilises an ultrashort pulse. This net-gain window also forms the minimal stability requirement, that is, the net loss immediately before and after the pulse approximately defines its extent. For solid-state lasers we can not apply slow saturable

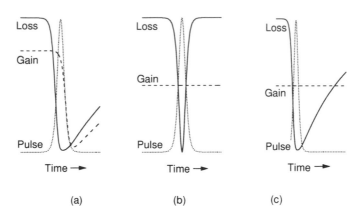

Figure 4. *Passive modelocking mechanisms explained by three fundamental models: (a) slow saturable absorber modelocking with dynamic gain saturation, (b) fast saturable absorber modelocking, (c) soliton modelocking.*

absorber modelocking as shown in Figure 4a, because no significant dynamic gain saturation is taking place as the the long upper state lifetime of the laser is long. Dynamic gain saturation means that the gain experiences a fast, pulse-induced saturation that then recovers again between consecutive pulses (Figure 4a). The upper state lifetime of solid-state lasers is typically in the microsecond to millisecond regime, much longer than the pulse repetition period which is typically in the nanosecond regime. We therefore do not observe any significant dynamic gain saturation and the gain is saturated to a constant level by the average intracavity intensity. This is not the case for dye or semiconductor lasers where the upper state lifetime is in the nanosecond regime. In Figure 4a, an ultra-short net-gain window can be formed by the combined saturation of absorber and gain for which the absorber has to saturate and recover faster than the gain, while the recovery time of the saturable absorber can be much longer than the pulse duration.

2.3.2 Fast saturable absorber (FSA) modelocking

In the fast saturable absorber model no dynamic gain saturation is required and the short net-gain window is formed by a fast recovering saturable absorber alone (Figure 4b). This was initially believed to be the only stable approach to passively modelock a solid-state laser such as Ti:sapphire.

The first demonstration of this method did not involve saturable absorption as such but intensity dependent (nonlinear) losses in the laser. Additive pulse modelocking (APM) (Blow and Nelson 1988, Kean *et al.* 1989, Ippen *et al.* 1989) with a nonlinear Kerr media, such as a fibre in a coupled cavity, was the first "fast saturable absorber" for solid-state lasers. However, APM required interferometric cavity length stabilisation because self-phase modulation in a fibre inside the coupled cavity generated a nonlinear phase shift that added constructively at the peak of the pulse in the main cavity and destructively in the wings, thus shortening the pulse duration inside the main cavity.

Kerr lens modelocking (KLM) (Spence *et al.* 1991) was the first useful demonstration of an intracavity fast saturable absorber for a solid-state laser and because of its simplicity,

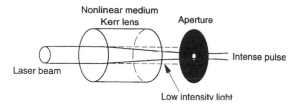

Figure 5. *Kerr lens modelocking (KLM) is very close to an ideal fast saturable absorber, where the saturable absorption is produced by the intensity dependent self-focusing through a hard aperture, for example.*

it replaced coupled cavity modelocking techniques. The principle of this scheme is shown if Figure 5. It relies on the intensity dependent refractive index of the gain (or other) medium. If the laser runs cw, there is little nonlinear effect and the broad beam is severely attenuated by an intracavity aperture. However, if the laser modelocks, the high instantaneous intensity of the laser in the gain medium will result in a significant change in refractive index through the Kerr effect. Because the beam has a (Gaussian) radially varying intensity profile, so does the change in refractive index. This refractive index profile acts as an intensity dependent lens. In short pulse operation the transmitted beam is self-focussed through the hard aperture (Keller *et al.* 1991, Salin *et al.* 1991, Negus *et al.* 1991). This mechanism is close to being an ideal fast saturable absorber with a significant modulation depth. Similar results have been seen as a result of increased gain in modelocked operation due to an increased overlap of the laser mode with the pump profile in the gain medium (Piché and Salin 1993). Only in the ultrashort pulse regime of a few optical cycles do more complicated space-time couplings occur and wavelength dependent effects start to limit further pulse reduction.

Besides the tremendous success of KLM, there are some significant limitations for practical or "real-world" ultrafast lasers. First, the cavity is typically operated near one end of its stability range, where the Kerr-lens-induced change of the beam diameter is large enough to sustain modelocking. This results in a requirement for critical cavity alignment where mirrors and laser crystal have to be positioned to an accuracy of several hundred microns typically. Once the cavity is correctly aligned, KLM can be very stable and under certain conditions even self-starting. However, KLM lasers in the sub-50fs regime have not yet been demonstrated without an additional starting mechanism. This is not surprising, since in a 10fs Ti:sapphire laser with a 100MHz repetition rate, the peak power changes by 6 orders of magnitude when the laser switches from cw to pulsed operation. Therefore, nonlinear effects that are still effective in the sub-10fs regime are typically too small to initiate modelocking in the cw-operation regime. In contrast, if self-starting is optimised, KLM effects tend to be less successful in the ultrashort pulse regime or the large self-phase modulation (SPM) will drive the laser unstable.

For an ideal fast saturable absorber, Haus *et al.* (1992) developed an analytical solution for the pulse width assuming a weakly saturated fast saturable absorber (Equation 13) which results in the Master equation:

$$\sum_i \Delta A_i = \left[g \left(1 + \frac{1}{\Omega_g^2} \frac{\partial^2}{\partial t^2} \right) - l + \gamma \left| A \right|^2 \right] A\left(T, t \right) = 0 \tag{34}$$

not taking into account GVD and SPM. The solution is an unchirped sech^2 pulse shape

$$I(t) = I_0 \operatorname{sech}^2\left(\frac{t}{\tau}\right) \tag{35}$$

with a FWHM pulse width:

$$\tau_p = 1.7627\tau = 1.7627\frac{4D_g}{\gamma F_{p,A}} \tag{36}$$

where I_0 is the peak intensity of the pulse, $F_{p,A}$ is the incident pulse energy fluence on the fast saturable absorber and D_g is the gain dispersion of the laser medium (Equation 10).

The shortest possible pulses can be obtained when we use the full modulation depth of the fast saturable absorber. We obtain an analytic solution only if we assume an ideal fast absorber that saturates linearly with the pulse intensity (Equation 13) over the full modulation depth. For a maximum modulation depth, we then can assume that $\gamma I_0 = q_0$. For a soliton shaped pulse (Equation 29), the pulse energy fluence is given by $F_{p,A} = \int I(t)dt = 2I_0\tau$. The minimum pulse width for a fully saturated ideal fast absorber then follows from Equation 36:

$$\tau_{p,\min} = \frac{1.7627}{\Omega_g}\sqrt{\frac{2g}{q_0}}. \tag{37}$$

This occurs right at the stability limit when the filter loss due to gain dispersion is equal to the residual loss a soliton undergoes in an ideal fast saturable absorber:

$$\text{Filter loss} = \frac{D_g}{3\tau^2} = \frac{q_0}{3} = q_s = \text{ residual saturable absorber loss.} \tag{38}$$

The residual saturable absorber loss q_s results from the fact that the soliton pulse initially experiences loss to fully saturate the absorber. This residual loss is exactly $q_0/3$ for a sech^2 pulse shape. This condition results in the minimum FWHM pulse duration given by Equation 37.

When GVD and SPM are also taken into account in the fast saturable absorber model, that is, in soliton formation, an additional pulse shortening of a factor of 2 was predicted. However, unchirped soliton pulses (*i.e.* ideal sech^2-shaped pulses, Equation 29) are only obtained for a certain negative dispersion value given by

$$\frac{A_L|D|}{\delta} = \frac{A_A D_g}{\gamma}. \tag{39}$$

where A_L and A_A are the laser mode areas in the gain medium and absorber respectively. This is also where we obtain the shortest pulses with a fast saturable absorber. Here we assume that the higher order dispersion terms are fully compensated or negligibly small. Computer simulations show that too much SPM can drive the laser unstable.

2.3.3   Soliton modelocking

In soliton modelocking, the pulse shaping is done solely by soliton formation, that is, the balance of group velocity dispersion (GVD) and self-phase modulation (SPM) at steady state, with no additional requirements on the cavity stability regime as for example in KLM. In contrast to KLM we use only the time dependent part of the Kerr effect at the peak intensity (*i.e.* $n(t) = n + n_2 I_0(t)$, Equation 20) and not the radially dependent part as well (*i.e.* $n(r,t) = n + n_2 I(r,t)$, Equation 20). The radially dependent part of the Kerr effect is responsible for KLM because it forms the nonlinear lens that reduces the beam diameter at an intracavity aperture (Figure 5) inside the gain medium. Thus, this nonlinear lens forms effectively a fast saturable absorber because the intensity dependent reduction of the beam diameter at an aperture introduces less loss at high intensity and more loss at low intensity. Such a radially dependent saturable absorber (e.g. KLM) couples the modelocking mechanism with the cavity mode. Therefore, soliton modelocking has the inherent advantage compared to KLM that the modelocking mechanism is decoupled from the cavity design and no critical cavity stability limit is required - it basically works over the full cavity stability regime. In soliton modelocking, an additional loss mechanism, such as a saturable absorber (Kärtner and Keller 1995b), or an acousto-optic modulator (Kärtner *et al.* 1995), is necessary to start the modelocking process and to stabilise the soliton pulse forming process. In soliton modelocking, we have shown that the net-gain window (Figure 4c) can remain open for more than 10 times the length of the ultrashort pulse, depending on the specific laser parameters (Jung *et al.* 1995a, Kärtner *et al.* 1996). This strongly relaxes the requirements on the saturable absorber and we can obtain ultrashort pulses even in the 10fs regime with semiconductor saturable absorbers that have much longer recovery times. With the same modulation depth, one can obtain almost the same minimum pulse duration as with a fast saturable absorber, as long as the absorber recovery time is less than about ten times longer than the final width of the optical pulse. In addition, autocorrelation measurements with high dynamic range have shown excellent pulse pedestal suppression over more than seven orders of magnitude in the 130fs pulses of a Nd:glass laser (Kopf *et al.* 1995). This is very similar to or even better than KLM pulses in the 10fs pulse width regime (Jung *et al.* 1997a). Even better performance can be expected if the saturable absorber also shows a negative refractive index change coupled with the absorption change as is the case for semiconductor materials (Kärtner *et al.* 1996).

We can describe soliton modelocking by Haus's master equation, where we take into account GVD, SPM and a slow saturable absorber $q(T,t)$ that recovers more slowly than the pulse duration (see Figure 4c) (Kärtner and Keller 1995, Kärtner *et al.* 1996):

$$\sum_i \Delta A_i = \left[-iD\frac{\partial^2}{\partial t^2} + i\delta |A(T,t)|^2 \right] A(T,t) + \left[g - l + D_g\frac{\partial^2}{\partial t^2} - q(T,t) \right] A(T,t) = 0.$$

(40)

This differential Equation can be solved analytically using soliton perturbation theory.

The first bracket term of this differential Equation (Equation 40) determines the nonlinear Schrödinger Equation for which the soliton pulse is a stable solution for negative GVD (*i.e.* $D < 0$) and positive SPM (*i.e.* $n_2 > 0$):

$$A(z,t) = A_0 \operatorname{sech}\left(\frac{t}{\tau}\right) e^{-i\phi(z)}.$$

(41)

The FWHM soliton pulse duration is given by $\tau_p = 1.7627\tau$ and the time-bandwidth product by $\Delta\nu_p\tau_p = 0.3148$. The phase shift of the soliton during propagation along the z-axis is given by $\phi(z)$:

$$\phi(z) = \frac{1}{2}k\,n_2 I_0 z. \tag{42}$$

Thus, the soliton pulse experiences during propagation a constant phase shift over the full beam profile in contrast to SPM alone. For a given negative dispersion and a pulse fluence F_p we obtain a pulse duration

$$\tau_p = 1.7627\frac{2\,|k_n''|}{k\,n_2 F_p} \tag{43}$$

for which the effects of SPM and GVD are balanced with a stable soliton pulse.

We treat the terms in the second bracket terms of Equation 40 as a perturbation to the soliton pulse which is a solution to the terms in the first bracket. We then predict a minimum pulse duration for a soliton-like pulse (Kärtner and Keller 1995, Kärtner *et al.* 1996):

$$\tau_{p,\mathrm{min}} = 1.7627\left(\frac{1}{\sqrt{6\Omega_g}}\right)^{3/4}\phi_0^{-1/8}\left(\frac{\tau_A g^{3/2}}{q_0}\right)^{1/4} \tag{44}$$

where ϕ_0 is the phase shift of the soliton per cavity round trip (Equation 42). Here, we assume a fully saturated absorber, a linear approximation for the exponential decay of the slow saturable absorber and that the absorber recovers on a time scale longer than the soliton pulse duration.

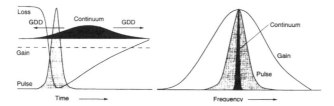

Figure 6. *Soliton modelocking in the time and the frequency domain. The continuum spreads in time due to group delay dispersion (GDD) and thus experiences more loss in the "slow" absorber, which is saturated by the shorter soliton pulse. However, the longer continuum pulse has a narrower spectrum and thus experiences more gain than the spectrally broader soliton pulse.*

Soliton modelocking can be explained in a more simple physical picture as follows (Figure 6). The soliton loses energy due to gain dispersion and losses in the cavity. Gain dispersion and losses can be treated as perturbation to the nonlinear Schrödinger equation for which a soliton is a stable solution. This lost energy, called continuum in soliton perturbation theory, is initially contained in a low intensity background pulse, which experiences negligible bandwidth broadening from SPM, but spreads in time due to GDD (Figure 6a). This continuum experiences a higher gain than the soliton, because it sees only the gain at line centre. In contrast, the soliton sees a lower average gain due to its larger bandwidth (Figure 6b). After a sufficient build-up time, the continuum would grow until it reached lasing threshold, de-stabilising the soliton. However, we can stabilise the soliton by introducing a "slow" saturable absorber into the cavity. This "slow" absorber adds additional loss to the continuum, preventing it from reaching its

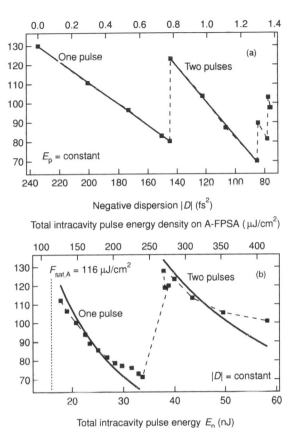

Figure 7. *Soliton modelocked diode-pumped Nd:glass laser using an intracavity SESAM (Figure 3). The solid lines are the theoretical predictions of an ideal soliton pulse.*

lasing threshold. The minimum pulse duration (Equation 44) is achieved right at the stability limit, where the continuum will start to see less loss than the soliton.

In soliton modelocking, the dominant pulse formation process is assumed to be soliton formation. Therefore, the pulse has to be approximately a soliton. In this case, the negative GVD is balanced with the SPM inside the laser cavity. The pulse duration is then given by the simple soliton solution (Equation 43), where F_p is the pulse fluence inside the SPM medium (*i.e.* usually the laser gain medium). This means that the pulse duration scales linearly with the negative group delay dispersion inside the laser cavity (*i.e.* $\tau_p \propto |D|$). In the case of an ideal fast saturable absorber, an unchirped soliton pulse is only obtained at a very specific dispersion setting (Equation 39), whereas for soliton modelocking an unchirped transform-limited soliton is obtained for all dispersion levels as long as the stability requirement against the continuum is fulfilled. This fact has also been used to confirm experimentally that soliton modelocking is the dominant pulse formation process and that a fast saturable absorber such as KLM is not the main mechanism (Aus der Au *et al.* 1997). We illustrate this with results (Figure 7) from a soliton modelocked Nd:glass laser pumped by laser diodes (Figure 3). Higher order dispersion only increases the pulse duration, therefore is undesirable and is assumed to be compensated.

The break-up into two or even three pulses in Figure 7 can be explained as follows: Beyond a certain pulse energy, two soliton pulses with lower power, longer duration, and narrower spectrum will be preferred, since their filter loss introduced by the limited gain bandwidth decreases so much that the slightly increased residual loss of the less saturated saturable absorber cannot compensate for it. This results in a lower total roundtrip loss and thus a reduced saturated or average gain for two pulses compared to one pulse. The threshold for multiple pulsing is lower for shorter pulses, that is, with spectra which are broad compared to the gain bandwidth of the laser. A more detailed description of multiple pulsing is given elsewhere (Kärtner *et al.* 1998).

Soliton formation is generally very important in femtosecond lasers; this was recognised in colliding pulse modelocked dye lasers. However, no analytical solution was presented for the soliton pulse shortening. It was always assumed that for a stable solution the modelocking mechanism without soliton effects had to generate a net gain window as short as the pulse (Figure 4a and b). In contrast to these cases, in soliton modelocking we present an analytical solution based on soliton perturbation theory, where soliton pulse shaping is clearly assumed to be the dominant pulse formation process, and the saturable absorber required for a stable solution is treated as a perturbation. Then, the net gain window can be much longer than the pulse (Figure 4c). Stability of the soliton against the continuum then determines the shortest possible pulse duration (Equation 44). This is a fundamentally different modelocking model from that previously described. We therefore refer to it as soliton modelocking, emphasising the fact that soliton pulse shaping is the dominant factor.

3   Saturable absorbers

The use of saturable absorbers in solid-state lasers is almost as old as the solid-state laser itself. Early examples are ruby lasers (Mocker and Collins 1965) using dye and colour filter glass saturable absorbers and Nd:glass lasers (DeMaria *et al.* 1966) using dye saturable absorbers. However, the first attempts to passively modelock solid-state lasers with long upper state lifetimes (*i.e.* significantly longer than the cavity round trip time) always resulted in Q-switched modelocking. This limitation was mostly due to the parameter ranges of available saturable absorbers (Haus 1976). For many years the success of passively modelocked dye lasers diverted the research interest away from solid-state lasers.

The first results of passively modelocked solid-state lasers with longer upper state lifetimes were reported in 1990, and used SESAMs in coupled cavities, a technique termed RPM (resonant passive modelocking) (Keller *et al.* 1990). This work was motivated by previously demonstrated coupled cavity modelocking techniques such as APM. Currently, however, most uses of coupled cavity techniques have been supplanted by intracavity saturable absorber techniques based on either Kerr lens modelocking (KLM) or intracavity SESAMs, due to their more inherent simplicity. In 1992, we demonstrated stable, purely cw-modelocked Nd:YLF and Nd:YAG lasers using an intracavity SESAM design, referred to as the anti-resonant Fabry-Perot saturable absorber (A-FPSA) (Keller *et al.* 1992). This was the first intracavity saturable absorber that self-started cw modelocking of a solid-state laser with long upper state lifetimes without any Q-switching. Since then, many new SESAM designs have been developed that provide stable pulse generation in the pico- and femtosecond regime for a variety of solid-state lasers. A more extended review of the SESAM designs is given by a recent book chapter (Keller 1999).

3.1   Important saturable absorber parameters

The macroscopic properties of a saturable absorber are the modulation depth ΔR , the non-saturable loss ΔR_{ns} , the saturation fluence $F_{\text{sat,A}}$, the saturation intensity $I_{\text{sat,A}}$ and the impulse response or recovery time τ_A . These parameters determine the operation of a passively modelocked or Q-switched laser. The modulation depth ΔR is the maximum amount of saturable loss of the absorber which can be bleached. This bleaching can be achieved by an incident pulse fluence (*i.e.* pulse energy density per unit area) much larger than the saturation fluence of the absorber. Normally the saturable absorber is integrated inside a mirror structure, thus the maximum nonlinear change in the reflectivity determines the modulation depth of the saturable absorber. The maximum absorber amplitude loss coefficient q_0 is then given by

$$\Delta R = 1 - e^{-2q_0} \approx 2q_0, \quad q_0 \ll 1 \,. \tag{45}$$

The non-saturable losses ΔR_{ns} of the SESAM are the residual losses for an incident pulse energy density much larger than the saturation fluence. Included in ΔR_{ns} are the less than 100% reflectivity of the bottom mirror, scattering losses due to impurities at the surface of the sample, residual absorption form defect states, and losses introduced by two-photon and free carrier absorption. Both the saturation fluence $F_{\text{sat,A}}$ and the absorber recovery time τ_A are determined experimentally. Standard pump-probe techniques determine the impulse response and therefore τ_A . In the femtosecond pulse regime we have to consider more than one absorber recovery time (Jung *et al.* 1997b). The saturation fluence $F_{\text{sat,A}}$ is determined and defined by the measurement of the nonlinear change in reflectivity of a pulse as a function of increased incident pulse energy (Keller 1999).

3.2   Semiconductor saturable absorber mirrors (SESAMs)

Semiconductor materials are well-established for electronic and optoelectronic devices. Remarkable advances in semiconductor growth and processing technology allow for the fabrication of artificial micro-structures approaching atomic dimensions. Epitaxial growth techniques, such as molecular beam epitaxy (MBE) and metal-organic chemical vapour deposition (MOCVD), can be used with different growth parameters to tailor linear and nonlinear optical properties in ultrathin layer structures. This precise control of optical nonlinearities combined with the large bandgap variations of different semiconductor materials from the visible to the infrared make semiconductors very attractive as saturable absorbers in solid-state lasers. This specialised application can benefit from a technology base that is driven by a large world-wide market in optoelectronics.

Why combine semiconductor materials with solid-state lasers? Solid-state lasers still provide some of the best laser qualities in terms of high-quality spatial modes, high output power, energy storage and large pulse energy when Q-switched, and the large optical bandwidths necessary for ultrashort pulse generation. Combined with the recent advances in semiconductor lasers (diode lasers) as pump sources, passive modelocking or Q-switching of solid-state lasers with semiconductor saturable absorbers forms a new and promising basis for an all-solid-state ultrafast laser technology with the potential for "turn-key" operation. Semiconductor saturable absorbers can offer significant improvements in performance, size, robustness, and cost of an ultrafast laser, compared to saturable absorbers used in the past such as organic dyes, colour filter glasses, dye-doped solids, or ion-doped crystals. Many of these absorbers only provided saturation for a limited range of wavelengths and with a limited range of recovery times and saturation fluences. In addition, organic dyes suffer from well-known problems such as short lifetimes and complicated handling (pumps, dye jets, flow tubes, and often toxic dyes).

Epitaxial growth techniques allow us to design devices at the optical wavelength level. We typically integrate a semiconductor saturable absorber directly into a mirror structure, resulting in a device whose reflectivity increases as the incident optical intensity increases. We call this general class of devices semiconductor saturable absorber mirrors (SESAMs) (Keller *et al.* 1992, 1996, 1999). SESAMs can be applied to mode-locked solid-state lasers from the sub-10fs range through the picosecond regime, then extending to Q-switched pulses beginning in the sub-100ps range and extending into the nanosecond range and even higher.

SESAMs offer a distinct range of operating parameters not available with other approaches. We can modify the absorber cross section of the saturable absorber if we integrate the absorber layer into a device structure. This allows us to modify the effective absorber cross section beyond its material value. We use various designs of SESAMs to achieve many of the desired properties. Figure 8 shows the different SESAM designs for a centre wavelength of $\approx$ 800nm. The first intracavity SESAM device was the anti-resonant Fabry-Perot saturable absorber (A-FPSA) (Keller *et al.* 1992, Brovelli *et al.* 1995a), initially used in a design regime with a rather high reflectivity top reflector, which we call now more specifically the high-finesse A-FPSA (Figures 8a,b). The Fabry-Perot is typically formed by the lower semiconductor Bragg mirror and a dielectric top mirror, with a saturable absorber and possibly transparent spacer layers in between. The thickness of the total absorber and spacer layers, the Fabry-Perot thickness, is adjusted such that the Fabry-Perot is operated at antiresonance. The Fabry-Perot thickness d (that can include non-absorbing transparent material to reduce the total absorption) has to fulfill the condition for antiresonance:

$$\varphi_{\mathrm{rt,a}} = \varphi_b + \varphi_t + 2k\bar{n}d = (2m+1)\,\pi \qquad (46)$$

where φ_{rt} is the round-trip phase inside the Fabry-Perot, $\varphi_{\mathrm{rt,a}}$ is φ_{rt} at antiresonance (*i.e.* destructive interference), φ_b and φ_t are the phase shifts of the reflectivity at the bottom and top mirrors, $\bar{n}$ is the average refractive index of the Fabry-Perot thickness layer, k is the wavevector, λ is the wavelength in vacuum and m is an integer. At antiresonance the interference of the partially reflected waves at the Fabry-Perot mirrors is destructive. Therefore, operation at antiresonance results in a device that is broadband and has minimal group velocity dispersion. The bandwidth of the A-FPSA is limited by either the free spectral range of the Fabry-Perot or the bandwidth of the mirrors.

The top reflector of the A-FPSA provides an adjustable parameter that determines the intensity entering the semiconductor saturable absorber and therefore the effective saturation fluence of the device. A high reflectivity top reflector results in a lower modulation

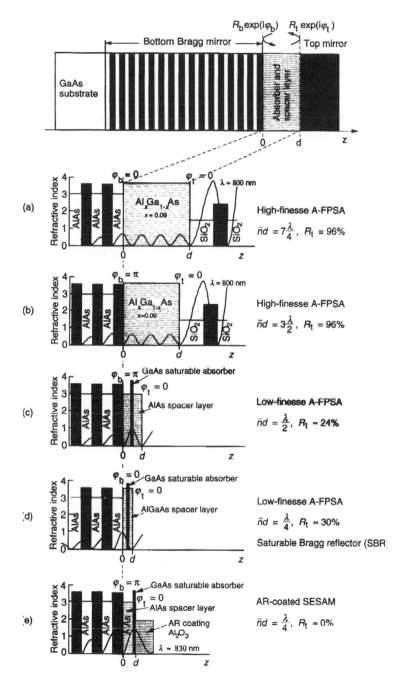

Figure 8. *Different SESAM designs for a centre wavelength of about 800nm which is suitable for a Ti:sapphire laser for example. The high-finesse A-FPSA was the first intracavity saturable absorber that stably cw modelocked solid-state lasers with long (> µs) upper state lifetimes. The saturable Bragg reflector is a special case of a low-finesse A-FPSA for which all layers are quarter-wave layers.*

depth for the same absorber thickness. However, with a thicker absorber layer the modulation depth can be increased again. A low reflectivity top reflector generally requires a thinner absorber layer to reduce the effects of the non-saturable insertion loss of the device. The two design limits of the A-FPSA are the high-finesse A-FPSA (Figures 8a,b) with a relatively high reflectivity top reflector and the AR-coated SESAM (Figure 8e) (Brovelli *et al.* 1995b). Using the incident laser mode area as an adjustable parameter, the incident pulse fluence can be adapted to the saturation fluence of both SESAMs for stable modelocking (see section 3.3)

A specific intermediate design is the low-finesse A-FPSA (Figures 8c,d) (Jung *et al.* 1995b, Hönninger *et al.* 1995), where the top reflector is formed by the $\approx 30\%$ Fresnel-reflection of the semiconductor/air interface. Reducing the top reflectance typically requires a thinner saturable absorber and a higher bottom reflectance to minimise non-saturable insertion loss. Because the absorber layers are typically rather thin (*i.e.* $< 30\mathrm{nm}$), the effective saturation fluence of the device can then be varied by changing the position of the buried absorber section within the Bragg reflector or simply within the last quarter-wave layer of the Bragg reflector, taking into account that an infinitely thin absorber layer at the node of a standing wave does not introduce any absorption. Different wavelengths have different positions for their nodes in the standing wave profile. This can be used to reduce the wavelength dependence of the absorber edge and obtain broadband performance (Jung *et al.* 1997). For a more detailed description of the different SESAM design we would like to refer the reader to a recently published book chapter (Keller 1999).

3.3   Stability requirements against self-Q-switching

The use of a saturable absorber as a passive modelocker in a solid-state laser can introduce a tendency for Q-switched modelocked operation. We have investigated the transition between the regimes of cw modelocking and Q-switched modelocking. We compared the predictions of an analytical model with the measurements from two lasers that were passively modelocked using saturable absorber mirrors (SESAMs). These were a Nd:YLF laser operating in the picosecond domain and a soliton modelocked Nd:glass laser operating in the femtosecond domain. The observed stability limits for the picosecond lasers agree well with a very simple criterion (Hönninger *et al.* 1999)

$$E_P^2 \; > \; E_{P,c}^2 = E_{\mathrm{sat,L}} \cdot E_{\mathrm{sat,A}} \cdot \Delta R. \tag{47}$$

The critical intracavity pulse energy $E_{P,c}$ is the minimum intracavity pulse energy which is required to obtain stable cw modelocking, that is, for $E_P > E_{P,c}$ we obtain stable cw modelocking and for $E_P < E_{P,c}$ we obtain Q-switched modelocking. For good stability of a modelocked laser against unwanted fluctuations of pulse energy, operation close to the stability limit (Equation 47) is not recommended. In Equation 47 we have assumed that the pulse duration τ_p is shorter than the absorber recovery time τ_A and τ_A is much shorter than the cavity round trip time T_R, (*i.e.* $\tau_p < \tau_A \ll T_R$).

We have shown that the experimental results for a Nd:YLF laser are in good agreement with the theory outlined above. When we checked experimental results from various other passively modelocked solid-state lasers, we found a similar agreement for other picosecond lasers. However, we discovered that Nd:glass and Yb:glass femtosecond soliton modelocked lasers show stable cw modelocking in a regime where they should actually be Q-switched modelocked according to the stability criterion of Equation 47. Thus, femtosecond soliton modelocked lasers showed a significant reduction of the tendency for QML. This can be explained by an extended theory, which takes into account soliton shaping effects and gain filtering (Hönninger *et al.* 1999) The basic idea is as follows: if the energy of an ultrashort pulse rises slightly due to relaxation oscillations, SPM and/or SAM broadens the pulse spectrum. A broader spectrum, however, reduces the effective gain due to the finite gain bandwidth, which provides some negative feedback, thus decreasing the critical pulse energy which is necessary for stable cw modelocking.

4 Approaching limits with Ti:sapphire lasers

4.1 Dispersion compensation

The challenge in ultrashort pulse generation is dispersion compensation over an extremely large bandwidth (Sutter *et al.* 1998, 1999). For optimum soliton formation of a sub-10fs pulse inside the laser, only a small amount of negative total intracavity dispersion is necessary. Higher order dispersion should be compensated as follows: (Table 3).

Phase Velocity v_p	$\dfrac{\omega}{k_n}$	$\dfrac{c}{n}$
Group Velocity v_g	$\dfrac{d\omega}{dk_n}$	$\dfrac{c}{n}\,\dfrac{1}{1-\dfrac{dn}{d\lambda}\dfrac{\lambda}{n}}$
Group Delay T_g	$T_g = \dfrac{z}{v_g} = \dfrac{d\phi}{d\omega}\ ,\ \phi \equiv k_n z$	$\dfrac{nz}{c}\left(1-\dfrac{dn}{d\lambda}\dfrac{\lambda}{n}\right)$
Dispersion: 1st order	$\dfrac{d\phi}{d\omega}$	$\dfrac{nz}{c}\left(1-\dfrac{dn}{d\lambda}\dfrac{\lambda}{n}\right)$
Dispersion: 2nd order	$\dfrac{d^2\phi}{d\omega^2}$	$\dfrac{\lambda^3 z}{2\pi c^2}\dfrac{d^2 n}{d\lambda^2}$
Dispersion: 3rd order	$\dfrac{d^3\phi}{d\omega^3}$	$\dfrac{-\lambda^4 z}{4\pi^2 c^3}\left(3\dfrac{d^2 n}{d\lambda^2}+\lambda\dfrac{d^3 n}{d\lambda^3}\right)$

Table 3. *Compensating terms for higher order dispersion. k_n the wavevector in the dispersive media i.e. $k_n = k\,n = n2\pi/\lambda$, and z is a certain propagation distance.*

For example, a 1mm thick Ti:sapphire crystal produces a second order dispersion of 58fs^2 and a third order dispersion of 42fs^3 at a centre wavelength of 800nm. In comparison, 1mm of fused quartz (SF10 glass) produces 36fs^2 (159fs^2) second order dispersion and 27fs^3 (104fs^3) third order dispersion. Dispersion compensation is critical. For example, a 10fs (1fs) Gaussian pulse given by

$$\frac{\tau_p(z)}{\tau_p(0)} = \sqrt{1 + \left(\frac{4\ln 2\ d^2\phi/d\omega^2}{\tau_p^2(0)}\right)^2}$$

is broadened to 100fs (1ps) after 1cm of fused quartz due to second order dispersion. This form of the pulse becomes

$$\tau_p(z) \approx \frac{d^2\phi}{d\omega^2}\Delta\omega_p \tag{48}$$

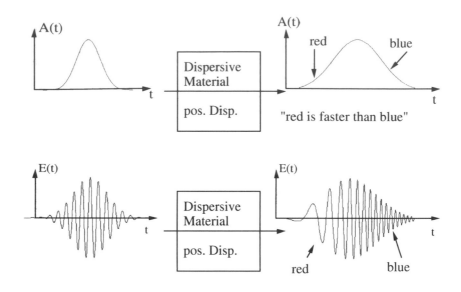

Figure 9. *Dispersive pulse broadening due to positive dispersion.*

since in the regime of strong pulse broadening we have $\dfrac{d^2\phi}{d\omega^2} \gg \tau_p^2\,(0)$ (Figure 9).

For optimum soliton formation of a sub-10fs pulse inside the laser, only a very small amount of negative total intracavity dispersion is necessary. This amount of negative second order dispersion can be determined from the balance of the chirp due to self-phase modulation with the chirp due to the negative dispersion. The resulting estimate for the dispersion with a given pulse duration follows from Equation 43. With a peak power of 11MW and an estimated self-phase modulation coefficient of 0.07/MW, we estimate the necessary dispersion to be -10fs^2. However, the self-phase modulation coefficient is not known precisely, because it is inversely proportional to the effective area of the laser mode in the laser crystal and, therefore, can only be estimated from ABCD-matrix calculations of the laser cavity. The strength of SPM also depends on the actual pulse width in the laser crystal. Ideally, this small amount of negative total intracavity dispersion, which is approximately zero on the scale shown in Figure 10, should be constant over a wavelength range as large as possible.

Prism pairs are well established for intracavity dispersion compensation (Fork *et al.* 1984) and offer two advantages. First, the pulsewidth can be varied by moving one of the prisms and second, the laser can be tuned in wavelength by simply moving a knife edge at a position where the beam's spectrum is spatially dispersed. Both properties are often desired for spectroscopic applications, for example. However, the prism pair suffers from higher order dispersion, which is the main limitation in pulse shortening. The higher order dispersion of the prism pairs is dominated by the prism spacing which is not changed when we adjust the dispersion. At a spacing between the fused quartz prisms of 40cm, the total dispersion that is produced by a double pass through the prism geometry amounts to -862fs^2 and a third order dispersion of $d^3\phi/d\omega^3 = -970\text{fs}^3$ at a centre wavelength of 800nm, assuming zero prism insertion into the beam. We can reduce

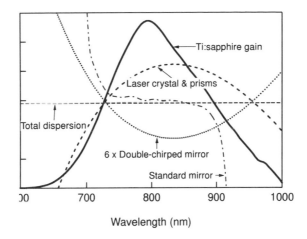

Wavelength (nm)

Figure 10. *Higher-order dispersion compensation requirements in sub-10fs Ti:sapphire lasers: Ti:sapphire gain spectrum and group delay dispersion (GDD) of fused quartz prism and 2.3mm thick Ti:sapphire crystal, GDD of six reflections from standard high-reflecting Bragg mirrors, and required GDD of six reflections from double-chirped mirrors for stable soliton pulses.*

the negative dispersion by moving the prisms into the laser cavity mode: Each additional millimetre of prism insertion produces a positive dispersion of $101fs^2$ but a third order dispersion of only $78fs^3$ per cavity round trip. Thus, the prism pairs can generally only be used to compensate for either second or third order dispersion. Another dispersion compensation is then required for the other order. Therefore, the double-chirped mirrors were designed to show the inverse dispersion in higher order group delay to that of the prism pair plus laser crystal. This would eliminate the higher order dispersion and obtain a slightly negative but constant group delay dispersion required for an ideal soliton pulse.

Chirped mirrors (Szipöcs *et al.* 1994, Kärtner *et al.* 1997) have been established in different set-ups for the generation of ultrashort laser pulses (Figure 11). According to Szipöcs *et al.* (1994), chirping means that the Bragg wavelength λ_B is gradually increased along the mirror, producing a negative group delay dispersion (GDD). However, no analytical explanation of the unwanted oscillations typically observed in the group delay and GDD of such a simple-chirped mirror was given. These oscillations were mostly minimised purely by computer optimisation. Recently, we developed a theory of chirped mirrors which is based on an exact coupled-mode analysis (Matuschek *et al.* 1997). Following our theory, one has to use the double-chirp technique in combination with a broadband AR coating, in order to avoid the oscillations in the GDD (Kärtner *et al.* 1997). Using our simple but very accurate analytical expressions for phase, group delay, and GDD, we designed double-chirped mirrors with a smooth and custom tailored GDD suitable for generating pulses in the two-cycle regime directly from a Ti:sapphire laser (Matuschek *et al.* 1999). According to Figure 11, a double chirped mirror is a multilayer interference coating that can be considered as a composition of at least two sections, each with a different task. The layer materials are SiO_2 and TiO_2. The first section is the AR-coating, typically composed of 10–14 layers. It is necessary because the theory is derived assuming an ideal

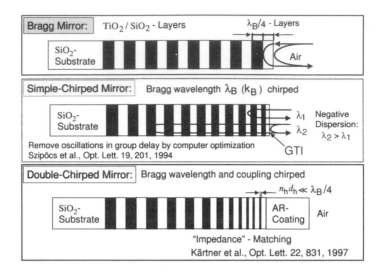

Figure 11. *A Bragg mirror is defined by alternating high and low refractive index quarter-wave layer materials. In the simple-chirp mirror, only the Bragg wavelength λ_B is chirped, whereas in the double-chirped mirror (DCM) both the Bragg wavelength and the coupling constant is chirped. Oscillations in the group delay and GDD are due to Gires-Tournois interference (GTI) effects.*

matching to air. The other section represents the actual DCM structure, as derived from theory. The double-chirp section is responsible for the elimination of the oscillations in the GDD. Double-chirping means that in addition to the local Bragg wavelength λ_B, the local coupling of the incident wave to the reflected wave is independently chirped as well. The local coupling is adjusted by slowly increasing the high-index layer thickness in every pair so that the total optical thickness remains approximately $\lambda_B/2$. This corresponds to an adiabatic matching of the impedance. The AR-coating, together with the rest of the mirror, is used as a starting design for a numerical optimisation program. Since we start from a theoretical design that is close to the desired design goal, a local optimisation using a standard gradient algorithm is sufficient.

4.2   New frontiers with pulses in the two-cycle regime

The shortest pulses directly from a laser oscillator, with durations of 6.5fs (Jung *et al.* 1997c) (Figure 12a), have been obtained from a KLM Ti:sapphire laser using double-chirped mirrors (DCMs) (Kärtner *et al.* 1997) and a prism pair for dispersion control, and a SESAM (Jung *et al.* 1997) to initiate modelocking. SESAM-assisted KLM relaxes the tight constraints on cavity alignment required for pure Kerr-lens modelocking. Moreover, SESAMs provide a reliable starting mechanism with modelocking build-up times down to $60\mu s$ for sub-10fs KLM lasers (Sutter *et al.* 1998).

Recent further improvements resulted in pulses of sub-6fs duration from a KLM Ti:sapphire laser at a repetition rate of 100MHz and an average power of 300mW (Sutter *et al.* 1999). Fitting an ideal $sech^2$ to the autocorrelation data yields 4.8fs pulse duration

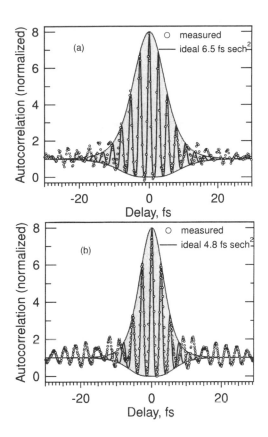

Figure 12. *Interferometric autocorrelation traces of pulses directly out of a cw mode-locked Ti:sapphire laser. An ideal sech²-shaped fit results in 6.5fs (a) and 4.8fs (b) pulse durations. These are the shortest pulses ever obtained directly from a laser without any further pulse compression.*

(Figure 12b). The pulse spectrum covers the wavelength range from 630nm to above 950nm, with some clearly visible spectral content in the yellow. This extreme spectral width has been achieved by optimised spectral shaping of the output coupling mirror. In an unoptimised version, we previously demonstrated similar spectral shaping (Sutter *et al.* 1998). The resonator is a standard 100MHz X-cavity with 10cm radius folding mirrors and a 2.3mm long Ti:sapphire crystal with 0.25wt.% doping (Figure 13). Except for the output coupling mirror, only double-chirped mirrors (DCMs) are used in the cavity. The DCMs compensate the intracavity dispersion together with a pair of fused silica Brewster prisms for adjustment of the group-delay dispersion. We used a custom designed output coupler for spectral shaping. A new broadband SESAM is used to provide a self-starting mechanism and to stabilise the KLM operation over a broader range of cavity parameters.

Proper measurement of sub-10fs pulses is still controversial and challenging. So far, the duration of world record ultrashort optical pulses has been estimated from interferometric autocorrelation (IAC) traces. Other techniques, such as spectral phase interferometry for direct electric-field reconstruction (SPIDER (Iaconis and Walmsley 1998)) and frequency-

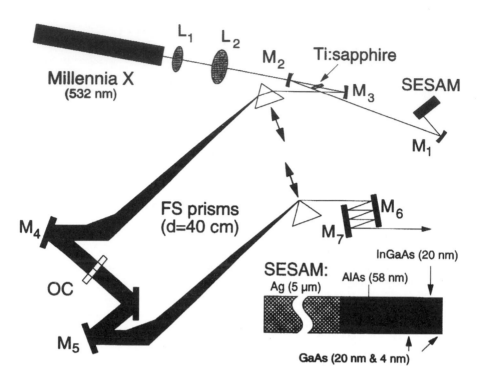

Figure 13. *Ti:sapphire laser set-up for sub-10fs pulse generation. M1–M7: double-chirped mirrors, OC: output coupler, FS: fused silica prisms with 40cm apex separation, SESAM: semiconductor saturable absorber mirror, bottom right: detailed broadband SESAM structure.*

resolved optical gating (FROG (Kane and Trebino 1993, Taft *et al.* 1996)) have been developed that retrieve both amplitude and phase of the pulse. However, further work will be necessary to evaluate the different trade-offs and limitations of SPIDER, FROG, and IAC for ultrashort pulses in the two-cycle regime. It is important to realise that deconvolution with an ideal sech^2 pulse shape slightly underestimates the true pulse width. In principle, only if the spectrum also shows an ideal sech^2-shape is it justified to make this ideal assumption. Therefore, it is always necessary to include the information on how the pulse duration was measured. Only the results of the same measurement techniques should be used as a basis for comparison with other publications. In the future, we hope that more accurate measurement techniques will be used to characterise new world record pulses.

5   Conclusion and future outlook

The SESAM helped to establish the feasibility of pure cw passive modelocking in solid-state lasers with upper state lifetimes (μs to ms) much longer than the cavity round trip time (typically ns). Earlier, most solid-state lasers also Q- switched during passive modelocking, a condition not desired for many applications. The combination of adjustable device structure and material parameters of the SESAM provides sufficient design freedom to choose the key macroscopic parameters such as modulation depth, recovery time, saturation intensity and saturation fluence to suppress Q-switched operation. The design freedom of SESAMs has also allowed us to systematically investigate the stability regimes of passive Q-switching and passive cw modelocking with an improved understanding and modelling of Q-switching (Hönninger *et al.* 1999) and multiple pulsing instabilities (Aus der Au *et al.* 1997, Kärtner *et al.* 1998). We consider this improved understanding and detailed experimental investigations as one of the main achievements of the SESAM technology. This means that if we want to look for or consider other saturable absorber materials in the future we then know their required properties.

On the practical side, modelocked diode-pumped solid-state lasers using SESAMs are becoming commercially available, providing more compact, reliable and "hands-off" ultrafast lasers. The pulse generation process is self-starting, and cavity stability adjustments are less critical when SESAMs are used compared to KLM. The design freedom of the laser cavity with SESAMs allows for good performance with diode-pumping. With optimised mode-matching, we have generated the highest reported average powers in diode-pumped mode-locked femtosecond Cr:LiSAF (Kopf *et al.* 1997) and Nd:glass (Aus der Au *et al.* 1998) lasers. Recently we have demonstrated more than 10W average output power and 16ps pulses from a passively modelocked diode-pumped Nd:YAG laser (Spühler *et al.* 1999). Scaling all-solid-state ultrafast lasers to the microjoule and millijoule pulse energy level will be an interesting future challenge. So far, picosecond and femtosecond pulses have been demonstrated from many solid-state lasers, so we can expect to see further applications of these lasers to commercial products at many different wavelengths and pulsewidth regimes.

SESAMs also have allowed us to push a number of key limits in laser performance. Presently, we have demonstrated pulses as short as 6.5fs (Jung *et al.* 1997) and more recently around 5fs (Sutter *et al.* 1999b) directly from a Ti:sapphire laser without any further pulse compression, the shortest cw modelocked pulses reported to date. In addition, we have also demonstrated the shortest Q-switched pulses of 37ps duration from a solid-state laser (Spühler *et al.* 1998).

However, we are approaching the fundamental limits in ultrashort pulse generation from a Ti:sapphire laser, having reached pulses lasting only two optical cycles centred at around 800nm. Substantially shorter pulses will only be achieved by switching to ultraviolet or even shorter wavelengths, and using nonlinear processes to increase the spectral bandwidth. Currently, the most promising path to attosecond pulse generation and attosecond spectroscopy is high-harmonic generation (HHG). Nearly single-cycle-excitation optical pulses can produce harmonic frequencies in gases extending even into the water window (2.3–4.4nm), and a continuum broad enough to support attosecond pulses. Key to this harmonic generation are optical pulses of a few cycles, since even slightly longer pulses result in reduced harmonic generation. Additionally, peak intensities of 10^{15}W/cm^2 are typically required at the focus for HHG. Presently, Ti:sapphire lasers can produce around 10^{13}W/cm^2, assuming a focused beam diameter of 3μm. Therefore, one of the goals is to extend sub-10fs pulse sources to higher powers. These high-intensity Ti:sapphire sources can extend the spectral range from the visible to the X-ray and open up the possibility of entering a new ultrafast era with attosecond pulse generation and spectroscopy - maybe in the future referred to as "hyperfast".

We would expect that with attosecond time resolution we open up a new world of physics with as much impact as has been demonstrated in the 1970s and 1980s with the transition from picosecond to femtosecond time resolution. In addition to this unsurpassed time resolution, we also obtain coherent sources from the visible into the X-ray regime which has been previously only reached by synchrotron radiation. At this point, however, we still have some significant problems to tackle before we have a chance to demonstrate useful attosecond pulse generation and X-ray radiation. Presently, only theoretical proposals for attosecond pulse generation exist. No experimental demonstrations have been made so far. There are some fundamental problems such as the synchronisation of the phase of the electric field with respect to the pulse envelope on a femtosecond timescale. Solving all of these problems will make the research in ultrashort pulse generation very exciting and rewarding for many years to come.

6 Acknowledgement

The author is greatly indebted to the following colleagues and graduate students who have made major contributions to this work: Luigi Brovelli, Kurt J Weingarten, Franz X Kärtner, Uwe Siegner, Rüdiger Paschotta, Günter Steinmeyer, Guodong Zhang, Francois Morier-Genoud, Michael Moser, Daniel Kopf, Bernd Braun, Isabella Jung, Regula Fluck, Clemens Hönninger, Nicolai Matuschek, Dirk Sutter, Jürg aus der Au, Gabriel Spühler, and Markus Haiml. The financial support of the Swiss National Science Foundation and the Swiss Priority Program in Optics is also acknowledged.

References

Aus der Au J, Kopf D, Morier-Genoud F, Moser M and Keller U, 1997, "60-fs pulses from a diode-pumped Nd:glass laser," *Opt. Lett.* **22** 307.

Aus der Au J, Loesel F H, Morier-Genoud F, Moser M and Keller U, 1998,"Femtosecond diode-pumped Nd:glass laser with more than 1-W average output power," *Opt. Lett.* **23**, 271

Baltuska A, Wei Z, Pshenichnikov M S, Wiersma D A and SzipöcsR, 1997, "All-solid-state cavity dumped sub-5-fs laser," *Appl. Phys. B* **65** 175.

Blow K J and Nelson B P, 1988, "Improved modelocking of an F-center laser with a nonlinear nonsoliton external cavity," *Opt. Lett.* **13** 1026.

Braun B, Weingarten K J, Kärtner F X and Keller U, 1995, "Continuous-wave mode-locked solid-state lasers with enhanced spatial hole-burning, Part I: Experiments," *App. Phys. B* **61** 429.

Brovelli L R, Jung I D, Kopf D, Kamp M, Moser M, Kärtner F X and Keller U, 1995a, "Self-starting soliton modelocked Ti:sapphire laser using a thin semiconductor saturable absorber," *Electron. Lett.* **31** 287.

Brovelli L R, Keller U and Chiu T H, 1995b, "Design and Operation of Antiresonant Fabry-Perot Saturable Semiconductor Absorbers for Mode-Locked Solid-State Lasers," *JOSA B* **12** 311.

DeMaria A J, Stetser D A and Heynau H, 1966, "Self mode-locking of lasers with saturable absorbers," *Appl. Phys. Lett.* **8** 174.

Fork R L, Martinez O E and Gordon J P, 1984, "Negative dispersion using pairs of prisms," *Opt. Lett.* **9** 150.

Haus H A and Silberberg Y, 1986, "Laser mode locking with addition of nonlinear index," *IEEE J. Quantum Electron.* **22** 325.

Haus H A, 1975a, "A Theory of Forced Mode Locking," *IEEE J. Quantum Electron.* **11** 323.

Haus H A, 1975b, "Theory of modelocking with a fast saturable absorber," *J. Appl. Phys.* **46** 3049.

Haus H A, 1976, "Parameter ranges for cw passive modelocking," *IEEE J.Quantum Electron.* **12** 169.

Haus H A, 1984, *Waves and fields in optoelectronics* (Prentice Hall,Inc., Englewood Cliffs, NJ).

Haus H A, Fujimoto J G and Ippen E P, 1992, "Analytic Theory of Additive Pulse and Kerr Lens Mode Locking," *IEEE J. Quantum Electron.* **28** 2086.

Hönninger C, Zhang G, Keller U and Giesen A, 1995, "Femtosecond Yb:YAG laser using semiconductor saturable absorbers," *Opt. Lett.* **20** 2402.

Hönninger C, Paschotta R, Morier-Genoud F, Moser M and Keller U, 1999, "Q-switching stability limits of continuous-wave passive mode locking," *JOSA B* **16** 46.

Iaconis C and Walmsley I A, 1998, "Spectral Phase Interferometry for Direct Electric Field Reconstruction of Ultrashort Optical Pulses," *Opt.Lett.* **23** 792.

Ippen E P, Haus H A and Liu L Y, 1989, "Additive Pulse Modelocking," *J.Opt. Soc. Am. B* **6** 1736.

Jung I D, Brovelli L R, Kamp M, Keller U and Moser M, 1995b, "Scaling of the antiresonant Fabry-Perot saturable absorber design toward a thin saturable absorber," *Opt. Lett.* **20** 1559.

Jung I D, Kärtner F X, Brovelli L R, Kamp M and Keller U, 1995a, "Experimental verification of soliton modelocking using only a slow saturable absorber," *Opt. Lett.* **20** 1892.

Jung I D, Kärtner F X, Henkmann J, Zhang G and Keller U, 1997a,"High-dynamic-range characterisation of ultrashort pulses," *Appl. Phys. B* **65** 307.

Jung I D, Kärtner F X, Matuschek N, Sutter D H, Morier-Genoud F, Shi Z, Scheuer V, Tilsch M, Tschudi T and Keller U, 1997b, "Semiconductor saturable absorber mirrors supporting sub-10fs pulses," *Appl. Phys. B: Special Issue on Ultrashort Pulse Generation* **65** 137.

Jung I D, Kärtner F X, Matuschek N, Sutter D H, Morier-Genoud F, Zhang G, Keller U, Scheuer V, Tilsch M and Tschudi T, 1997c, "Self-starting 6.5fs pulses from a Ti:sapphire laser," *Opt. Lett.* **22** 1009.

Kärtner F X and Keller U, 1995, "Stabilisation of soliton-like pulses with a slow saturable absorber," *Opt. Lett.* **20** 16.

Kärtner F X, Kopf D and Keller U, 1995, "Solitary pulse stabilization and shortening in actively mode-locked lasers," *JOSA B* **12** 486.

Kärtner F X, Jung I D and Keller U, 1996, "Soliton Modelocking with Saturable Absorbers," *Special Issue on Ultrafast Electronics, Photonics and Optoelectronics, IEEE J. Selected Topics in Quantum Electronics (JSTQE)* **2** 540.

Kärtner F X, Matuschek N, Schibli T, Keller U, Haus H A, Heine C, Morf R, Scheuer V, Tilsch M and Tschudi T, 1997, "Design and fabrication of double-chirped mirrors," *Opt. Lett.* **22** 831.

Kärtner F X, d. Au J A and Keller U, 1998, "Modelocking with slow and fast saturable absorbers - What's the difference?," *IEEE J. Selected Topics in Quantum Electronics (JSTQE)* **4** 159.

Kane D J and Trebino R, 1993, "Characterization of Arbitrary Femtosecond Pulses Using Frequency-Resolved Optical Gating," *IEEE J. Quantum Electron.* **29** 571.

Kean P N, Zhu X, Crust D W, Grant R S, Landford N and Sibbett W, 1989, "Enhanced modelocking of color center lasers," *Opt. Lett.* **14** 39.

Keller U, Knox W H and Roskos H, 1990, "Coupled-Cavity Resonant Passive Modelocked (RPM) Ti:Sapphire Laser," *Opt. Lett.* **15** 1377.

Keller U, 'tHooft G W, Knox W H and Cunningham J E, 1991, "Femtosecond Pulses from a Continuously Self-Starting Passively Mode-Locked Ti:Sapphire Laser,"*Opt. Lett.* **16** 1022.

Keller U, Miller D A B, Boyd D G, Chiu T H, Ferguson J F and Asom M T, 1992, "Solid-state low-loss intracavity saturable absorber for Nd:YLF lasers: an antiresonant semiconductor Fabry-Perot saturable absorber," *Opt. Lett.* **17** 505.

Keller U, Weingarten K J, Kärtner F X, Kopf D, Braun B, Jung I D, Fluck R, Hönninger C, Matuschek N and Aus der Au J, 1996, "Semiconductor saturable absorber mirrors (SESAMs) for femtosecond to nanosecond pulse generation in solid-state lasers," *IEEE J. Selected Topics in Quantum Electronics (JSTQE)* **2** 435.

Keller U, 1999, "Semiconductor nonlinearities for solid-state laser modelocking and Q-switching," in *Nonlinear Optics in Semiconductors* **59** 211, E. Garmire, A. Kost, Eds. (Academic Press, Inc., Boston).

Kim B G, Garmire E, Hummel S G and Dapkus P D, 1989, "Nonlinear Bragg reflector based on saturable absorption," *Appl. Phys. Lett.* **54** 1095.

Kopf D, Kärtner F, Weingarten K J and Keller U, 1994, "Pulse shortening in a Nd:glass laser by gain reshaping and soliton formation," *Opt. Lett.* **19** 2146.

Kopf D, Kärtner F X, Weingarten K J and Keller U, 1995, "Diode-pumped modelocked Nd:glass lasers using an A-FPSA," *Opt. Lett.* **20** 1169.

Kopf D, Weingarten K J, Zhang G, Moser M, Emanuel M A, Beach R J, Skidmore J A and Keller U, 1997, "High-average-power diode-pumped femtosecond Cr:LiSAF lasers," *Appl. Phys. B* **65** 235.

Matuschek N, Kärtner F X and Keller U, 1997, "Exact Coupled-Mode Theories for Multi-layer Interference Coatings with Arbitrary Strong Index Modulations," *IEEE J. Quantum Electron.* **33** 295

Matuschek N, Kärtner F X, and Keller U, 1998, "Analytical Design of Double-Chirped Mirrors with Custom-Tailored Dispersion Characteristics,"*IEEE J. Quantum Electron.* **35** 129.

Mocker H W and Collins R J, 1965, "Mode competition and self-locking effects in a Q-switched ruby laser," *Appl. Phys. Lett.* **7** 270.

Morgner U, Kärtner F X, Cho S H, Chen Y, Haus H A Fujimoto J G, Ippen E P, Scheuer V, Angelow G and Tschudi T, 1999, "Sub-two-cycle pulses from a Kerr lens mode-locked Ti:sapphire laser", *Opt. Lett.* **24** 411

Negus D K, Spinelli L, Goldblatt N and Feugnet G, 1991, "Sub-100 femtosecond pulse generation by Kerr lens modelocking in Ti:Sapphire," in *Advanced Solid-State Lasers* G. Dubé, L. Chase, Eds. (Optical Society of America, Washington, D.C.) **10** 120.

New G H C, 1974, "Pulse evolution in mode-locked quasi-continuous lasers," *IEEE J. Quantum Electron.* **10** 115.

Nisoli M, Stagira S, Silvestri S D, Svelto O, Sartania S, Cheng Z, Lenzner M, Spielmann C and Krausz F, 1997, "A novel-high energy pulse compression system: generation of multi-gigawatt sub-5-fs pulses," *Appl. Phys. B*, **65** 189.

Perry M D, Stuart B C, Tietbohl G, Miller J, Britten J A, Boyd R, Everett M, Herman S, Nguyen H, Powell H T and Shore B W, 1996, *Tech. Digest CLEO'96* Paper CW14, 307

Piché M and Salin F, 1993, "Self-mode locking of solid-state lasers without apertures," *Opt. Lett.* **18** 1041.

Salin F, Squier J and Piché M, 1991, "Modelocking of Ti:Sapphire lasers and self-focusing: a Gaussian approximation," *Opt. Lett.* **16** 1674.

Shank C V, 1988, "Generation of Ultrashort Optical Pulses," in *Ultrashort Laser Pulses and Applications* W. Kaiser, Ed. (Springer Verlag, Heidelberg), Chapter 2.

Shirakawa A, Sakane I, Takasaka M and Kobayashi T, 1999, Sub-5-fs visible pulse generation by pulse-front-matched noncollinear optical parametric amplification" *Appl.Phys, Lett.* **74** 2268

Spence D E, Kean P N and Sibbett W, 1991, "60-fs pulse generation from a self-mode-locked Ti:Sapphire laser," *Optics Lett.* **16** 42.

Spühler G J, Paschotta R, Fluck R, Braun B, Moser M, Gini G, Zhang G and Keller U, 1999, "Experimentally confirmed design guidelines for passively Q-switched microchip lasers using semiconductor saturable absorbers," *J. Opt. Soc. Am. B* **16** 376.

Spühler G J, Paschotta R, Fluck R, Braun B, Moser M, Zhang G, Gini E and Keller U, 1999, " Diode-pumped passively modelocked Nd:YAG laser with 10W average power in a diffractio-limited beam," *Opt. Lett.* **24** 528

Strickland D and Mourou G, 1985, *Opt. Comm.* **56** 219

Sutter D H, Jung I D, Kärtner F X, Matuschek N, Morier-Genoud F, Scheuer V, Tilsch M, Tschudi T and Keller U, 1998, "Self-starting 6.5-fs pulses from a Ti:sapphire laser using a semiconductor saturable absorber and double-chirped mirrors," *IEEE J. of Sel. Topics in Quantum Electronics* **4** 169.

Sutter D H, Steinmeyer G, Gallmann L, Matuschek N, Morier-Genoud F, Keller U, Scheuer V, Angelow G and Tschudi T, 1999a, "Ultrabroad band pulses in the two-cycle regime by SESAM-assisted Kerr-lens modelocking of an all-solid-state Ti:Sapphire laser" *ASSL*

Sutter D H, Steinmeyer G, Gallmann L, Matuschek N, Morier-Genoud F, Keller U, Scheuer V, Angelow G and Tschudi T, 1999b, "SESAM-assisted Kerr-lens modelocked Ti:sapphire laer producing pulses in the two-cycle regime," *Opt. Lett.* **24** 631

Szipöcs R, Ferencz K, Spielmann C and Krausz F, 1994, "Chirped multilayer coatings for broad-band dispersion control in femtosecond lasers," *Opt. Lett.* **19** 201.

Taft G, Rundquist A, Murnane M M, Christov I P, Kapteyn H C, DeLong K W, Fittinghoff D N, Krumbügel M A, Sweetser J N and Trebino R, 1996, "Measurement of 10-fs Laser Pulses," *IEEE Journal of Selected Topics in Quantum Electronics* **2** 575.

Tsuda S, Knox W H, d. Souza E A, Jan W Y and Cunningham J E, 1995, "Low-loss intracavity AlAs/AlGaAs saturable Bragg reflector for femtosecond modelocking in solid-state lasers," *Optics Letters* **20** 1406.

Umstadter DP, Barty C Perry M and Mourou G A, 1998, "Tabletop, Ultrahigh-Intensity Lasers: Dawn of Nonlinear Relativistic Optics," *Optics and Photonics News,* **9** 41

Weingarten K J, Shannon D C, Wallace R W and Keller U, 1990 "Two gigahertz repetition rate, diode-pumped, mode-locked, Nd:yttrium lithium fluoride (YLF) laser," *Opt. Lett.* **15** 962.

Materials for lasers and nonlinear optics

William F Krupke

Lawrence Livermore National Laboratory (LLNL), USA

1 Introduction

Selecting from but a handful of commercially available laser and nonlinear optical (NLO) crystals, laser designers have produced an impressive array of lasers able to perform a wide range of scientific, commercial, industrial, and military applications. Laser design possibilities were enormously expanded with the development in the late-1980s of efficient and powerful semiconductor laser diodes as pump sources for solid state lasers. Despite this, it sometimes proves difficult, if not impossible to design lasers meeting some application requirements, when drawing only on presently available laser and NLO materials, especially when particularly demanding size, weight, efficiency, and cost requirement are imposed. Thus, there continues to be a need to identify, characterise, and develop novel laser and NLO materials possessing characteristics that significantly broaden design options of future laser devices and systems. Because the cost is rather large to transition a newly discovered optical material from laboratory experiments to commercial readiness, it is necessary that such materials possess significantly distinct and enabling characteristics, relative to available materials. Fortunately, since the discovery of the laser 40 years ago, an extensive array of modelling tools, design experiences, and materials data bases have evolved, which can be exploited to guide the search for, and development of, enabling new laser and NLO materials.

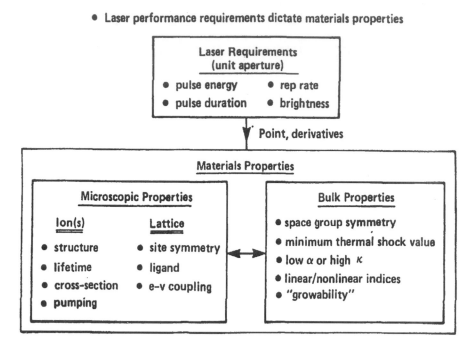

Figure 1. *Laser performance requirements flow down to the microscopic and bulk properties of the implementing laser gain material.*

Figure 1 shows schematically the relationship between an intended laser application and the properties of a laser gain medium responsive to that application. An analysis of an intended application provides a detailed set of laser performance requirements (e.g., output waveform, power or pulse energy, repetition rate, wavelength(s), tunability, beam-quality, efficiency, size, weight, cost, etc.). Knowing the desired laser system performance, a candidate laser architecture and implementation of active and passive components can be postulated (e.g., diode or lamp pumped; end or transverse pumped; CW or repetitively-pulsed; Q-switched, cavity-dumped, or mode-locked; power oscillator or master-oscillator-power-amplifier; etc.). Simplified models for the operation of such laser architectures and component arrangements are now well described in the literature, and computer design and simulation models often can be purchased. By assuming parameter values for the constituent active and passive components (pump source characteristics, resonator cavity optics and mirrors, and laser gain and NLO materials characteristics) one can estimate laser performance. By altering the values of assumed parameters, the impact of these changes on laser performance can be assessed quantitatively. Through such a process, one can identify desirable ranges of values for key parameters of novel laser and/or NLO crystals which, if available, would lead to superior laser performance compared to lasers utilising only currently available crystals. In general terms, two characteristic sets of material parameters are needed: (1) laser-spectroscopic parameters , and (2) thermal, mechanical, and optical bulk parameters. Key laser-spectroscopic parameters include: laser transition cross-section, wavelength, and bandwidth; metastable upper laser level

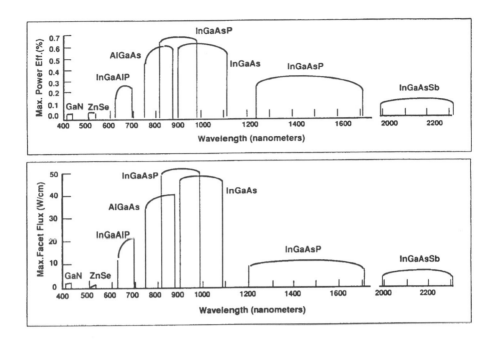

Figure 2. *Current status of the power and efficiency of semiconductor diode laser bars.*

quantum yield and lifetime; pump transition wavelengths, cross-sections, and bandwidths. Key bulk crystal parameters include: thermal conductivity, linear-expansion coefficient, Young's modulus, Poisson's ratio, tensile rupture strength, index of refraction and its temperature derivative.

Using various *ab initio* and semi-empirical methods, propensity-rules, and data bases, researchers at the Lawrence Livermore National Laboratory have conducted several "directed searches" for novel laser crystals in recent years, with some positive outcomes: (1) frequency-doubled diode-pumped Nd:YSiO₅ laser (Beach *et al.* 1990), whose output was intended to match precisely the 455nm wavelength of the caesium atomic resonance fluorescence detector (Marling *et al.* 1979), and to scale appropriately in power and pulse energy for submarine to satellite communications; (2) Cr:LiCAF (Payne *et al.* 1988) and Cr:LiSAF (Payne *et al.* 1989) lasers as efficient pulsed sources of ~750–850nm radiation to pump Nd:glass fusion lasers; (3) InGaAs diode-pumped Yb:FAP and Yb:S-FAP lasers (DeLoach *et al.* 1993) for use in inertial fusion laser drivers (Orth *et al.* 1996); and (4) diode-pumped, mid-IR tunable, divalent transition metal ions lasers in chalcogenide crystals (DeLoach *et al.* 1996; Page *et al.* 1997; Krupke *et al.* 1997a). After briefly indicating some search tools for key laser-spectroscopic and bulk properties, and other novel material considerations, two of these directed searches will be outlined in more detail.

The development of semiconductor laser diode pump sources throughout the 650–2000nm spectral band that are efficient, powerful and narrowband, has revolutionised the field of lasers based on rare earth and transition-metal doped solid state materials. Figure 2 shows the operating spectral bands of laser diodes produced using various III/V

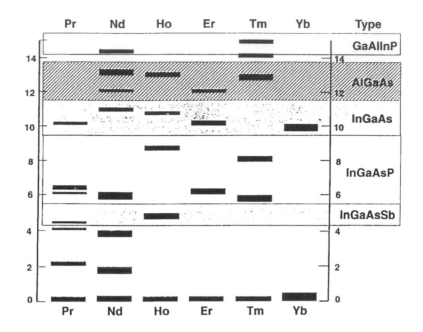

Figure 3. *Spectral overlay of laser diodes with the low-lying energy levels of Pr, Nd, H, Er, Tm, and Yb trivalent rare earth ions.*

semiconductor materials, and their associated maximum powers and efficiencies obtained from high power laser bars. The operating spectral bands of laser diodes are overlaid on the characteristic distributions of electronic levels of rare earth and transition metals in Figures 3 and 4, respectively. We see that laser-diode pump-sources allow for the selective excitation of many levels of these ions, and thus the possibility of many efficient new laser schemes that were not practical with broadband lamp pump sources. To help identify practical novel candidate laser schemes suggested by Figures 3 and 4, it is particularly beneficial to assess radiative and nonradiative processes in potentially interesting laser ion/host combinations. In the case of rare earth laser ions, one can use the semi-empirical theory for crystal-field-induced radiative transition probabilities developed by Judd (1992) and by Ofelt (1962). Using three empirically determined (Judd-Ofelt) parameters for a given laser-ion/host-crystal combination, which are obtained from simple absorption measurements, the radiative transition probability between *any pair* of electronic levels of the rare earth ion can be calculated, along with the radiative lifetimes of all electronic levels. This procedure was first applied in relation to laser materials by Krupke (1966, 1971, 1974), and has been widely used in the decades since, especially in the fields of upconversion lasers and erbium doped fibre amplifiers (EDFAs).

The nonradiative decay of electronic levels in laser crystals also strongly influences the performance of solid state lasers. The "energy-gap" law, elucidated by empirical studies performed by Riseberg and Moos (1966) and by Riseberg and Weber (1976), is especially useful in the screening of possible novel diode-pumped laser schemes. They showed that the probability of nonradiative multiphonon mediated relaxation of a given

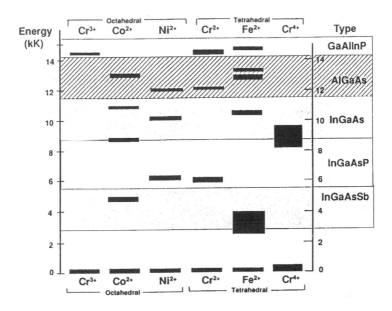

Figure 4. *Spectral overlay of laser diodes with the low-lying energy levels of some transition metal ions in octahedral and tetrahedral sites.*

electronic level of a rare earth ion scales exponentially with the energy gap to the next lower lying electronic energy level. The absolute relaxation rate in a given crystal depends on an ion-lattice coupling strength (approximately constant for all rare earth ions) and the maximum energy of optical phonons in the crystal. The latter declines from a high of ~1400 wavenumbers in materials with oxygen anions, to a low of ~200 wavenumbers in materials with heavy anions such a chlorine, sulphur, selenium, and tellurium. Because the non-radiative decay rate scales exponentially with this parameter, one can expect orders of magnitude changes in level lifetimes in the different host materials. This effect is therefore extremely important in identifying novel new laser materials, especially for lasers emitting at longer wavelengths with necessarily smaller energy gaps.

In addition to the "single ion" processes mentioned above, multi-ion interactions also come into play in practical solid state lasers. Important processes may include (1) migration of excitation energy among laser ions, (2) self-quenching of excitation energy by laser ions, (3) transfer (and possibly loss) of excitation energy to co-dopants and unintended impurities, and (4) interaction of two excited laser ions, resulting in the further excitation of one ion and de-excitation of the other ion (the so-called Auger process). Quantitative descriptions of these and other processes, in terms of spectroscopic and lattice properties, have been developed in recent years (for example, Zharikov *et al.* 1983, Payne *et al.* 1992). In general, energy transfer probabilities are highly dependent on the concentrations of interacting ions and on the spectroscopic details of the participating electronic transitions. Such studies are therefore not especially helpful in screening potential laser candidates, other than to make order-of-magnitude estimates of concentrations at which the various energy transfer processes may become important, relative to single-ion processes.

2   Bulk properties of laser crystals

Bulk thermal, mechanical, and optical properties of crystals also figure centrally in the performance of solid state lasers, as they govern the average output power and beam quality of the laser. To facilitate achieving high average output power, crystals with high values of the thermal shock resistance parameter, R_T, (Marion 1985, Marion *et al.* 1987a, Marion 1987b, Krupke *et al.* 1996) are preferred. Usually some minimum value of R_T is required to produce a given average output power from a laser. R_T is defined as

$$R_T = (1 - v)KS_T/Ea \tag{1}$$

where v is Poisson's ratio, K is the thermal conductivity, S_T is the tensile breaking strength, E is Young's modulus, and a is the coefficient of linear expansion. As described by Marion (1985), the breaking strength S_T of a brittle crystal is usually not determined by the intrinsic bond strength of the lattice ($\sim$10GPa), but is a much smaller value ($\sim$1/100) determined by microscopic flaws created just below the crystal surface during machining and polishing. A better measure of the intrinsic strength of a crystal is the fracture toughness, K_{IC}, which characterises the resistance of a brittle crystal to propagation of cracks. Its value is measured using diamond indentation to make a microcrack in the crystal surfaces using a known force. In terms of this parameter, one can write a modified form of Equation (1) as:

$$R_T' = (1 - v)KK_{IC}/Ea \tag{2}$$

Equations (1) and (2) show that crystals are preferred with small Poisson's ratio, high thermal conductivity, high tensile breaking strength or fracture toughness, and low coefficient(s) of thermal expansion and Young's modulus. Measured breaking strengths of conventionally finished laser crystals, when correlated with fracture toughness measurements, suggest that characteristic flaw sizes are about 50μm in length (and are probably oriented perpendicular to the surface as they are not normally observed optically). Assuming this typical flaw size, R_T' and R_T can be correlated. Tables 1 and 2 give compilations of R_T values for many crystals of interest for laser applications. Inspection of Table 1 reveals that covalently bonded oxide crystals generally possess R_T values that are order of magnitude higher than those of ionically bonded fluoride crystals. Table 2 indicates that many semiconductor crystals which are not normally thought of as host crystals for impurity doped solid state lasers, possess R_T values considerably higher than most oxide (laser) crystals. They may therefore prove to be of considerable interest as host crystals for novel transition metal and rare earth laser ions (see below). As mentioned above, the working tensile breaking strength can be increased, as shown by Marion *et al.* (1987a), by developing machining and polishing methods that reduce residual subsurface crack flaws to sizes below the typical 50μm size, while preserving optically flat crystal surfaces.

Fortunately, for generally directing attention to superior classes of crystals for use in lasers and nonlinear optics, there are numerous structure-property propensity rules that link bonding characteristics of the constituent cations and anions in the crystal to crystal properties such as electronic band structure, optical refractivity, mechanical stiffness, and various other mechanical and thermal characteristics. These correlation trends can be utilised as broad guides in identifying types and families of crystals likely to possess values of physical characteristics lying in the ranges being sought.

Crystal	K (W/m-K)	a $(10^{-6}/°C)$	E (GPa)	K_{IC} $(MPa[m]^{1/2})$	S_T (MPa)#	R_T (W/m)
Al_2O_3	28	6.7	405	2.2	440	3,400
MgO	59	13.5	260	0.8	160	2,000
$BeAl_2O_4$	23	6.5	520	2.6	520	2,650
$MgAl_2O_4$	13	9.	280	1.2	240	940
$LaMgAl_{11}O_{19}$	10	10.	180	1.1	220	920
$Y_5Al_3O_{12}$	13	6.7	280	1.4	280	1,450
$Gd_5(ScGa)_3O_{12}$	7	7.5	210	1.2	240	800
$Be_2Al_2(SiO_3)_6$	5	1.2	210	0.8	160	2,350
$YSiO_5$	5	~3	180	0.4	80	~1000
MgF_2	21	15	140	0.9	180	1,350
$LiYF_4$	6	14	80	0.3	60	240
$LiCaAlF_6$	5.5	19	110	0.43	85	170
$Ca_5(PO_4)_3F$	2	10	120	0.48	95	120
$Sr_5(PO_4)_3F$	2	91	20	0.51	100	140
$Sr_5(VO_4)_3F$	1.7	10	120	0.36	70	75
CaF_2	10	26	110	0.33	65	170
SrF_2	8	20	85	0.45	90	310
LiF	11	37	90	0.36	70	160
ZnSe	17	8	76	0.32	65	1,420
ZnSe	18	7	85	0.25	50	1,380

Table 1. *Thermal shock resistance, R_T of oxide, fluoride and sulphide crystals, along with the thermal conductivity K, coefficient of thermal expansion , a, Young's modulus E, fracture toughness, K_{ic} and tensile breaking stress, S_T.*

Parameter	Si	C	SiC	GaN	AlN	GaAs	GaP	ZnS	ZnSe
Conductivity K (W/m °C)	163	2000	500	130	230	53	97	17	19
Young's modulus E (GPa)	130	1050	460	308	440	85	103	75	66
Linear expansion $a(10^{-6}/°C)$	2.6	0.8	2.2	4.3	4.4	6.4	5.8	6.0	7.7
Poisson's ratio	0.28	0.1	0.21	0.23	0.29	0.31	0.31	0.29	0.28
Toughness $K_{IC}(m^{1/2}MPa)$	0.95	5.3	3.3	0.79	2.8	0.43	0.8	0.8	0.5
Rupture strength* $S_T(MPa)$	135	800	470	113	350	60	115	115	70
Thermal shock, $R'_T(W/m^{1/2})$	330	11360	1290	60	236	29	90	18	13
Thermal shock, R_T (W/cm)	471	16130	1850	86	337	41	130	26	19

R_T(YAG) ~10W/cm	R_T(TiS) ~34W/cm	R_T(YLF) ~1.6W/cm

Table 2. *Thermal, optical, and mechanical properties of semiconductor crystals, and a comparison of their thermal shock resistance with three well-established laser crystals.(* The rupture strength assumes a 50µm flaw size.)*

To facilitate achieving a laser output with high beam quality (i.e, near the diffraction limit) laser crystals with relatively small thermal focusing are desirable. Thermal focusing arises when a temperature gradient is formed transverse to the output laser beam axis (due to pumping and cooling of the laser gain element). For example, in a transverse

Crystal	Symmetry	n (av)	a_a $(10^{-6}/°C)$	a_c $(10^{-6}/°C)$	dn/dT_a $(10^{-6}/°C)$	dn/dT_c
Al_2O_3	trigonal	1.76	6.7	7.2	13.6	19.7
YVO_4	tetragonal	2.1	11.4	4.4	3	8.5
$LiYF_4$	tetragonal	1.46	13.3	8.3	-2	-4.3
$LiCaAlF_6$	trigonal	1.39	22	3.6	-4.2	-4.6
$Sr_5(PO_4)_3F$	hexagonal	1.63	9.5	8.4	-10	-8
$Y_3Al_5O_{12}$	cubic	1.82	7.7	7.7	9.1	9.1
nSe	cubic	2.59	7.1	7.1	~59	~59

Table 3. *Index of refraction (n), index-temperature derivative (dn/dT) and linear expansion coefficients (a) of some laser crystals*

pumped and cooled laser rod, the thermally induced focal length varies inversely with the waste heat deposited in the rod, and inversely with the quantity:

$$\frac{1}{2}\frac{dn}{dT} + aC_{r,0}n_0^3 + \frac{ar_0(n_0 - 1)}{L} \tag{3}$$

where dn/dT is the temperature derivative of the refractive index, n_0, $C_{r,0}$ are the radial and tangential photoelastic coefficients, a is the coefficient of linear thermal expansion, r_0 is the rod radius, and L is the rod length. The terms in this expression describe lensing effects due to (1) a variation of refractive index with the spatially varying temperature, (2) photo-elastic effects and (3) non-uniform thermal expansion. For many materials the photoelastic contribution to thermal focusing is small compared to the other two terms in Equation (3). Table 3 provides values for n_0, dn/dT, and a for several well known laser crystals. Note that all of the fluorine containing crystals possess negative dn/dT values, giving rise to a partial cancellation of the thermal focusing produced by the linear thermal expansion term of Equation (3). Such materials are designated partially or fully a-thermal, depending on the degree of cancellation. This feature of negative dn/dT is of significant importance in designing high average power lasers with excellent output beam quality, as demonstrated by the commercial Nd:YLF lasers produced by Q-peak (Finch *et al.* 1998). The athermal character of S-FAP is also an essential property in being able to design a diode-pumped inertial fusion energy driver with the necessary beam quality (see below).

3   Crystal growth

To be practical, a laser or NLO crystal must be growable in adequately large size, in adequately high optical quality, and be available at an affordable price (valued in terms of intended applications). While there are many methods for growing single crystals of optical quality (Laudise 1968), the prospect for the growth is greatly enhanced if the crystal chemistry allows the crystal to be grown from a "congruent" melt . Figure 5 shows the temperature-composition "phase diagram" (Nassau 1971) of a melt consisting of two chemical components, A and B. The heavy line of curved segments marks the boundary between liquid and solid, for all mixtures of A and B. The vertical line at AB_2 denotes a congruently melting compound, for which the solid and liquid at the melting point have

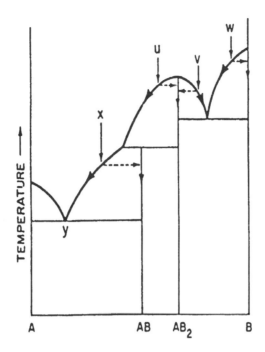

Figure 5. *Temperature-composition phase diagram, for a two component (A and B) system.*

the same composition. Thus a congruently melting crystal can be pulled from a melt of congruent composition (using the Czochralski technique) without experiencing significant shifts in composition. Such crystals can have extremely high chemical homogeneity and a correspondingly high homogeneity of optical refractive index, even when the pulled crystals are very large, and a large fraction of the melt volume is pulled. For these reasons, today, most laser crystals are congruently melting compounds and are pulled from a congruent melt by the Czochralski method.

The situation is rather different for crystals which do not melt congruently, such as compound AB in Figure 5. As the temperature increases, the solid crystal of composition AB reaches the temperature at which it melts into a solid of composition differing from AB, and a liquid specified by the point of intersection with the liquid-solid curve defined by projecting a horizontal line at this melting temperature. A single crystal of composition AB can be pulled from a liquid "solution" of composition different from the crystal composition, e.g. say point x in Figure 5, which lies below the melting point of compound AB. However, because the composition of the solution will necessarily shift as the crystal is pulled, the fraction of the melt that can be pulled, while yielding the desired crystal, is greatly reduced. Thus, in general terms, it is highly desirable if the laser or NLO crystal to be developed is congruently melting, as this will allow for the widest range of possible growth methods, and permit the growth of large high quality crystals, usually at the lowest unit cost.

4   Search for mid-IR tunable lasers for remote sensing

Growing environmental concerns are creating expanding needs for ever more capable laser sources for remote sensing of trace atomic and/or molecular species in the atmosphere. Such sensing is usually carried out using narrowband tunable radiation lying within one of the atmosphere's infrared transmission windows, and at wavelengths for which the target species possesses strong spectroscopic absorption features. At LLNL, a sensing application presented itself calling for a capable laser source in the 2000–2500nm atmospheric window (Krupke *et al.* 1997a). Analysis of the sensing scenario (detection range, anticipated target concentrations and spectral signatures, detector sensitivity, atmospheric scintillation noise, data acquisition, deployment platform utilities, etc.) indicated the need for a compact laser with the following performance characteristics:

- tunable wavelength range: ~2300–2500nm;
- pulse-repetition-rate: ~50kHz;
- efficiency: >5%;
- compactness: diode pumping.

- energy per pulse: ~0.2mJ;
- average output power: ~10W;
- operating temperature: ~300K;

There are several ways of progressively combining these performance values to delimit the spectroscopic parameters which a laser gain material must possess in order for the laser to deliver the required output. For example, the 50kHz pulse repetition rate infers that the upper laser level should have a fluorescence lifetime equal to or less than the inverse of the repetition rate, that is less than $20\mu s$. Of course, for efficiency reasons, we would like the upper laser level to be purely radiative, with this lifetime. Using the Einstein relationship linking the stimulated emission cross-section to emission wavelength (~2500nm), radiative decay rate ($<20\mu s$) , and transition bandwidth (~1800cm^{-1}, to meet the tuning requirement), we can estimate that the stimulated emission cross-section should equal $\sim 10^{-18}$cm^2, with a corresponding saturation fluence of 0.1mJ/cm^2. These numbers suggest that the laser transition must have a rather large f-number (about 10^{-3}), characteristic of a 3d transition metal ion in a crystal field site with strong odd parity electric field components. The 3d electrons of transition metal ions are known to interact strongly with lattice anions, resulting in strong and broad absorption bands. Given this background analysis, our attention was directed to the following class of ions and host crystals: divalent transition metal ions (TM^{2+}) placed in the asymmetric (T$_d$) lattice sites of the wide bandgap chalcogenide crystals , MX, where M=Cd, Zn; X = S, Se, Te. The heavy anions in these crystals ensure that the optical phonon cutoff occurs at very low energy, thus maximising the prospects for radiative decay of mid-IR luminescence in these crystals (Renz and Schulz 1983). Surprisingly, from Table 1 we see that the chalcogenides ZnSe and ZnS possess excellent bulk thermo-mechanical properties, having thermal shock resistance R_T values comparable to the thermo-mechanically robust garnets, such as YAG. From Table 3 we see that ZnSe also has a rather large value of dn/dT; its effect on thermal focusing will have to be mitigated through the utilisation of a thermal lens compensating gain element/cavity geometry, such as the "zig-zag" slab configuration. The divalent transition metal ions were selected as a potential class of laser ions because they are known (Fazzio *et al.* 1984) to be rich in energy levels in the mid IR spectral region (see Figure 6), and because their divalent charge state allows for substitution in T$_d$ lattice sites of MX crystals without the need for charge compensation. The T$_d$ site symmetry assures strong configurational mixing, and thereby strong induced electric dipoles between 3d

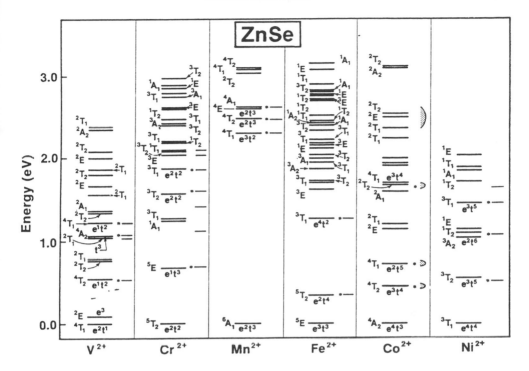

Figure 6. *Low-lying energy levels of divalent transition metal ions in ZnSe (after Fazzio 1994).*

levels, as called for by the above analysis of requirements for laser performance.

Inspection of Figure 6 shows why our attention for the present application was directed to Cr^{2+} as the most likely target for initial laser studies: (1) its first excited level lies at about the right energy ($\sim$4200 wavenumbers) to generate $\sim$2400nm emission; (2) the ground and first excited levels have the same spin, and therefore will have a relatively high transition f-number; (3) higher lying levels have spins that are lower than the ground and first excited levels, greatly mitigating the potential for significant excited state absorption at the pump or laser transition wavelengths; (4) the orbital characteristics of the ground and first excited levels are different, and will experience a significant Franck-Condon shift between absorption and emission, resulting in broadband "dye-like" absorption and emission characteristics, suitable for a broadly tunable laser.

On the basis of the foregoing analysis, experimental samples of Cr^{2+}:ZnSe Laser-spectroscopic parameters were measured by DeLoach *et al.* 1996. The room temperature absorption and emission spectra between the ground $^5T_2$ and first excited ^5E levels of Cr^{2+}:ZnSe, and the temperature dependent fluorescence lifetime of the ^5E upper laser level are shown in Figure 7. The Cr^{2+} ion shows a broad, strong absorption band peaked at $\sim$1800nm, and a broad emission band peaked at a wavelength of $\sim$2400nm. The spectral width of the emission band is about 1800cm^{-1}. The low temperature lifetime is about 4.8μs, increases to about 7.5μs at room temperature, and declines above room temperature. This temperature behaviour suggests that the ^5E level decays radiatively up to room temperature, and begins to decay non-radiatively above room temperature. The

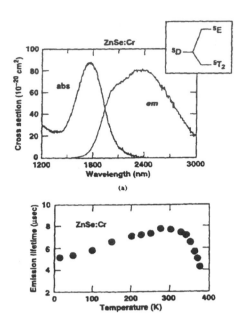

Figure 7. *Properties of Cr^{2+}:ZnSe after Page 1997a. top: room temperature absorption and emission spectra; bottom: fluorescence lifetime as function of temperature.*

relatively short lifetime of $7.5\mu s$ (at room temperature) is expected for a spin-allowed 3d-3d transition of an ion in tetrahedral coordination (as mentioned above). The calculated stimulated emission cross-section for Cr^{2+} in ZnSe is $0.8 \times 10^{-18} cm^2$, and the corresponding saturation fluence and flux are, respectively, $0.11 J/cm^2$ and $13 kW/cm^2$. Initial laser experiments (Page *et al.* 1997a) were carried out using a Co:MgF$_2$ laser as a pump source at 1860nm. For the resonator utilised, a threshold energy of 0.5mJ was observed, along with a slope efficiency of 27% relative to absorbed pump energy. Laser emission occurred at a wavelength of 2350nm (near the peak of the fluorescence spectrum, suggesting the presence of little or no excited state absorption). Using a grating as a tuning element, tuning over the spectral region from ~2180 to 2660nm was observed. More recently Page *et al.* (1997b) pumped a Cr^{2+} :ZnSe laser using a 2-D InGaAsP diode array emitting at ~1650nm. These demonstrations all support the initial intention of providing a novel, diode-pumped, tunable laser gain in the mid-IR to enable efficient and compact remote sensing applications to be performed from utility-constrained platforms.

While performing the Cr^{2+}:ZnSe search campaign, it was noted (Krupke *et al.* 1997a) that the tuning range of Cr^{2+} could probably be extended to wavelengths as long 2900nm, corresponding to the peak of a strong water absorption band. Thus, it seemed clear that the Cr^{2+} laser, when tuned from ~2200–2900nm might serve as a very versatile surgical "laser scalpel", varying the depth of energy deposition in tissue from about a micrometre near 2900nm to about a millimetre near 2200nm.

The laser results reported by Page *et al.* (1997a,b) prompted other groups to explore analogous chalcogenide crystal hosts for tunable Cr^{2+} lasers. Hoemmerich *et al.* (1997)

explored the mixed cation telluride, $Cd_{0.85}Mn_{0.15}Te$. CdMnTe crystals doped with Cr^{2+} ions can be grown with optical quality by the Bridgman method (Laudise 1968). Hoemmerich *et al.* (1997) found absorption and emission bands analogous to those of Cr:ZnSe, with high quantum efficiency at room temperature, a room temperature lifetime of $1.4\mu s$, and a stimulated emission cross-section of 2.7×10^{-18} cm^2. Room temperature laser action was demonstrated by Hoemmerich *et al.* (1997) using a Raman-shifted Nd:YAG laser as a pump source at $\sim 1.9\mu m$. More recently, Seo *et al.* (1998) demonstrated improved laser action using a Tm:Ho:YAG laser at $2.01\mu m$ as a pump, and reported a maximum slope efficiency of 44.2% with respect to absorbed pump energy. Their measurements also implied that little or no excited state absorption was present, and that the pump quantum yield was nearly unity. They demonstrated tuning from 2300 to 2600nm.

Schepler *et al.* (1997) explored CdSe as a host for Cr^{2+} lasers and found encouraging spectroscopic characteristics: a room temperature fluorescence lifetime of $6\mu s$, high quantum efficiency of emission, and a stimulated emission cross-section of 1.8×10^{-18}cm^2. Laser action has just been reported (McKay *et al.* 1998) for this ion/host combination. The laser was pumped at 2090nm and laser action occurred at 2520nm. The threshold absorbed pump energy was $10\mu J$, and the maximum output energy obtained was $70\mu J$, with a slope efficiency of 60%.

That three different chalcogenide host crystals doped with Cr^{2+} ions have been demonstrated as effective tunable lasers, suggests that the Cr^{2+} ion can be utilised in a variety crystal hosts, which can be selected for development on the basis of possessing superior bulk crystal properties (ease of growth, dopability, optical quality, residual background loss, etc.). It is also interesting to note that the Cr^{2+} laser is in many ways the mid-infrared analogue of the titanium-doped sapphire (TiS) laser which has operated with diverse output waveforms: CW, free-running long-pulse, repetitively-Q-switched, mode-locked, and chirp-pulse-compressed. It is expected that, in the near future, Cr:MX lasers will also be realised with all of these output waveforms, but with the added advantage of being directly pumpable with InGaAsP laser diodes.

A great number of hydrocarbon molecules possess strong spectral features in the $3.6\mu m$ spectral region that are suitable for remote sensing at trace levels. Inspection of Figure 6 suggests that the Fe^{2+} ion is a likely candidate for a diode-pumped, tunable laser in the 3400–3800nm region. The Fe^{2+} ion has six 3d electrons that (in T_d symmetry) give rise to a ^5E ground level and a $^5T_2$ first excited level (the inverse of the Cr^{2+} ion with four 3d electrons). As in the Cr^{2+} ion, in the spectral region of interest (pump and laser wavelengths) there are no spin allowed transitions from either of these levels to higher lying levels, so that one would not expect the presence of deleterious excited state absorption processes (within the Fe^{2+} ion). Laser action in Fe^{2+} has already been reported (Klein *et al.* 1983) in the InP semiconductor host crystal. In InP, the Fe^{2+} ion ground level lies relatively close to the conduction band, and laser action could be observed only at cryogenic temperatures, through a complex pumping process involving hole-electron pair production followed by conduction-band electron recombination on Fe^{3+} centres, and population inversion on the resulting excited Fe^{2+} centre. In the case of Fe^{2+} in ZnSe, it is expected that the $^5T_2$ upper laser level of Fe^{2+} near 3000nm can be optically pumped directly without parasitic processes involving either the valence or conduction bands. Recently, Adams *et al.* (1998) has realised the Fe^{2+}:ZnSe laser using an Er:YAG pump laser near 2700nm. The peak emission wavelength is ~ 4000nm with a fluorescence lifetime at low temperature of $\sim 40\mu s$. The lifetime increases to a maximum value of $115\mu s$ at $\sim 150K$, and decreases to a few microseconds at $T > 220K$. The peak stimulated emission cross-section at 3982nm is estimated to be $\sim 3 \times 10^{-19} cm^2$. Laser action was observed at a free-running wavelength of 3982nm, near the peak of the fluorescence emission spectrum. As in the case of Cr^{2+}, this suggests that there is little or no significant excited state absorption, consistent with the electronic structure of the Fe^{2+} ion in T_d symmetry. These initial laser results suggest that the prospect for a compact tunable laser in the 3600–4000nm region is quite good. The further characterisation and development of this, the second example of the new class of solid state lasers, TM^{2+}:MX, is in progress; the search for Fe^{2+}:MX crystals with radiative emission at room temperature also seems warranted.

From a broader perspective, given the attractive thermo-mechanical properties of III/V and II/VI wide bandgap semiconductor crystals, and these demonstrations of excellent laser performance in such crystals, bulk semiconductor crystals doped with rare earth and/or transition metal ions should be viewed as fertile grounds for future solid state laser development.

5   An inertial-fusion-energy laser-driver gain crystal

During the past 25 years, Nd:glass lasers have been successfully developed and utilised worldwide for conducting single-shot inertial confinement fusion (ICF) experiments of ever increasing energy (>100kJ) and power (>200TW). In looking beyond the laboratory demonstration of scientific energy break-even and energy gain, studies have projected the performance requirements for a laser driver of $\sim 1GW$ inertial fusion energy (IFE) power reactor for the production of central electric power: (1) average output power $\sim 10MW$; (2) efficiency >10%; (3)pulse energy $\sim 4MJ$; (4) pulse duration $\sim$few ns; (5) wavelength $\sim 350nm$; (6) beam quality $M^2 \leq 5$.

While flashlamp pumped Nd:glass laser systems are viewed as excellent for laboratory research, solid state lasers historically have been regarded as unlikely to be utilised in the IFE application for several technical reasons: (1) they cannot meet the efficiency requirement (flashlamp pumped Nd:glass lasers are < 1% efficient; (2) they cannot reasonably be scaled to $\sim MW$ of average output power (because the crystals will break); (3) they cannot achieve the required high beam quality at >MW power levels (due to unmanageable thermally-induced birefringence and focusing). All proposed IFE drivers (heavy and light ion accelerators, and KrF lasers) are also seen as too expensive. All of these perceptions deserved to be reassessed (Krupke 1997b) with the development of efficient high power semiconductor laser diodes as pump sources for solid state lasers. Diode pumped solid state lasers (DPSSLs) with efficiencies in excess of 10% have already been demonstrated (for selected configurations), and such efficiencies can be anticipated for the DPSSL IFE architecture (parallel master-oscillator, power-amplifier (MOPA) chains) described below. One technical flow-down pathway to identify an acceptable architecture for a DPSSL IFE driver is presented in Table 4. Laser system requirements are given in the first column, the controlling technical considerations are given in the middle column, and the implications for the system architecture are given in the third column. Starting in the upper left hand corner, pulse energy, pulse duration, and peak power are given. For pulses of a few nanoseconds duration, optical damage of robust optical mate-

Requirement	Consideration	Architecture Implication
Pulse Energy ~4MJ Pulse length ~3ns	Optical damage ~<20J/cm^2	Total output beam area ~20m^2
output area ~20m^2	ASE	Segment into multiple apertures Maximum aperture diameter ~20cm
	Self-focusing B-Integral	High gain coefficient of ~0.1cm^{-1} Low n_2 gain material (<1.3x10^{-13} esu)
Pulse rep-rate ~3–10Hz Average Power ~10MW	Thermal stress Beam quality	Segment gain medium axially (disks) Gas cool faces of disk gain elements
Efficiency ~10%	Pump brightness	Use laser diode pump arrays
Cost <1B$	Diode utilization	>ms energy storage lifetime

Table 4. *IFE DPSSL driver requirements - architecture flow-down sequence*

rials occurs above ~20J/cm^2. Dividing the total system energy of ~4MJ by the working damage fluence of 20J/cm^2 yields a total output beam area of ~20m^2. Taking this total beam area as a derived requirement, we ask how this area must be subdivided due to the combined constraints of amplified spontaneous emission transverse to the output beam axis, and nonlinear self-focusing at high beam intensities. Experience with Nd:glass lasers instructs us to design to small-signal gain coefficient values of about >0.05cm^{-1} (assuming a gain-material nonlinear refractive index of ~1.3×10^{-13}esu). This strikes a reasonable balance between the total gain path of solid state material necessary to amplify the small input pulse energy to the desired output pulse energy in a given amplifier aperture, and the diameter of that aperture, as limited by the reduction of gain due to transverse amplified spontaneous emission (aperture × gain coefficient <2). Efficient volumetric utilisation of the gain material suggests designing to the somewhat higher a gain coefficient of ~0.1cm^{-1}. These considerations tell us that we need to segment the total output area of ~20m^2 into several hundred "beamlets", each with an optical diameter of about ~20cm. The output energy from an aperture of this size can be adequately delivered to a millimetre-sized IFE fuel pellet using a mirror system of 50m focal length, if the beam quality is $M^2 \leq 5$. The next system requirement is ~10MW of average power at a pulse repetition rate of 3–10Hz.

Thermally-induced stress and birefringence considerations dictate that the gain medium be segmented along the amplifier axis into thin disks (or slabs), whose broad faces are both pumped with diode light and cooled with transversely flowing helium gas (Krupke 1997b). Of course, the system efficiency of >10% dictates that efficient laser diode pump arrays be used. But because the pump radiation must be propagated through the faces of the axially-segmented gain disks over a rather long distance, the entendue or brightness of the pump arrays must be extraordinarily high in this application. To mitigate the issues of pump brightness and diode cost central to this approach, one is forced to find a gain medium with unusually low pump and laser saturation fluxes and an energy storage lifetime that is many times that of Nd:glass lasers (typically ~350μs).

Several years ago researchers at LLNL set out to search systematically for a suitable laser gain medium that responds to these IFE DPSSL architecture/gain-material trade-offs. The then recent development of very reliable InGaAs laser pump diodes emitting in the 900–980nm spectral region (mainly for pumping EDFAs) opened up the interesting

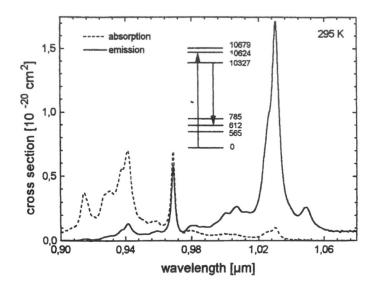

Figure 8. *Energy levels, absorption and emission cross-sections of Yb:YAG.*

possibility of using ytterbium as the active ion in an IFE DPSSL (Lacovara *et al.* 1991, Krupke and Chase 1990). Figure 8 shows the simple two-manifold energy level structure of the Yb^{3+} ion and the absorption and emission spectrum of Yb:YAG (Lacovara *et al.* 1991, DeLoach *et al.* 1993). Local crystal fields split the two Yb^{3+} ion manifolds into two groups of Stark levels. Diode pumping takes place from the ground Stark level of the ground $^2F_{7/2}$ manifold to a high lying Stark level of the excited (upper laser) $^2F_{5/2}$ manifold. Laser emission occurs from the lowest lying Stark level of the $^2F_{5/2}$ manifold to a high lying Stark level of the ground $^2F_{7/2}$ manifold. Because manifold splittings are typically 500–1000cm^{-1}, the terminal Stark levels of both pump and laser transitions will contain some thermal population at room temperature. For this reason, Yb^{3+} lasers are said to be "quasi-four-level" lasers. Attractive features of Yb^{3+} for the IFE DPSSL application are: (1) an energy storage lifetime in the 1–2ms range (~4x that of Nd doped laser materials; (2) a low quantum energy defect between laser and pump photon energies (0.09 for Yb vs 0.24 for Nd); (3) absence of ion-ion self quenching (since there are no intermediate electronic manifolds between the excited $^2F_{5/2}$ and $^2F_{7/2}$ ground manifolds; and (4) no possibility for Auger radiationless decay, again since there are no other excited manifolds in the Yb^{3+} ion.

Given all of these very positive features, we need to ask what are the negative features, or challenges for this laser ion. The main problems are two-fold: (1) the relatively long radiative lifetime of the excited $^2F_{5/2}$ manifold, and the distribution of emission transition moment among many possible inter-Stark emission transitions, which usually results in a small laser transition cross-section. Similarly, the pump transition cross section is also small; (2) the thermal population in the laser transition terminal Stark level represents a resonance loss that must be overcome by pumping to transparency before laser action can occur. The generally small pump and laser transition cross-sections result in much

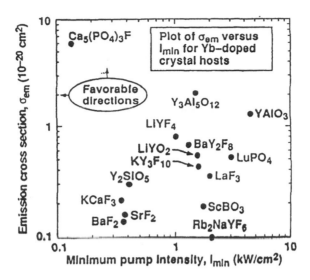

Figure 9. *Peak laser emission cross-section and minimum pump intensity map of Yb^{3+} doped crystals (DeLoach 1993).*

higher values for the pump and laser saturation fluxes than those characterising Nd doped laser crystals. These high saturation parameters, in turn, place very stressing flux and brightness demands on diode pump arrays for Yb lasers.

To identify and develop Yb doped laser crystals possessing unusually high pump and laser emission cross-sections, and correspondingly low pump and laser saturation fluxes, DeLoach *et al.* (1993) conducted a broad study of the spectra of a wide range of Yb doped crystals. They used as key figures of merit, the peak laser emission cross-section, and the minimum pump flux, I_{min}, needed to reach transparency. Figure 9 shows the results of this survey on a log-log plot. As a high value of stimulated emission cross-section and a low value of I_{min} are preferred, promising Yb doped materials are located toward the upper left hand region of the plot. Note that all but one of the crystals studied fall in the lower right hand half of the figure. In contrast, Yb doped fluorapatites (defined by the mineral, $Ca_5(PO_4)_3F$, or FAP) exhibit remarkably high stimulated emission cross-sections and very low I_{min} values. This is because Yb-FAP exhibits an unusually large Stark splitting of both ^2F manifolds, and allocates the available transition moment only to the pump and laser transitions (a very sparse absorption and emission spectra). Table 5 lists the key laser spectroscopic parameters for the well known Yb:YAG crystal and for Yb:FAP. Note that the pump and laser saturation flux and fluences differ between these two gain crystals by significant factors. To get a quantitative practical feel for the impact of these differing parameter values, we need to calculate the excited fractions of Yb ions, and the maximum extractable fraction of excited ions in Yb lasers (Krupke and Chase, 1990).

Figure 10 shows the Stark levels of the two manifold Yb^{3+} ion, and necessary definitions: j-th Stark level population, n_j and degeneracy, g_j; total Yb concentration, N_T;

Parameter	YAG	FAP
pump wavelength,nm	941	900
pump transition width, nm	18	2.4
pump cross-section, 10^{20} cm^2	0.8	10
pump saturation intensity, kW/cm^2	28	1.9
laser wavelength, nm	1030	1043
laser transition width, nm	9	4
laser cross-section, 10^{-20} cm^2	2.0	5.9
laser saturation intensity, kW/cm^2	10	3
radiative lifetime, msec.	0.94	1.1
minimum pump intensity, kW/cm^2	1.54	0.13

Table 5. *Yb:YAG and FAP laser spectroscopic parameters.*

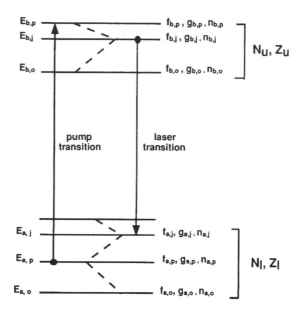

Figure 10. *Stark energy levels in the Yb^{3+}ion with pump and laser transitions shown. Z_l and Z_u are the partition functions for the lower and upper manifolds. The relationships between Stark level and manifold populations are as follows:*
(1) $n_{a,p} = f_{a,p} N_l$; (2) $n_{b,l} = f_{b,l} N_u$; (3) $n_{a,l} = f_{a,l} N_l$; (4) $n_{b,p} = f_{b,p} N_u$; (5) $N_T = N_l + N_u$.

upper and lower manifold populations, N_u and N_l, respectively; upper and lower partition functions, Z_u, and Z_l, respectively; and initial and terminal Stark levels subscript indices for the pump transition, ap, bp, and for the laser transition, bl, al, respectively. It is reasonably assumed that, for all pump and laser transition rates, the population distribution within each manifold is given by a Boltzmann distribution at the lattice temperature. Then, using the principle of detailed balancing in the presence of pump and laser radiation

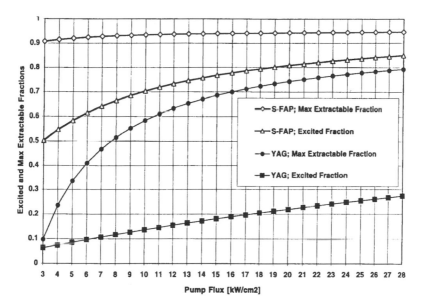

Figure 11. *Maximum extractable fraction and excited fraction of Yb^{3+} ions in YAG and FAP, as functions of pump flux.*

fields with intensities, I_p and I_l respectively, one can write down the fraction of Yb ions excited, $F'_{excited}\,(I_p)$ as a function of the pump intensity:

$$F_{excited}(I_p) = F_{excited,max}[1 + (I_{sat}/I_p)]^{-1} \qquad (4)$$

due to bleaching of the ground manifold population. I_{sat} is the pump saturation flux, defined as the pump photon energy divided by the product of pump transition cross-section and $^2F_{5/2}$ manifold fluorescence lifetime. The maximum fraction of Yb ions that can be excited to the upper manifold, $F'_{excited,max}$ due to an infinitely high pump intensity is given by:

$$F'_{excited,max} = \left[1 + \frac{Z_l}{Z_u} \exp\left(\frac{E_{a,p} - E_{b,p}}{kT}\right)\right]^{-1} \qquad (5)$$

The maximum extractable fraction, $F'_{extracted,max}$ of the fraction of excited Yb ions, for an infinitely high laser intensity, is given by:

$$F'_{extracted,max} = 1 - \frac{1}{F_{excited}(I_p)}\left[1 + \frac{Z_l}{Z_u} \exp\left(\frac{E_{b,l} - E_{a,l}}{kT}\right)\right]^{-1} \qquad (6)$$

In these expressions, the Stark level energies, $E_{a,j}$ and $E_{b,j}$ are measured from the lowest lying Stark level in their respective manifolds. The calculated values for $F_{excited}(I_p)$ and $F_{extracted,max}$ as functions of pump intensity, I'_p, for Yb:YAG and Yb:FAP, are shown in Figure 11.

The calculations were carried out using the spectroscopic values for Yb:YAG and Yb:FAP given in Table 5. The upper two curves for Figure 11 are for Yb:FAP, and the lower two curves are for Yb:YAG. We see that even for the relatively small pump intensity of 3kW/cm^2, the maximum extractable fraction for Yb:FAP is over 90% while that for Yb:YAG is only 10%. Similarly, the excited fraction at this pump intensity is about 50% for Yb:FAP while it is only about 8% for Yb:YAG (i.e., Yb:YAG is barely over the transparency condition). From this Figure we can see that the Yb:FAP gain material greatly relieves the demand for array pump intensity, compared to Yb:YAG. Also, because Yb:FAP has a laser transition cross-section that is about three times bigger that that of Yb:YAG, the small signal gain coefficient of Yb:FAP will be three times larger than for Yb:YAG at the same excited manifold population density. It was for these reasons that Yb:FAP (actually the strontium-substituted fluorapatite, $Sr_5(PO_4)_3F$) was adopted for the development of a scalable IFE DPSSL.

Of course, it was necessary for this crystal to meet other requirements as well. Detailed modeling (Orth *et al.* 1996) showed that Yb:S-FAP has adequate thermo-mechanical properties to scale in the gas-cooled slab architecture to the average power levels needed (about 100kW in a single coherent 10×16cm^2 aperture). Its partial a-thermal focusing property mitigates against excessive beam focusing in the design as well. This crystal also possesses the all important property of melting congruently, so that with proper development (Schaffers *et al.* 1998), it is expected that adequately large, high optical quality crystals of Yb:S-FAP can be grown successfully. High quality experimental Yb:S-FAP crystals have been grown, completely characterised spectroscopically and physically, and functionally demonstrated in the face gas-cooled geometry, at small scale (Marshall *et al.* 1996, 1998). On the basis of satisfactory results from these efforts, LLNL has committed to design, develop, and construct the Yb:S-FAP Mercury laser system producing a kilowatt of average power (100J/10Hz) at a 10% efficiency. To implement this system in the year 2000, 4×6cm^2 Yb:S-FAP gain crystals, cost-effective 2-D InGaAs diode pump arrays, and gas-cooled disk technologies are in current development. It is hoped and expected, that once developed, the Yb:S-FAP laser crystal will find many other, more conventional applications because of its ability to store diode array energy in a very cost competitive manner.

6   NLO crystals for high average power applications

Presently, an active area of development in nonlinear optics is scaling the average power (Eimerl 1987) of 2nd, 3rd, and 4th harmonics of $\sim$1060nm Nd lasers, for applications in micro-machining, isotope separation, and micro-lithography (Velsko and Krupke 1996). For these harmonically up-converted lasers to be economically viable, the relatively expensive nonlinear harmonic crystal must operate without significant degradation for a long period of time (thousands of hours). The most robust and effective nonlinear harmonic converters and frequency-mixers have proven to be BBO and LBO (Velsko *et al.* 1991). Both of these borate crystals are characterised by high transparency into the ultraviolet, low scatter loss, high optical damage threshold, moderate nonlinear coefficients, and moderate angular and thermal sensitivities (Chen *et al.* 1989, Becker 1998). As the demand for ever higher average powers of visible and UV radiation increase, now moving from the tens of watts to hundreds of watts in the green spectral region, and to the few watt level in the UV, there is an increasing need for larger BBO and LBO crystals. The main problem in scaling the size of BBO and LBO crystals, while retaining the necessary high optical

Crystal (Process)	PM Angle(°) (θ, ϕ)	Angular Sensitivity. B_O (cm-rad)$^{-1}$	Angular Accept. $\Delta\theta.L$ (cm-deg)	Thermal Sensitiv. B_T (cm-°C)$^{-1}$	Thermal Accept. $\Delta T.L$ (cm-°C)
KTP (II)	(90, 23.5)	618	0.52	0.22	25
LBO (II); 23 °C	(90, 11.6)	1336	0.24	0.87	6.4
LBO (I) 158 °C	(90, 0) NCPM	0	–	1.42	3.9
LBO (II)	(21.4, 90)	571	0.56	0.69	8.1
GdCOB (I)	(90, 70.4)	2530	0.12	0.05-0.15	38-105
CLBO (I)	(29.3)	1130	0.28	0.18	31.1

Table 6. *Phase-matching angles and angular and thermal sensitivities and acceptances of some borate NLO crystals. (Frequency doubling of 1064nm radiation)*

quality, is that these crystals melt incongruently, and are grown from a high temperature flux (solution). The technical NLO properties of borates in general are so attractive that the search for additional borate NLO crystals has continued in recent years. Two new borate NLO crystals have been reported: $CsLiB_6O_{10}$ or CLBO by Mori *et al.* (1995), Yap *et al.* (1996), and $Ca_4GdO(BO_3)_3$ or GdCOB , by Aka *et al.* (1997). These NLO crystals share the same NLO qualities as the other borates mentioned above (see Table 6), but they also possess the additional feature of melting congruently. As we have noted in the section on crystal growth, congruent melting allows CLBO (Mori *et al.* 1995, Ryu *et al.* 1998) and GdCOB (Aka *et al.*. 1997) crystals to be grown in large size, with high optical quality, using the Czochralski method. CLBO is now finding its way into high average power UV laser systems (Finch *et al.* 1998).

The GdCOB NLO crystal is also interesting because it represents a family of congruently melting rare-earth and yttrium substituted oxoborates with complimentary NLO properties. For example, Yoshimura *et al.* (1998) reported that a 25% yttrium substituted GdCOB crystal was noncritically phase-matched for producing 355nm radiation by sum-frequency mixing 1064 and 532nm radiation from a Nd:YAG laser. This can have considerable importance for IFE lasers that require the generation of 355nm light at high pulse energies (e.g., requiring large crystal sizes). Mougel *et al.* (1997) showed that GdCOB could be doped with Nd ions because of the presence of Gd lattice sites, allowing for effective self-frequency-doubling to be demonstrated. Jang *et al.* (1998) showed the analogous case of self-frequency-doubling in Yb doped YCOB. Thus, it seems that borate-based NLO crystals continue to provide new design options and capabilities for high performance NLO devices from the near infrared to the deep ultraviolet.

7 Conclusions

The advent of efficient high power laser diodes has spawned a renaissance in the field of solid state lasers in the past decade or so, not only by replacing flashlamps in well known lasers (rendering them more compact, efficient, and utility benign), but also opening the possibility for entirely novel lasers. Novel impurity doped crystals likewise have much to contribute to this revolution in the field of diode-pumped solid state lasers. Fortunately, the laser community is in possession of an array of capable tools to help us identify and develop such novel enabling laser and NLO materials.

Acknowledgements

The author is greatly indebted to many colleagues at LLNL for their major contributions to the guided searches for novel laser materials outlined here. Indeed, determination of the specific paths that such searches should take is not so much a proscriptive process at present, but rather derives from the "expert system" established at LLNL, comprised of the following principle members: Steve Payne, Ray Beach, Eric Honea, Ralph Page, Camile Bibeau, Chris Marshall, Laura DeLoach, and Kathleen Schaffers. We are all indebted to Prof. Bruce Chai of CREOL, Univ. of Central Florida, an external member of this expert system. The author also is pleased to acknowledge the expertise in semiconductor laser diodes shared by LLNL colleagues Mark Emanuel, Jay Skidmore, and Rich Solarz.

References

Adams J J, Bibeau C, Payne, S A and Page R H, 1999, "Tunable laser action at 4.0 microns from Fe:ZnSe", *OSA Advanced Solid State Lasers*, February, 1999.

Aka G, Kahn-Harari A, Mougel F, Vivien D, Salin F, Coquelin P, Colin P, Pelenc D and Damelet JP, 1997, "Linear- and nonlinear optical properties of a new gadolinium calcium oxoborate crystal, $Ca_4GdO(BO_3)_3$", *J Opt Soc Am B* **14**, 2238

Beach R, Albrecht G, Krupke W, Comaskey B, Mitchell S, Brandle C, and Berkstresser G, 1990, "Q-switched laser at 912 nm using ground-state-depleted neodymium in yttrium orthosilicate", *Opt Lett*, **15**, 1020.

Becker P, 1998, "Borate materials in nonlinear optics", *Adv Mater*, **10**, 979.

Chen C, Wu Y, Jiang B, Wu B, You G, Li R and Lin S, 1989, "New nonlinear optical crystal: LiB_3O_5", *J Opt Soc Am*, **6**, 616.

DeLoach L D, Payne S A, Chase L L, Smith L K, Kway W L and Krupke W F, 1993, "Evaluation of absorption and emission properties of Yb^{3+} doped crystals for laser applications", *IEEE J Quantum Electron*, **29**, 1179.

DeLoach L D, Page R H, Wilke G D, Payne S A, and Krupke W F, 1996, "Transition metal doped zinc chalcogenides: spectroscopy and laser demonstration of a new class of gain media", *IEEE J Quantum Electron*, **32**, 885.

Eimerl D, 1987, ""High average power harmonic generation", *IEEE J Quantum Electron*, **23**, 575.

Fazzio A, Caldus M J and Zunger A, 1984, "Many-electronmultiplet effects in the spectra of 3d impurities in heteropolar semiconductors", *Phys Rev B*, **30**, 3430.

Finch A, Obsako Y, Sakuma J, Deki K, Horiguchi M, Mori Y, Sasaki T, Wall K, Harrison J and Moulton P J, 1998, "Development of a high-power, high repetition rate, diode-pumped, deep UV laser system", *OSA TOPS, Adv. Solid State Lasers*, **19**, 16.

Hoemmerich U, Wu X, Davis V R, Trivedi S B, Grasza K, Chen R J and Kutcher S, 1997, "Demonstration of room-temperature laser action at 2.5μm from $Cr^{2+}:Cd_{0.85}Mn_{0.15}Te$", *Opt Lett*, **22**, 1180.

Jang W K, Ye Q, Eichenholz J, Richardson M C and Chai B H T, 1998, "Second harmonic generation in Yb doped $YCa_4O(BO_3)_3$", *Opt Commun*, **155**, 332.

Judd B R 1962, "Optical absorption intensities of rare-earth ions", *Phys Rev*, **127**, 750.

Klein P B, Furneaux J E and Henry R L, 1983, "Laser oscillation at 3.53 microns from Fe^{2+} in n-InP:Fe", *Appl Phys Lett*, **42**, 638.

Krupke W F 1966, "Optical absorption and fluorescence intensities in several rare-earth-doped Y_2O_3 and LaF_3 single crystals", *Phys Rev*, **145**, 325.

Krupke W F 1971, "Radiative transition probabilities within the $4f^n$ ground electronic configuration of Nd:YAG", *IEEE J Quantum Electron*, **7**, 153.

Krupke W F, 1974, "Induced-emission cross-sections in neodymium laser glasses", *IEEE J Quantum Electron*, **10**, 450.

Krupke W F and Chase L L, 1990, "Ground-state depleted solid state lasers: principles, characteristics, and scaling", *Opt and Quantum Electron*, **22**, S1.

Krupke W F, Shinn M D, Marion J E, Caird J A, and Stokowski S E, 1996, "Spectroscopic, optical, and thermomechanical properties of neodymium- and chromium-doped gadolinium scandium gallium garnet", *J Opt Soc Am B*, **3**, 102.

Krupke W F, Page R H, Schaffers K I, Payne S A, Beach R J, Skidmore J A and Emanuel M A, 1997a, "A new class of diode-pumped, mid-IR, broadly-tunable lasers based on TM^{2+} ions in T_d coordination: Cr2+:ZnX (X=S,Se)", *Tunable Solid State Lasers, SPIE* , **3176**, 47.

Krupke W F, 1997b, "Diode-pumped solid state lasers (DPSSLs) for inertial fusion energy", *Solid State Laser for application to Inertial Confinement Fusion*. SPIE **3047**, 73.

Lacovara P, Choi H K, Wang C A, Aggarwal R L and Fan T Y, 1991, "Room- temperature diode-pumped Yb:YAG laser", *Opt Lett*, **16**, 1089.

Laudise R A, 1968, "Techniques in Crystal Growth", Chapter 6, *Appl. Solid State Physics*, Plenum Press, NY; W. Low and M. Schieber, eds.

Marion J, 1985, "Strengthened solid-state laser materials", *Appl Phys Lett*, **47** 694.

Marion J, Devlin M, Gualtieri B and Morris R C, 1987a, "Compressive epitactic layers on single-crystal components for improved mechanical durability and strength", *J Appl Phys*, **62**, 2065.

Marion J, 1987b, "Appropriate use of the strength parameter in solid state slab laser design", *J Appl Phys*, **62**, 1595.

Marling J B, Nilsen J, West L C and Wood L L, 1979, "An ultrahigh Q, isotropically sensitive optical filter employing atomic resonance transitions", *Appl Phys*, **50**, 610.

Marshall C D, Smith L K, Beach R J, Emanuel M A, Schaffers K I, Skidmore J, Payne S A and Chai B H T, 1996, "Diode-pumped ytterbium doped $Sr_5(PO_4)_3F$ laser performance", *IEEE Quantum Electron*, **32**, 650.

Marshall C D, Beach R J, Bibeau C, Ebbers C A, Emanuel M A, Honea E C, Krupke W F, Payne S A, Powell H T, Schaffers K I, Skidmore J A and Sutton S B, 1998, "Next-generation laser for inertial confinement fusion", *Laser Physics*, **8**, 741.

McKay J, Schepler K and Kueck S, 1998, "Chromium (II):cadmium selenide laser", Technical abstract ThGG1, OSA Annual Meeting, October 8, 1998.

Mori Y, Kurode I, Nakajima S, Sasaki T and Nakai S, 1995, "New nonlinear optical crystal: cesium lithium borate", *Appl Phys Lett*, **67**, 1818.

Mougel F, Aka A, Kahn-Harari A, Hubert H, Benitez J M and Vivien D, 1997, "Infrared laser performance and self-frequency doubling of $Nd^{3+}:Ca_4GdO(BO_3)_3$ (Nd:GdCOB)", *Opt Mater*, **8**, 161.

Nassau K, 1971, *The Chemistry of Laser Crystals*, Applied Solid State Science, **2**, 174; Academic Press, NY.

Ofelt G S 1992, "Intensities of crystal spectra of rare earth ions", *J Chem Phys*, **37**, 511

Orth C D, Payne S A and Krupke W F 1996, "A diode pumped solidstate laser driver for inertial fusion energy", *Nuclear Fusion*, **36**, 75.

Page R H, Schaffers K I, DeLoach L D, Wilke G D, Patel F D, Tassano J B, Payne S A, Krupke W F, Chen K T, and Burger A, 1997a, "Cr^{2+} doped zinc chalcogenides as efficient widely-tunable mid-infrared lasers", *IEEE J Quantum Electron*, **33**, 609.

Page R H, Skidmore J A, Schaffers K I, Beach R J, Payne S A and Krupke W F, 1997b, "Demonstration of diode-pumped and grating-tuned ZnSe:Cr^{2+} lasers", *OSA TOPS, Advanced Solid State Lasers,* **19**.

Payne S A, Chase L L, Newkirk H W, Smith L K and Krupke W F, 1988, "LiCaAlF$_6$:Cr^{3+}: a promising new solid state laser material", *IEEE J Quantum Electron*, **24**, 2243.

Payne S A, Chase L L, Smith L K, Kway W L and Newkirk H W, 1989, "Laser performance of LiSrAlF$_6$:Cr^{3+}", *J Appl Phys*, **66**, 1051.

Payne S A, Smith L K, Kway W L, Tassano J B and Krupke W F, 1992, "The mechanism of Tm-Ho energy transfer in LiYF$_4$", *J Phys: Condens Matter*, **4**, 8525.

Renz R and Schulz H J, 1983, "The decay of infrared luminescence in II-VI compound semiconductor doped by 3d transition elements", *J Phys: Solid State C;* **16**, 4917.

Riseberg L A and Moos H W, 1966, "Multiphonon orbit-lattice relaxation of excited states of rare-earth ions in crystals", *Phys Rev*, **174**, 429

Riseberg L A and Weber M J, 1976, "Relaxation phenomena in rare-earth luminescence", *Progress in Optics* **XIV**, 91, North-Holland, E. Wolf, ed.

Ryu G, Yoon C Y, Han T P and Gallagher H G, 1998, "Growth and characterization of CsLiB$_6$O$_{10}$ (CLBO) crystals", *J Crys Growth*, **191**, 492.

Schaffers K I, Tassano J B, Payne S A, Hutcheson R L, Equall R W and Chai B H T, 1998, "Crystal growth of Yb:Sr$_5$(PO$_4$)$_3$F for 1.047 micron laser operation", *OSA TOPS, Adv. Solid State Lasers*, **19**, 437.

Schepler, Kueck S and Shiozawa L, 1997, "Cr^{2+} emission spectroscopy in CdSe", *J Lumines*, **72**, 116.

Seo J T, Hoemmerich U, Trivedi S B, Chen R J and Kutcher S, 1998, "Slope efficiency and tunability of a Cr^{2+}:Cd$_{0.85}$Mn$_{0.015}$Te mid-infrared laser", *Opt Commun*, **153**, 267.

Velsko S P, Webb M, Davis L and Huang C, 1991, "Phase-matched harmonic generation in lithium triborate (LBO)", *IEEE J Quantum Electron*, **27**, 2182.

Velsko S P and Krupke W F, 1996, "Applications of high-average-power nonlinear optics, *Nonlinear Frequency Generation and Conversion*, SPIE Proceedings, **2700**, 6.

Yap Y K, Inagaki M, Nakajima S, Mori Y and Sasaki T, 1996, "High-power fourth and fifth-harmonic generation of a Nd:YAG laser by means of a CsLiB$_6$O$_{10}$ crystal", *Opt Lett*, **21**, 1348.

Yoshimura M, Kobayashi T, Furuya H, Murase K, Mori Y and Sasaki T, 1998, "Crystal growth and optical properties of yttrium calcium oxoborate YCa$_4$O(BO$_3$)$_3$", *OSA TOPS Adv Solid State Lasers*, **19**, 561.

Zharikov E V, Iichev N N, Laptev V V, Malyutin A A, Ostroumov G, Pashinin P P, Pimenov A S, Smirnov V A and Shcherbakov I A, 1983, "Spectral, luminescence, and lasing properties of gadolinium scandium gallium garnet crystals activated with neodymium and chromium ions", *Sov J Quantum Electron*, **13**, 82.

Periodically poled materials for nonlinear optics

Lawrence E Myers

Lightwave Electronics, California, U.S.A.

1   Introduction

Quasi-phasematched (QPM) frequency conversion has been an area of practical importance over the past five years with the development of techniques for fabricating periodically poled ferroelectric materials. This chapter reviews these materials and their applications to QPM devices. This first section contains background information on the role of phasematching in nonlinear frequency conversion. Section 2 presents a theoretical analysis of QPM structures and fabrication tolerances. Section 3 describes methods of fabricating periodically poled materials. Finally, Section 4 discusses QPM devices that have been demonstrated and that are envisioned.

1.1   The need for phase-matching

Nonlinear frequency conversion is continually limited by the availability of suitable materials. The requirements for a suitable nonlinear material include nonlinearity, transmission, phasematching, homogeneity, damage, mechanical and thermal properties, lifetime, cost and availability. All of these requirements must be simultaneously satisfied for a given crystal to be of practical importance. Once the requirements for transmission at the wavelengths of interest and basic nonlinearity have been satisfied, phasematching is the most difficult criterion to meet. Professor Byer stated in his presentation at a previous Scottish Universities Summer School in Physics that "phasematching is the most restrictive requirement placed on a crystal and reduces the number of potential crystals to only a few hundred out of over 13,000 known crystals" (Byer 1977). The need to satisfy simultaneously other requirements restricts this set further, so that only a handful of crystals is actually used in practical nonlinear optical devices. The method of quasi-phasematching is important for overcoming the restriction of satisfying the conventional phasematching requirement and extending the range of utility of existing crystals.

The need for phasematching comes about because of dispersion in the refractive index of nonlinear crystals (Armstrong *et al.* 1962, Zernicke and Midwinter 1973). In a nonlinear

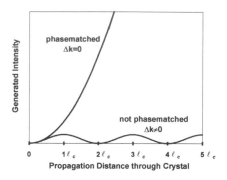

Figure 1. *Generated intensity vs. propagation distance though the crystal in a nonlinear frequency conversion interaction.*

frequency conversion interaction, the driving waves force a nonlinear polarisation in the material. This generates free waves, and the direction of power flow is determined by the relative phases of these interacting waves. Dispersion in the phase velocities causes a relative phase change as the interacting waves propagate through the crystal which changes the direction of power flow. Assuming fields of the form $E(z)e^{-ikz}$ with wave vector $k=2\pi n/\lambda$, the change in field strength for field E_3 in a three-wave interaction is expressed by a nonlinear wave equation of the form

$$\frac{dE_3}{dz} \propto E_1 E_2 e^{-i\Delta kz} \tag{1}$$

where $\Delta k = k_3 - k_2 - k_1$ is the wave-vector mismatch , assuming small signals and plane waves. As shown in Figure 1, the wave equation has two forms of solution depending on the value of Δk. When $\Delta k=0$, the interaction is phasematched and the generated intensity grows exponentially as the waves propagate through the crystal. In contrast, when $\Delta \neq 0$, the solution is oscillatory. The coherence length ℓ_c is defined as that distance which results in a phase mismatch of $\Delta k \cdot \ell_c = \pi$.

As specific examples, consider the cases of second harmonic generation (SHG) and optical parametric oscillators (OPO) in lithium niobate (LN). The coherence length and wave-vector mismatch are defined as

$$\ell_c = \frac{\pi}{\Delta k}, \qquad \Delta k = 2\pi \frac{2}{\lambda} \left(n_{2\omega} - n_\omega \right) \tag{2}$$

The dispersion of LN is shown in Figure 2 (Edwards and Lawrence 1984, Jundt *et al.* 1991). For the representative interactions of SHG of 532nm from 1064nm and a degenerate OPO pumped by 1064nm generating 2128nm, the coherence lengths are 6μm and 30μm respectively at 25°C. These lengths are similar to comparable non-phasematched interactions in other materials. If the interaction is not phasematched, then the oscillatory nature of the solution means that the generated power is at most that obtained within one coherence length. Generally, efficient conversion requires the use of crystal lengths of the order of 1cm. Since the typical crystal is many coherence lengths long, phasematching is essential for efficient conversion.

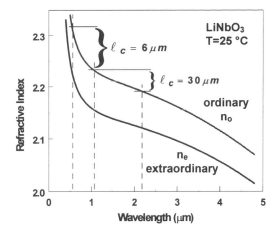

Figure 2. *Dispersion of LN, showing the coherence lengths of 6μm for SHG of 532nm from 1064nm, and 30μm for a 1064nm pumped degenerate OPO. Similar values are obtained for the extraordinary index.*

1.2   Phase-matching using birefringence

The traditional method of obtaining phase matching is through use of the birefringence of the nonlinear crystal (Zernicke and Midwinter 1973). In LN, the allowable birefringent interactions are of Type I, where the two longer wavelengths in a three-wave interaction have the ordinary polarisation while the shortest wavelength has the extraordinary polarisation. For the case of SHG, or equivalently a degenerate OPO, phasematching requires that both waves have the same refractive index as seen in Equation 2. Figure 3 shows that for LN at 25°C there are two solutions for phase matching along the crystal axes. Such solutions have $\Delta k = 0$ and satisfy the energy conservation condition for SHG that the shorter wavelength is half the value of the longer wavelength. Unfortunately these solutions are not of practical use since they do not match the wavelengths of common laser lines.

Fortunately some degree of tunability in the phasematching condition can be obtained by adjusting the angle of the crystal relative to the propagation direction. Light polarised in the plane containing the optic axis and the propagation vector experiences a refractive index that varies between the ordinary and extraordinary index values, n_0 and n_e respectively (Zernicke and Midwinter 1973). At 25°C, a phasematching solution can be found for the 1064nm pumped degenerate OPO at $\theta = 45°$. However, no solution is possible for the important interaction of 532nm SHG.

Although angle tuning helps offer more phasematching possibilities, it is at the cost of introducing walk-off of the Poynting vector for the extraordinary wave which propagates at an angle to the optic axis in the anisotropic crystal. Walk-off reduces the spatial overlap between the interacting waves, which limits the interaction length and the ability to focus tightly, thus lowering efficiency. Another manifestation of propagation at non-

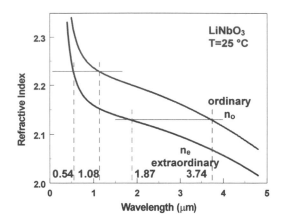

Figure 3. *Type I birefringent phasematching solutions for SHG (or degenerate OPO) in LN at 25C. The horizontal lines indicate those interactions which simultaneously satisfy phasematching ($n_{2\omega}=n_\omega$) and energy conservation ($\lambda_{2\omega}=\lambda_\omega/2$). Solutions are obtained for SHG of 0.54μm from 1.08 μm and 1.87μm from 3.74μm.*

normal incidence is that the acceptance angle is limited since Δk is linear with angle, a condition referred to as critical phasematching. For these reasons, propagation at normal incidence is preferred, in which case there is no first-order dependence of Δk with angle (noncritical phasematching) and there is no walk-off.

Noncritical phasematched solutions can be found by employing the temperature dependence of the refractive indices and this allows tunability in the phasematching conditions without the drawbacks of angle tuning. In LN, the rate of change of the index with temperature is much greater for the extraordinary index than for the ordinary index, so the birefringence is also a function of temperature (Edwards and Lawrence 1984, Jundt *et al.* 1991). At normal incidence, a phasematching solution for the 1064nm pumped degenerate OPO can be found at 526°C, and 532nm SHG can be phasematched at –20°C. Unfortunately these are not convenient temperatures for practical devices.

A final possibility for tuning birefringent-phasematching is to adjust the material composition. This offers the possibility of providing noncritical phasematching at convenient operating temperatures; however, it has the serious disadvantage of requiring almost a complete redevelopment of the crystal growth to achieve the requisite quality and uniformity. Most such work has been motivated by the 532nm SHG interaction. In the case of LN, it has been found that the exact ratio of Li to Nb in the crystal influences the temperature for noncritical phasematching (Bordui and Fejer 1993). LiNbO$_3$ can exist over a range of compositions of the Li$_2$O and Nb$_2$O$_5$ components, resulting in a solid solution of Li$_x$Nb$_{1-x}$O$_3$ where $x = 45$–50%. The standard commercial material has the value $x = 48.4\%$ since this is the congruent composition (*i.e.* at the solid/liquid phase transition temperature, the material exists in both solid and liquid form with the same composition). The important advantage of congruent material is that it can be grown with no compositional fluctuation since its composition does not change as it is pulled from the melt. As stated previously, the congruent material has noncritical phasematching at −20° C; however, material with stoichiometric composition ($x = 50\%$) phasematches at 234°C. This would be a useful temperature for practical devices since it also happens to exceed the operating temperature required to avoid photorefractive damage. However, the stoichiometric crystal is very difficult to grow, although research is underway in this area (Kitamura *et al.* 1997). Alternatively, doping the congruent crystal with MgO has been found to improve the resistance to photorefractive damage, and it has the fortuitous benefit of raising the noncritical phasematching temperature to 110°C. Thus in some situations MgO:LN is a practical choice for 532nm SHG and in fact it is used successfully in at least one commercial product (Lightwave Electronics model 142). Development is also underway in the KTP family using the composition of $K_{1-x}Na_xTiOPO_4$ to obtain noncritical phasematching at 25°C. While compositional alteration of crystals can improve the phasematching properties for given interactions, the long and costly development makes this a difficult approach for general application.

To summarise, birefringent phasematching has been the predominant technique for phasematching nonlinear optical frequency conversion interactions for the past thirty years. Given that birefringent phasematching requires a fortuitous intersection of inherent dispersion conditions, combined with the simultaneous satisfaction of a list of other requirements for a nonlinear crystal, it is remarkable that phasematching solutions can be found at all, let alone at interactions that best match available laser sources. Some relief is possible since the birefringence can be tuned by several methods, although each has drawbacks. Angle tuning introduces walk-off which limits the interaction length, spot size, and acceptance angle. Tuning with temperature may help obtain noncritical phasematching conditions, but possibly at unacceptably inconvenient operating temperatures. Compositional tuning may permit noncritical phasematching at more convenient temperatures, but the development is long and costly. Even in the best cases, there are only limited solutions available from these tuning methods and hence the method of birefringent phasematching requires the search for new materials to expand the available pool of useful materials. The development process for new materials is slow and expensive, so alternative phasematching techniques are valuable.

1.3 Phasematching by periodic phase shift

Quasi-phasematching is an alternative approach to conventional birefringent phasematching that does not require finding a coincidence of material properties to obtain $\Delta k=0$. Instead the oscillatory solution with $\Delta k \neq 0$ is allowed, but whenever the relative phase mismatch of the interacting waves slips by π (*i.e.* after one coherence length), the mismatch is reset to zero so that overall the process proceeds with power flowing in the desired direction. As shown in Figure 4, quasi-phasematching produces efficient frequency conversion even with non-phasematched crystals.

Quasi-phasematching is most commonly implemented by reversing the sign of the nonlinear coefficient. The nonlinear coefficient is a signed value related to the crystal structure. Reversing the sign of the nonlinear coefficient has the effect of adding a π phase shift to the relative phase of the interacting waves:

$$\frac{dE_3}{dz} \propto -d_{\text{eff}} E_1 E_2 e^{-i\Delta kz} = d_{\text{eff}} E_1 E_2 e^{-i(\Delta kz+\pi)} \tag{3}$$

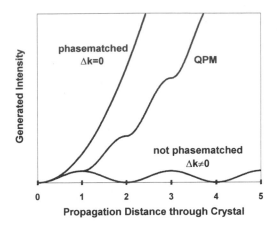

Figure 4. *Quasi-phasematching produces efficient frequency conversion with non-phasematched crystals by resetting the phase of the interaction whenever the phase mismatch slips by* π.

where d_{eff} is the effective nonlinear coefficient. When $z = \ell_c$, then $\Delta k \cdot \ell_c = \pi$ which is exactly cancelled by the sign reversal of d_{eff}. Thus quasi-phasematching can be implemented by building into the crystal a nonlinear coefficient with a grating pattern having a period of $\Lambda_g = 2\ell_c$.

Quasi-phasematching has several important advantages over conventional birefringent phasematching. Quasi-phasematching allows an arbitrary selection of operating temperature and angle, so that noncritical phasematching at convenient temperatures is possible at any wavelength within the transparency range of the material. The interaction can also make use of the highest component of the nonlinear susceptibility tensor of the material. In many cases this means that all the waves have the same polarisation which makes birefringent phasematching impossible. Thus quasi-phasematching extends the utility of existing materials, and enables the use of materials with no birefringence and even optically isotropic materials. Tuning of quasi-phasematching is possible with angle, temperature, and grating period. This is an important additional design parameter that makes possible a variety of novel devices.

The use of quasi-phasematching replaces the search for new materials having adequate birefringence with the search for fabrication processes to implement the necessary microstructured modulation. The form of an ideal crystal for QPM has sections with reversed sign of the nonlinear coefficient as shown in Figure 5. The method of implementing QPM first proposed was to stack crystal plates with alternating layers rotated (Armstrong *et al.* 1962). However, implementing this with many slices, tens of micrometres thick, is not readily achievable. Instead in a ferroelectric material such as LN and KTP, the sign of the nonlinear coefficient can be reversed by reversal of the ferroelectric domain polarity, thus making it possible to build the desired structure into a monolithic crystal. The process of reversing the ferroelectric domain polarity is referred to as periodic poling because the prototype pattern for QPM nonlinear frequency conversion is a

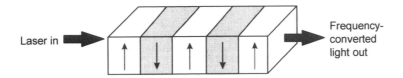

Figure 5. *Form of an ideal crystal for QPM frequency conversion. The sign of the nonlinear coefficient is reversed with period a multiple of the coherence length.*

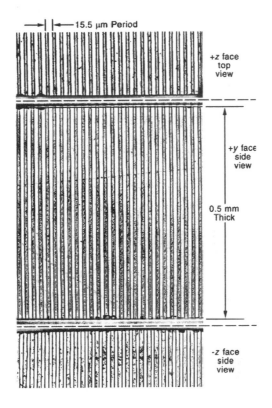

Figure 6. *Example of a bulk periodically poled lithium niobate (PPLN) crystal fabricated by electric-field poling with lithographic electrodes (Myers, 1995b). This process produces a domain structure that is nearly ideal for efficient quasi-phasematching and has led to the current interest in this technology.*

periodic grating structure. The recent surge of interest in QPM material is attributable to the development of the method of electric-field poling with lithographic electrodes which produces material having nearly ideal domain structures for efficient quasi-phasematching, as shown in Figure 6.

Research in quasi-phasematching materials and devices is currently of strong worldwide interest. Most of the activity involves the material LN (and its isomorph lithium tantalate (LT)), and KTP (and its isomorphs RTA, and KTA), although other ferroelectrics have been considered for operating further into the IR and UV. For non-ferroelectrics, poling for domain reversal does not apply, so periodic patterning of crystallographic twins is being investigated. With semiconductors, the process of diffusion bonding is used to fuse together a rotated stack of plates into a monolithic patterned crystal. In the case of glass and fibres, nonlinearity is induced by an electric-field poling process (not to be confused with domain reversal), and quasi-phasematching can be implemented by selectively creating and erasing the nonlinearity.

Much of the research on periodic poling of ferroelectrics has been focused on generating the shorter periods of the domain grating that are needed to match the coherence lengths of visible and UV interactions. Also, as the materials become more widely used, more attention is being paid to their linear properties such as absorption and thermal effects, which limit their applicability even when ideal domain patterns are possible. QPM device research has often been motivated by the desire for blue sources based on frequency doubling of diode lasers, and recently the area of QPM OPOs has received much attention. QPM devices used to be closely associated with waveguide frequency conversion, but QPM devices using bulk crystals are now possible with the development of electric-field poled material. Some of the leading current work in QPM devices involves power scaling and the investigation of novel device designs such as cascaded processes, temporal/spatial wave shaping, and tailored phasematching bandwidths.

2   Quasi-phasematching theory

The theory of quasi-phasematching has analogies to other applications involving the coherent superposition of electromagnetic fields, such as diffraction gratings, Fourier optics, binary optics, phased array radar, and optical thin film design. Quasi-phasematching can be analysed from the perspectives of both space domain (*i.e.* propagation distance through the crystal) and mismatch domain (*i.e.* Δk) with each approach offering different insights. QPM analysis is important for understanding the implications of the tuning and acceptance bandwidths on QPM design so that it is possible to specify tolerances on fabrication of materials for QPM devices.

2.1   Nonlinear conversion efficiency and phase mismatch

A primary tool for analysis and material characterisation is the tuning curve, which is a description of conversion efficiency as a function of a parameter which affects the phase mismatch. The tuning curve function can be developed by integrating the nonlinear wave Equation 1

$$\frac{dE_3}{dz} = \Gamma d_{\text{eff}} e^{-i\Delta k z} \tag{4}$$

over the crystal length L to get

$$P_3\left(L\right) = \Gamma^2 d_{\text{eff}}^2 L^2 \operatorname{sinc}^2\left(\frac{\Delta k L}{2}\right) \tag{5}$$

where P_3 is the generated power and Γ is the nonlinear gain parameter. This parameter includes the driving fields E_1 and E_2 from Equation 1 as well as other proportionality

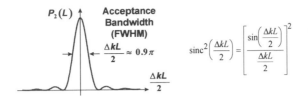

Figure 7. *The ideal tuning curve for nonlinear frequency conversion has the form of* $sinc^2 (\Delta kL/2)$.

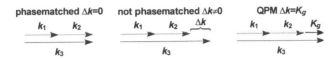

Figure 8. *Wavevector diagrams show that tuning the grating vector allows quasi-phasematching even when* $\Delta k \neq 0$.

terms for the particular material and interaction (Zernicke and Midwinter 1973). The form of the ideal tuning curve is shown in Figure 7. The aspects of the tuning curve that are most important for use in material characterisation are the height of the central peak, which gives the maximum conversion efficiency, the acceptance bandwidth defined as the full width at half maximum (FWHM) of the central peak, and the shape and height of the sidelobes. The tuning curve is explicitly a function of the phase mismatch Δk, and ultimately depends on a variable experimental parameter such as wavelength, temperature, or refractive index implicit in the definition of Δk.

2.2 Nonlinear conversion efficiency with quasi-phasematching

The analysis can be extended to quasi-phasematching by including the periodic nature of the nonlinear coefficient in the nonlinear wave equation

$$\frac{dE_3}{dz} = \Gamma d_{eff} e^{-i\Delta kz} = \Gamma \left| d_{eff} \right| e^{iK_g z} e^{-i\Delta kz} = \Gamma \left| d_{eff} \right| e^{-i\Delta k_Q z} \qquad (6)$$

where the grating vector is defined analogously to the wavevector as $K_g = 2\pi/\Lambda_g$ with grating period Λ_g. Most analytical results obtained for conventional phasematching can be extended to quasi-phasematching by substituting $\Delta k_Q = \Delta k - K_g$ for Δk. The impact of the grating vector on phasematching can be seen by considering the wavevector diagram shown in Figure 8. When the interaction is perfectly phasematched, the sum of the wavevectors of the driving waves equals the wavevector of the generated wave. When not phasematched, there is a gap between the driving and generated wavevectors, which is the value of Δk. In quasi-phasematching, this gap is filled by K_g. Thus the grating vector K_g is an essential additional tuning parameter for quasi-phasematching.

Wavevector diagrams, such as Figure 9, are especially helpful for visualising angle tuning in QPM interactions. When propagating at an angle relative to the grating vector,

Lawrence E Myers

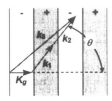

Figure 9. *Angle tuning in quasi-phasematching.*

the interaction experiences an effectively longer grating period is given by (Fejer 1992b)

$$\Lambda_{\text{eff}} \approx \Lambda_g \left[\cos\theta \ \left(1 + \frac{n_{2\omega} - n_\omega}{2n_{2\omega}} \tan^2\theta \right) \right]^{-1} \approx \frac{\Lambda_g}{\cos\theta}$$

Thus it is possible to tune a QPM interaction by crystal rotation. However, rotation relative to the grating vector leads to walk-off similar to that experienced in angle tuning of birefringent phasematching. This walk-off is associated with the wavevector rather than the Poynting vector, but the effects on conversion efficiency and angular acceptance are similar. The wavevector walk-off is a consequence of propagation at an angle relative to the grating structure and thus occurs even if propagation is perpendicular to the crystal axes and even if the crystal is isotropic. Quasi-phasematching is noncritical if the interacting waves are aligned with the grating vector, so tuning of the grating period is preferred to angle tuning in most QPM devices.

The periodic nature of quasi-phasematching lends itself to a Fourier series analysis. The periodic nonlinear coefficient can be expanded as

$$d(z) = d_{\text{eff}} \sum_{m=-\infty}^{\infty} G_m e^{imK_g z} \tag{7}$$

where G_m is the Fourier coefficient of the m-th harmonic (Fejer *et al.* 1992b). For $d(z)$ described by a rectangular wave with duty cycle $D = a/\Lambda_g$ (a is the width of a domain section), the Fourier coefficients are $G_m = (2/\pi m) \sin \pi m D$. Hence for optimum choice of D, the QPM effective nonlinear coefficient is reduced by $2/\pi m$ compared to the value for true phasematching of the same interaction.

The Fourier analysis approach makes it evident that higher order phasematching is possible, as shown in Figure 10. Both the effective nonlinear coefficient and the duty cycle which optimises its value depend on the order m. For the case of first-order, the optimum duty cycle is $D = 0.50$. This also optimises all odd-order cases, though for higher order, other choices also work (*i.e.* whenever $\sin \pi m D = 1$). For even order, a choice of $D = 0.50$ is not correct. Rather than optimising d_{eff}, this causes $d_{\text{eff}} = 0$ since it makes domain reversal occur at a null rather than a peak of the generated field. In the case $m = 2$, the optimum duty cycle is $D = 0.25$.

The Fourier approach can be further generalised to an analysis in a transform space defined in terms of the phase mismatch Δk. The mismatch domain analysis arises from reconsidering the integral of the wave equation Equation 6 to determine the power gen-

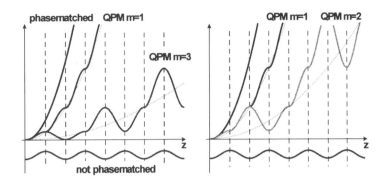

Figure 10. *Higher order quasi-phasematching. The optimised value of the effective nonlinear coefficient is $d_{eff} = 2/\pi m$, which is shown as dotted lines on the plots. The duty cycle that optimises d_{eff} is D=0.50 for m = 1, D=0.25 for m = 2, and D=0.50 (or 0.167) for m = 3.*

erated in a crystal of length L:

$$P_3(L) = \Gamma^2 d_{\text{eff}}^2 L^2 \left| \frac{1}{L} \int_L g(z) e^{-i\Delta kz} dz \right|^2 \tag{8}$$

The quantity in brackets has the form of the Fourier transform of the normalised nonlinear coefficient $g(z)$ with kernel $e^{-i\Delta kz}$ and transform variable Δk. For a conventional crystal, there is no modulation of the nonlinear coefficient, so $g(z) = 1$ for $0 < z < L$ (*i.e.* a rectangle function), and the Fourier transform becomes the previously discussed tuning curve of the form $\text{sinc}^2(\Delta kL/2)$. For a QPM crystal, the normalised nonlinear coefficient has a modulation pattern that is a square wave within the crystal length L taking on values of ± 1, as shown in Figure 11. This square wave also represents the relative phase of the nonlinear coefficient modulo 2π. By unfolding the 2π ambiguity, the phase can be represented as a stair-step function tilted along the line $\exp(iz\pi/\ell_c)$. By applying the shift theorem of Fourier transforms, the phase tilt of $\exp(iz\pi/\ell_c)$ in the space domain corresponds to a shift of π/ℓ_c in the Δk mismatch domain. Thus the domain grating in effect steers the phasematching from $\Delta k = 0$ to $\Delta k - \pi/\ell_c = 0$. The bandwidth of the peak is unaffected since it is still determined by the crystal length L, but the amplitude is reduced by an envelope function related to the Fourier transform of the individual domain lengths ℓ_c which make up the grating. This is a consequence of the phase being a stair-step rather than a smooth tilt. The domain sections appear as convolution in the space-domain expression for the square wave, and hence the envelope function appears as a multiplicative factor in the mismatch domain. Higher orders also appear at integer multiples of the grating period, reduced further by the increasingly lower amplitude of the envelope function at higher values of Δk. The power of the mismatch domain analysis is that the familiar tools of Fourier transforms can be applied to designing QPM structures with more complicated phasematching response (Fejer *et al.* 1992b).

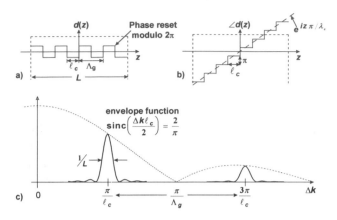

Figure 11. *Mismatch domain interpretation of quasi-phasematching. a) The phase modulation of the nonlinear coefficient across the crystal length L is a square wave. b) Removing the 2π ambiguity, the modulation is seen to be a phase tilt. c) Applying the shift theorem of Fourier transforms, the phasematching peak is steered from $\Delta k = 0$ to $\Delta k - \pi/\ell c = 0$, and the amplitude is reduced by the envelope function which is the Fourier transform of the individual domain sections of length ℓc.*

2.3   Tolerances on grating errors

The approaches to QPM analysis discussed above are appropriate for treating average behaviours of structures consisting of many domain periods. Obtaining tolerances on the structure fabrication requires analysis of the details of local domain wall position. A space-domain analysis treats the overall generated field as the coherent superposition of the contributions from each domain section, explicitly including the phase errors in each domain compared to ideal quasi-phasematching. This approach accommodates analysis of errors in the placement of domain walls and leads to exact expressions for tuning and tolerances due to deviations from ideal structures. The results derived by Fejer *et al.* (1992b) are summarised in Table 1. Two types of errors are considered: period error and duty cycle error. Both errors arise from domain widths that differ from those in an ideal structure, but in the case of duty cycle errors, the distribution of domain widths has an average value that is the same as the ideal case, while period errors have a different average value. Both period and duty cycle types occur as either constant mean error or randomly distributed error about the mean. Note that the expressions for period error contain the term $N = L/\Lambda g$ which is the number of grating periods in the crystal. Because of this large factor (N∼1000 for typical crystals, *e.g.* $L = 1$cm , $\Lambda g = 6$ μm), the efficiency is quite sensitive to this type of error. This is expected since period deviation results in a phase error that accumulates over the grating structure. In effect, a grating with mean period error steers the phasematching peak to a different value in Δk space, whereas a random period error smears out the tuning curve. On the other hand, the efficiency is less sensitive to duty cycle deviations because the phase error does not accumulate. Thus duty cycle error results in a lower peak value but the peak is still located at the correct point in Δk space.

	Period	Duty Cycle
Constant Error	$\eta = \mathrm{sinc}^2 \left(\frac{\delta \Lambda_g}{\Lambda_g} \frac{N}{2} \right)$	$\eta = \mathrm{sinc}^2 \left(1 - \frac{\delta D}{D} \right) \frac{\pi}{2}$
Random Error	$\langle \eta \rangle = 1 - \frac{\pi^2}{6} N \left(\frac{\sigma_\ell}{\ell_c} \right)^2$	$\langle \eta \rangle = 1 - \frac{\pi^2}{2} \left(\frac{\sigma_\ell}{\ell_c} \right)^2$

$N = L/\Lambda_g$ = number of domain grating periods in the crystal.
$\delta D/D$ = error in duty cycle relative to the design value of $D = 0.5$.
σ_ℓ/ℓ_c = standard deviation of domain width relative to the coherence length.

Table 1. *Expressions for the reduction in QPM efficiency caused by errors in the grating pattern (Fejer, 1992).*

The analysis of the errors in Table 1 leads to the specification of tolerances for device fabrication summarised in Table 2. It is evident that period errors are far more detrimental to device performance than duty cycle errors. Thus a suitable fabrication method must have a high level of period control to meet the tight tolerances for high efficiency, but the tolerances for duty cycle errors are relatively low. The method of electric-field poling meets these criteria since the period of the grating electrode is precisely controlled by a lithographic mask while the duty cycle is less well controlled because of fringing fields between the grating lines, as will be described in Section 3. Nonetheless there are other reasons for specifying tolerances for duty cycle control tighter than those indicated by Table 2, for example, to reduce photorefractive susceptibility or to avoid parasitic processes, as will be discussed in Section 4.

		Tolerance	
		Period	Duty Cycle
Constant Error	$\frac{\delta \Lambda_g}{\Lambda_g}$, $\frac{\delta D}{D}$	0.03%	50%
Random Error	$\frac{\sigma_\ell}{\ell_c}$	1%	38%

Table 2. *Tolerances on material fabrication for efficiency >50% of ideal (from Table 1 with $N = 1000$).*

One final form of error that will be considered is that of missing domains, which is an error to which the electric-field poling technique is also susceptible. A missing domain is where domain reversal should occur but does not, but the pattern picks up again correctly after this glitch. Thus some grating sections are merged together, but otherwise all domain walls are in their correct positions and the overall structure has the correct period. The effect of missing domains depends on their distribution within the grating pattern. Two limiting cases are considered here as shown in Figure 12. In the first case, the domains are all missing from the ends of the crystal. This is equivalent to having a shorter crystal of length L', so the bandwidth is broadened and the efficiency is reduced to $\eta = (L'/L)^2$. In the other case, all of the missing domains are clustered together in the center of the crystal. If the length of the section with missing domains is the same as in the previous case, the result is that the bandwidth is still determined by the crystal length L, but the

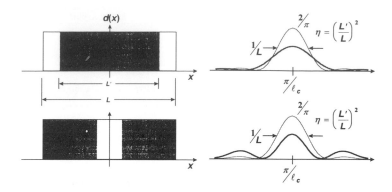

Figure 12. *Two cases of missing domains produce different tuning curves, with the same reduction in peak efficiency.*

sidelobes are higher and the peak efficiency is again reduced to the same value. Since no period error is involved, the structure is tolerant of missing domains such that the acceptance for less than 50% efficiency reduction is $\delta L/L = 30\%$. Missing domains can occur in electric-field-poled materials because of breaks in the electrode lines, for example, or excessive spreading of domain width due to microdomains or crystal defects.

To summarise the quasi-phasematching theory presented in this section, the addition of the grating vector as another adjustable phasematching parameter under the control of the crystal designer allows a great amount of flexibility for QPM devices. Fourier analysis provides design tools for developing QPM structures. The key analytical results that have implications on material fabrication tolerances are that QPM crystals are highly sensitive to period error, but relatively tolerant of duty cycle variations. As opposed to conventional approaches, quasi-phasematching replaces the search for new materials with the development of processing methods for existing materials, which is the subject of the next section.

3   Material fabrication

After the initial discovery of nonlinear laser frequency conversion, quasi-phasematching was the first technique invented for overcoming the limiting effects of phase velocity dispersion (Armstrong *et al.* 1962). However, it was mostly dropped in favour of the birefringent phasematching approach which was demonstrated shortly thereafter. The early history of quasi-phasematching was limited to observation of SHG enhancement in multi-domain ferroelectrics and rotationally twinned crystals. In 1976, the first demonstrations of quasi-phasematching with stacked rotated thin crystal plates were performed. However, it was not until the mid-1980s that research into quasi-phasematching began to grow with the development of techniques for implementation of monolithic QPM materials such as poled polymers and glass fibres. Much research focused on the processing of the ferroelectric materials LN, LT, and KTP, with bulk QPM crystals obtained by growth modulation, and waveguide devices obtained using methods employed in integrated optics. Citations describing these early devices can be found in references reviewed in Fejer *et al.* (1992b).

3.1 Ferroelectric Materials

Ferroelectric materials are suited to implementation of quasi-phasematching because the nonlinear coefficient changes sign with reversal of the ferroelectric domain polarity. Ferroelectricity can be regarded as a phase transition such that below a critical temperature (referred to as the Curie temperature T_c) the material develops a built-in polarisation due to displacement of ions relative to their neutral positions (Kittel 1986, Lines and Glass 1977). Depending on the direction of the displacement, the built-in polarisation can have different polarity values, and a region with a given polarity value is referred to as a domain. In the common QPM ferroelectrics, polarities of different domains are oriented at 180° relative to each other, but other orientations are possible such as 90° domains in $KNbO_3$. Piezoelectricity and pyroelectricity are similar phenomena, but the distinguishing feature of ferroelectricity is that the polarity of a domain can be reversed by application of an electric field.

Two useful parameters for characterising a ferroelectric are the coercive field and the spontaneous polarisation. The coercive field E_c is the strength of the external field that must be applied to cause domain reversal to occur (typical units are kV/mm). E_c is related to the potential barrier that the ions must overcome in order to re-orient into a domain of the opposite polarity. The spontaneous polarisation P_s is the electric dipole moment per unit volume (typical units are C-m/m^3 or $\mu C/cm^2$). P_s manifests itself as a surface charge which serves to compensate the field due to the built-in dipoles. The reversability of P_s and the reversal threshold E_c result in a hysteresis loop given by the constitutive relation $D = \varepsilon E + P_s$ where D is the electric flux density, E is the applied field, and ε is the dielectric permittivity. When a field exceeding E_c is applied in the direction to oppose the built-in dipole orientation, the polarity of the domain reverses and the surface charge that previously compensated the dipole field is no longer appropriate for the new domain orientation, so charge flows to screen the new field. The flow of screening charge can be either rearrangement of charges within the crystal itself or current through the circuit which is used to apply the external field, as is usually the case when the ferroelectric material is a good insulator. When the screening charge is supplied by the external circuit, the poling current exhibits a switching pulse which ends when the area of the sample A has completely reversed as shown in Figure 13. The amount of charge transferred during the switching pulse is $Q = 2P_s A$, where the factor of two comes in because the screening charge associated with the original domain polarity must first be cancelled, then the same amount again must be supplied to compensate the new opposite polarity.

3.2 Older techniques for patterned domain reversal

Prior to the advent of the current electric-field poling method, two classes of techniques had been investigated for accomplishing patterned domain reversal of ferroelectrics, differentiated by whether they produce bulk crystals or waveguide structures. The methods to make bulk crystals rely on modulation of parameters during the crystal growth (Byer 1994, Feng *et al.* 1989). Approaches include passing current through the crystal while pulling it from the melt to cause a dopant concentration gradient, differential laser heat-

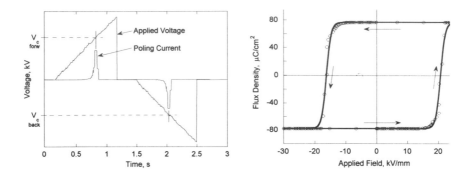

Figure 13. *Ferroelectric domain reversal occurs when the applied field reaches the coercive voltage. Poling current flows through the external circuit to compensate the new domain orientation. The integral of the poling current versus the applied field results in a hysteresis loop. For LN, the coercive field is asymmetric for poling in the forward versus the backward directions.*

ing with a fibre pedestal growth station, and temperature gradient created by off-centre rotation of the seed in the melt. The advantage of the bulk crystals fabricated by these methods is the possibility of large apertures, but they are subject to period errors due to inaccuracies in the modulation and pull rates. As discussed in Section 2, the conversion efficiency has stringent tolerances for this type of error, so typically the useful interaction length of crystals fabricated by these methods is limited to a few millimetres. Another problem is that the modulation process itself introduces conditions which are generally counter to the production of the best crystallinity. Thus while this approach seems like a natural way to produce bulk periodically poled crystals, its problems have prevented their development into commercial products.

The other class of techniques involves altering the surface of already grown crystals, which tends to produce domain-patterned regions in shallow layers that are suitable for use in waveguide devices (Byer 1994, Eger *et al.* 1994, Fejer 1992a, Webjörn *et al.* 1989). In LN, the methods include chemical in-diffusion of Ti or MgO plus thermal gradients, Li out-diffusion, and proton exchange plus thermal shock, and in KTP ion exchange. While these methods form only shallow surface layers making them unsuitable for applications requiring bulk crystals (such as high power), the high conversion efficiency due to waveguide confinement (*e.g.* 600 %/Wcm^2 vs. 4%/Wcm for bulk) has assured that much research effort has been devoted to their development. These methods also overcome the chief limitation of bulk growth poling by using lithographic patterning to achieve precise period control. The major difficulties are that the domain shapes may not be rectangular which reduces the QPM conversion efficiency, and that the processing may affect material properties (*e.g.* reducing the nonlinearity or the lifetime). Waveguide frequency conversion in general has proven difficult to commercialise.

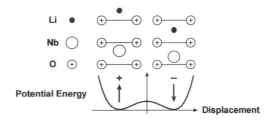

Figure 14. *Picture of ferroelectricity in LN.*

3.3 Electric-field poling

The method of electric-field poling has been the breakthrough that has promoted QPM ferroelectrics to their current importance. In fact, all of the methods described above accomplish poling with electric fields when the fundamental physics is examined. For instance, methods involving differential heating produce thermal gradients which couple to electric fields through the pyroelectric effect, and chemical composition gradients can produce electric fields through a change in T_c which affects the temperature-dependent P_s. However, the common usage of the phrase electric-field poling refers to the direct application of an electric field with an external circuit. For the case of LN, the ferroelectric domain polarity relates to the direction of offset of the Li and Nb ions relative to the oxygen lattice as shown in Figure 14 (Räuber 1978). Application of a field with strength exceeding E_c in the direction that opposes the built-in dipole allows the ions to overcome the potential energy barrier to re-orient to the opposite dipole positions. Ti and K ions play similar roles in determining domain polarity in the case of KTP (Laurell *et al.* 1992).

Early investigators considered LN to be a "frozen ferroelectric" at room temperature because the domain polarity could not be reversed by application of an external field before the sample suffered dielectric breakdown. The coercive field of LN is quite high by the standards of most ferroelectrics (21kV/mm vs. 2kV/mm for KTP or 0.2kV/mm for SBN), so early crystal quality may explain the inability to achieve room-temperature electric-field domain reversal. At high temperature, domain reversal was successful, requiring only a few volts near T_c. This technique is employed to make single-domain crystals from the large boules possible with Czochralski growth, which are critical to the modern LN industry. Electric-field domain reversal at room temperature was first accomplished using a pulsed field allowing a direct measurement of P_s (Camlibel 1969). In this work, 30 μm thick plates were used and differences between liquid and metal electrodes were studied. Later research on polarisation reversal involved poling of thicker plates (1mm) with low DC fields (poling time >1min) at slightly elevated temperatures (100–200°C) using metal electrodes (Prokhorov and Kuz'minov 1990).

Periodic poling of both LN and LT using direct e-beam and ion-beam writing is a form of electric-field poling that produces impressive results with bulk crystals (Kurimura *et al.* 1996, Mizuuchi and Yamamoto 1993). In this approach, the electric field is created by the build-up of charge deposited by the writing beam. The advantages of this method are that the crystals can be obtained with high aspect ratio patterns (up to 1mm thick with

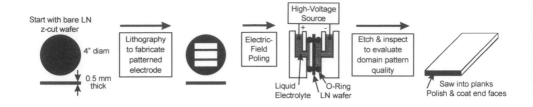

Figure 15. *Basic process of electric-field poling.*

periods as small 4μm demonstrated) and can have good periodicity due to the precise positional control of beam-writing equipment. Disadvantages of this approach are that there are problems with pattern uniformity because of charge diffusion and with pattern discontinuity because of repeated cycles of charge build-up followed by domain reversal with charge cancellation; hence, the interaction lengths of these crystals are limited. Also the process is slow and expensive since it requires that each line be individually written using costly beam-writing equipment. Thus this technique has not made the transition to commercial crystal fabrication.

The current success of periodically poled ferroelectrics began with the demonstration of domain patterning using an electric-field applied through a lithographically patterned electrode on the crystal surface and an external high-voltage circuit. In the first demonstration, 100μs long voltage pulses were used to fabricate a 2.8μm period, 3mm long grating in a 100μm thick LN substrate, and the sample was used to produce efficient SHG in a waveguide (Yamada *et al.* 1993). The process was extended to PPLN crystals 250μm thick and to SHG in the bulk crystals (Burns *et al.* 1994). The following year two groups introduced the method of electric-field poling with liquid electrodes (Myers *et al.* 1995b, Webjörn *et al.* 1994), which allowed the scaling of PPLN thickness to 0.5mm and the first demonstration of a QPM OPO. The 0.5mm thickness proved to be an important milestone because it enabled practical fabrication of bulk devices, it was a commonly available substrate thickness, and it allowed compatibility with processing equipment for semiconductor wafers. This technique was adopted by several companies, and PPLN was offered for sale commercially. The most widely used process of electric-field poling of PPLN is described in references (Myers *et al.* 1995b) and shown schematically in Figure 15. The state of the art of commercial PPLN fabrication is now 6μm period in 100mm diameter, 0.5mm thick wafers, and 17μm period in 1mm thick wafers.

Domain quality can be evaluated by etching poled samples followed by visual examination. In the case of LN, different crystal faces etch at dramatically different rates in HF acid. The different etch rates result in surface textures for the different domains that appear as a frosted pattern on unaided visual inspection and allow the detailed domain edges to be clearly seen through a microscope as shown in Figure 6. Similar etch techniques to distinguish domains are available for KTP. There is excellent agreement between the domain pattern quality determined by this inspection and the conversion efficiency and tuning bandwidth in nonlinear optical experiments. The etching step does not affect the use of the crystal because only a submicron surface layer of material is removed during the etch and the inspection is performed on the *z*- and *y*-surfaces rather than the *x*-faces which must be polished as entrance and exit apertures.

One concern about periodically poled materials is that domain interfaces could contribute loss above the bulk value; however, excess loss is not observed in practice. Ferroelectric domain reversal changes the sign of the second-order susceptibility without changing the refractive index; therefore, no Fresnel losses are encountered at the interfaces. Also the domain walls are no more than a few lattice constants thick so the walls themselves have negligible impact. After poling, there are residual fields at domain walls that cause index perturbations through the electro-optic effect which can be seen as depolarisation of cross-polarised illumination; however, these fields are relaxed by annealing, for example, at 100°C for 1 hour. Loss of <0.4% has been measured in 15mm long PPLN samples. This was dominated by the loss from the coated surfaces (Myers *et al*. 1995b). Since there were 600–1000 domain walls in these samples, losses attributable to the domain interfaces and propagation through the crystal were negligible.

3.4 The poling technique with LN

The parameter space for electric-field poling is multi-valued and there are different choices which can be used to produce high-quality periodically poled crystals. Some of the parameters that are influential are the circuit values (current, field strength, duration of voltage pulse, repetition rate of pulses), circuit topology (high-voltage switch versus amplifier, diode and other nonlinear elements), substrate state (temperature, thickness, crystal homogeneity, history of handling), and the electrode (substrate surface preparation, conductor material, insulator material, processing methods, pattern layout). Typical parameters used in commercially fabricated PPLN are: 0.5mm thick congruent LN substrates, with 21kV/mm voltage applied at room temperature, current 10μA–10mA, pulse length one second, and domain periods >6μm. Variations reported in the literature include shorter pulses, multiple pulses, higher fields, and higher or lower temperature. Commercial material is mostly fabricated using a liquid electrode fixture as shown in Figure 15. In this approach, the sample is connected to the circuit by contact with a conducting liquid such as LiCl solution. The fixture is clamped together with o-rings to hold the sample and contain the electrolyte, and it allows many samples to be processed easily. The soft termination of the liquid at the O-ring helps reduce the fringe fields at the outer perimeter where they are most intense and can lead to catastrophic dielectric breakdown. This is particularly important in the case of PPLN where the poling field is very high, but the liquid electrode structure is also used in the case of KTP poling for convenience.

The electrode structure is important for controlling the fields and supplying the current for domain reversal. Two methods of fabricating the electrode structure are employed as shown in Figure 16. The insulator that is part of the patterned surface electrode helps shape the fields and blocks the flow of charge to the region between the grating lines. Due to the high aspect ratio of the crystal thickness to the grating period, there is significant fringing of the fields around the electrode lines. This results in the field modulation dying away after a distance approximately Λ_g below the crystal surface. Thus the fields in most of the crystal are essentially uniform, so pattern formation is dominated by the field modulation and current blocking at the surface. The fringing fields at the edges of the grating lines contribute to the formation of domain nuclei which grow beyond the width of the electrode pattern. This spreading of the domain width beyond the pattern defined by the electrode makes it difficult to produce domain gratings with short periods. If the liquid

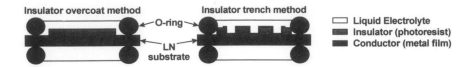

Figure 16. *Two designs for electrode structures used in producing periodically poled materials.*

electrolyte covers the electrode insulator, the electrolyte/insulator interface becomes an isopotential surface which helps control the fringing fields around the electrode edges, at the expense of reducing field modulation. This electrode design can be used to manage the field distribution for optimal patterning. In practice, the insulator trench method is easier to implement since it requires only a single lithography step, and it is not affected by broken grating lines since the current is fed through the surface. However, it is difficult to make high-resolution structures because the need to maintain the insulator thickness means that deep, narrow trenches are required. Also the exact shape of the trench (*e.g.* sloped vs. vertical sidewalls) can have a significant effect on the field profile, but the details of the shape can change widely as a function of minor processing variations. The insulator overcoat method avoids these problems because the overcoat itself is not patterned (except for a crude opening to allow contact of the metal electrode to the circuit). Short-period gratings can be more readily fabricated because the thin metal film is easier to pattern with high resolution. In addition, the smooth surface of the insulator overcoat makes for a uniform and reproducible field pattern and allows the use of insulators that can not be readily patterned. The insulator is usually chosen to be photoresist because it is easily applied and patterned; however, other choices such as SiO_2 film or spin-on glass have also been used. Selecting an insulator with the right dielectric and chemical properties may be important for optimising the process to achieve the best domain patterns.

The basic poling circuit, shown in Figure 17, uses a high-voltage amplifier, a series resistor to regulate current, a voltage monitor to measure the actual voltage applied to the sample after the series resistor, and a current monitor to measure charge applied to the sample. A rectangular voltage pulse is applied by the voltage source to achieve a voltage on the sample exceeding its coercive voltage V_c (21kV/mm for LN) after allowing for the voltage drop through the series resistor due to its loading by the voltage monitor. During poling, the voltage across the sample clamps at V_c because the domain wall velocity rises steeply with applied voltage. If the voltage should rise above V_c, the resulting faster domain velocity causes a large increase in the poling current leading to a larger voltage drop across the resistor which cancels the voltage rise. Similarly if the voltage should fall below V_c, the slower domain wall velocity causes a large decrease in the current leading to a smaller voltage drop across the resistor which cancels the voltage fall. Since the voltage across the sample clamps at V_c, the poling current is regulated by the driving voltage and the series resistor. With these values set, the current is essentially constant during the poling and so the duration of the poling time is proportional to the area of the sample. Integrating the current over time provides a convenient monitor of the poling progress by measuring the charge transferred.

When the poling model described above is applied to a sample with a patterned

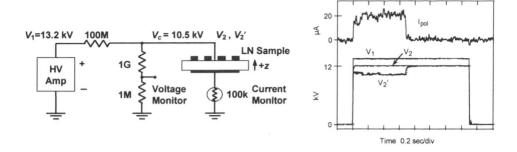

Figure 17. *Basic circuit and typical waveforms for electric-field poling. The circuit parameters are those used in the standard PPLN fabrication process. The voltage and current traces are from an LN sample 3mm diameter by 0.5mm thick. Trace V_1 is the voltage pulse applied by the voltage source. Trace V_2 is the voltage after the series resistor when there is no sample in the circuit, or the voltage that would be on the sample if there were no domain reversal. Trace V_2 is the actual voltage on the sample during poling.*

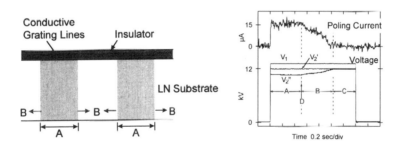

Figure 18. *Poling waveforms to produce a pattered crystal. Region A is poling under the conductor, similar to the type of poling shown in Figure 17. Region B shows a drop in poling current as the domains are driven under the insulator. Region C is the completion of poling when the entire LN substrate has been completely domain reversed. The stopping point D is when the poling under the electrode pattern is complete and before the domains spread into the region between the grating lines.*

electrode, the outcome is shown in Figure 18. The first stage of poling occurs by domain reversal under the conductive portion of the electrode, similar to that previously shown in Figure 17 for an unpatterned piece. The second stage of poling occurs after domain reversal under the conductor is complete and the domains start to spread into the region covered by the insulator. A higher field is required to drive the domains into this region effectively raising V_c, so the clamped voltage on the sample climbs to higher values and the poling current decreases because of the current regulation of the series resistor. Finally poling is complete when domain reversal has occurred in the entire area covered by electrode, and the poling current then ceases. Of course if poling proceeds this far,

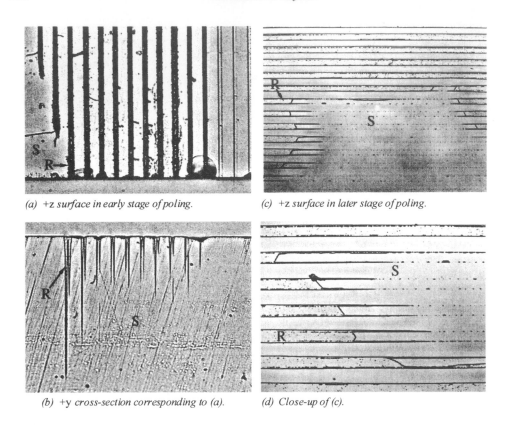

(a) +z surface in early stage of poling.

(c) +z surface in later stage of poling.

(b) +y cross-section corresponding to (a).

(d) Close-up of (c).

Figure 19. *Stages of domain growth in the PPLN poling process. (a) and (b) show the early stage when nucleation occurs at the edges of the electrode lines (13μm period, 3.5μm lines). (c) and (d) show a later stage when domains are filling in the electrode lines (31μm period, 11.5μm lines). S denotes a region with the original domain orientation of the substrate and R denotes a region where the domain has been reversed*

the pattern is obliterated. The strategy for producing a patterned crystal is to stop the poling when domain reversal has occurred under the electrode pattern but not under the insulator, as indicated by the down-turn in the poling current. An integrator on the current monitor is useful for controlling the poling event since the stopping point is reached when the charge transferred is $Q = \int I_{\mathrm{pol}} dt = 2P_s A$ where A is the area of the conductor portion of the electrode pattern where domain reversal is desired. A picture of poling in the time before the stopping point is shown in Figure 19. Domain reversal begins at nuclei formed by high fringe fields at the edges of the electrode lines. These nuclei extend slightly beyond the edges of the metal lines (typically $\sim 2\mu m$) following the fringing fields. Further domain growth beyond the electrode at this point in the process is prevented by the field modulation and current blocking of the insulator. Instead the domains fill in the conductor region of the electrode while the voltage stays clamped at V_c. Nucleation and filling continues until the entire conductor area is poled.

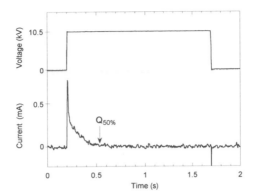

Figure 20. *Voltage and current traces for poling with a low impedance source, for a 0.5mm thick PPLN sample with 15.5μm grating period and 3.25 mm diameter poled area. Around the charge transfer for a 50/50 duty cycle grating design ($Q_{50\%} \approx 8\mu C$), the switching current self terminates, but this is sensitive to the exact setting of the drive voltage.*

A variant of the electric-field poling circuit eliminates the series resistor. For this low-impedance source, the poling current has more of the appearance of the ferroelectric switching pulse without the current clamping effects, as shown in Figure 20. The strategy for periodic patterning in this case is to set the poling voltage just enough above V_c to get active domain reversal so that the voltage never rises to a higher value and domains are not driven under the insulator. In this case, the poling is self terminating rather than charge controlled as it is with the high-impedance source. This approach is best for ferroelectrics that have relatively low resistivity, but it has the disadvantage that it requires a precise setting of the driving voltage. The voltage set point that maximises the effect of the field modulation, giving highest contrast between reversed and unreversed regions, is where the change in domain wall velocity is greatest for a given change in applied voltage, but this is also the regime with the most sensitivity to the field setting (Miller *et al.* 1997). Since the exact function relating domain wall velocity to applied field can vary from sample to sample, the high-impedance approach has the advantage that it is not very sensitive to the actual driving voltage because it automatically picks out the near-optimal setpoint by the clamping action of V_c. The high-impedance source also allows poling at low currents which makes the poling time long enough to permit real-time computer control to monitor the charge and shut off the voltage at the correct stopping point.

The choice of the materials in the electrode structure is an important consideration. Typical insulators are photoresist, SiO_2, and spin-on glass, while typical conductors are metal films (Al, Ta, Cr, and Ni–Cr) and liquid electrolyte (LiCl or NaCl solutions). An example of the issues concerning this selection is shown in Figure 21. When Al is used as the conductive electrode, domain reversal behaves differently on different crystal surfaces

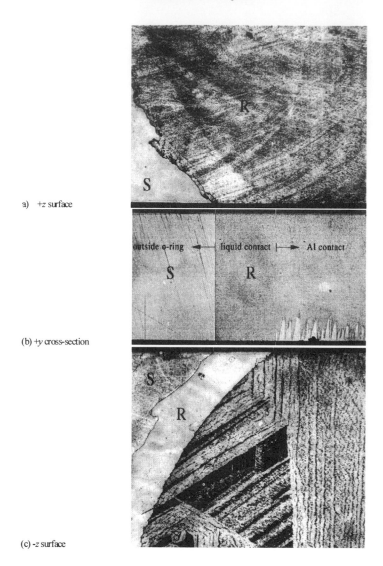

a)   +z surface

(b) +y cross-section

(c) -z surface

Figure 21. *Poling is different for Al and liquid contacts. The aluminium contact leaves an unreversed layer under the -z surface, the depth of which depends on the poling current. S and R are described in Figure 19.*

(Myers 1995a). On the -z surface, there is an unreversed layer that is absent on the +z surface, and it does not occur on either surface with LiCl contact. The explanation of this phenomenon is that the metal/semiconductor interface forms a blocking potential in a way somewhat analogous to a Schottky contact potential. As another example of the importance of material selection, it has been reported that domain nucleation density is increased with the use of sputtered Ni–Cr film on the +z surface of LN (Miller *et al.* 1997). Thus choice of electrode material is open to optimisation.

If the voltage is abruptly removed after poling a sample of LN, the domains will revert back to their original polarity. However the domains are stable after approximately 50ms, so after applying a poling pulse, the domain pattern can be frozen by applying a field slightly below the coercive field for ~50ms to prevent the domain from switching back while also avoiding further forward growth (Myers 1995a). Once stable domain reversal has been accomplished, the domain pattern is permanent in the crystal, even up to temperatures of 800°C (Missey *et al.* 1997). Thus a periodically poled crystal can be sawed, polished, coated, and handled like any ordinary single-domain nonlinear optical crystal. The tendency for domains to switch back may be caused by a built-in field in the crystal, which could also be related to the phenomenon of asymmetric hysteresis (Myers 1995a). After a piece of LN piece has been reversed (and the domain stabilised), if poling is attempted with the opposite field polarity to force the domain back to its original orientation, the coercive field that will be observed will be ~16kV/mm as opposed to 21kV/mm for the original poling. If another voltage pulse of the original polarity is applied to again cause domain reversal in the forward direction, E_c will again be its original value of 21kV/mm. This asymmetry in the coercive field will be repeated if cycles of domain reversal are continued. If the sample with domain reversed regions is stored for a long time before the second switch-back pulse is applied (*e.g.* months), the lower coercive field 16kV/mm will still be observed. However if the sample is heated after the forward reversal (*e.g.* 100°C for 1 hour), the second switch-back pulse will have the original E_c of 21kV/mm. Thus the built-in field that exists for the original domain polarity appears to have a long decay time at room temperature, but moderate heating allows the crystal to relax and charges to rearrange to accommodate the new domain structure. The statement that it takes 50ms for domain reversal to stabilise seems to contradict the literature reports of poling with shorter voltage pulses (*e.g.* 100µs). The discrepancy is resolved by recognising that the circuits may contain elements that block the back flow of charge such as a diode or a less obvious element with a similar current blocking effect such as a high-voltage switch that shuts off as an open circuit.

3.5   Poling other materials

Other materials related to LN have also been periodically poled. A potentially interesting crystal is MgO:LN since MgO doping reduces the photorefractive susceptibility. MgO:LN substrates are not as readily available as undoped LN, but the crystals are used commercially in at least one product. MgO:LN has a coercive field of ~7kV/mm versus 21kV/mm for undoped LN, but the domain spreading is much greater making periodic patterning difficult. Waveguide devices have been fabricated using surface poling methods, and a variant of the electric-field poling technique using a corona electrode has yielded the impressive result of ~ 5µm periods in 0.4mm thick, 5mm long substrates (Harada and Nihei 1996). Other dopants that are of interest are Nd and Er which can possibly be used to make quasi-phasematched nonlinear frequency conversion in combination with active lasing in the same crystal.

LT is an isomorph of LN which has been an important acoustic material but previously of little use in nonlinear optics because of its small birefringence. The melting temperature of 1530°C versus 1250°C for LN makes it harder to grow, and commercial substrates are not as well developed. However it has deeper UV transparency than LN (280nm

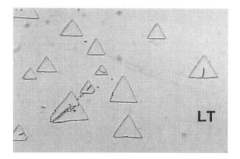 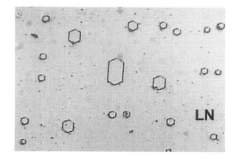

Figure 22. *Pictures of etched +z faces after applying a brief voltage pulse. The natural shape of domains in LT is a triangle; in LN it is a hexagon.*

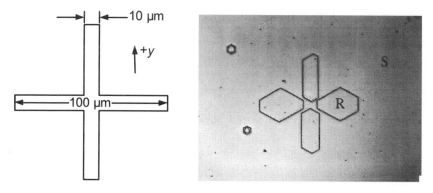

Figure 23. *Effects of pattern orientation relative to crystal axes in PPLN. (a) Test pattern formed in photoresist trench. (b) Etched +z surface of PPLN crystal. Note that poling occurs last in the centre of the cross because of lowered fringe fields at the conductive inside corners.*

vs. 350nm), and appears to pole with short periods more easily (Mizuuchi *et al.* 1997). Thus periodically poled LT (PPLT) could be an important material for short wavelength interactions. Poling parameters are similar to LN, with the same coercive field even though T_c is only about half that of LN (610°C vs. 1200°C). P_s is slightly less ($55\mu C/cm^2$), but the built-in field is apparently much stronger (domain set-up time $\sim$2s for LT vs. 50ms for LN, and E_c for switch back $\sim$7kV/mm vs. 16kV/mm). The major difficulty with PPLT is in maintaining the domain pattern away from the surface electrode. The natural domain shape of LN is a hexagon while that of LT is a triangle, as shown in Figure 22. Thus in LN the edges of the hexagonal domain can be aligned with the grating lines and good fidelity can be obtained as shown in Figure 23. In LT, it is not possible to align the two parallel edges of the grating lines with the edges of the triangular domains. While LT appears to be easier to pole with short periods when examined near the surface, inside the crystal the domains tend to take on their natural triangular shape so that degradation of pattern fidelity below the surface is worse than LN.

Besides LN and its relatives, the other important family of ferroelectrics for periodic poling is that of KTP and its isomorphs RTA, KTA, and CTA (Chen and Risk 1994, Karlsson *et al.* 1996). The coercive field of these materials is an order of magnitude lower than that of LN, and the anisotropy of the ionic channels in the crystal make domain patterns of high aspect ratio easier to achieve. These properties allow the fabrication of thicker periodically poled crystals, and combined with high optical damage threshold, they are useful for applications involving frequency conversion of higher pulse energy lasers. The main problems with these materials is their variability and uniformity relative to LN. The development of periodically poled versions has been hampered by their high conductivity; however, great progress has been made and important device demonstrations are now being performed with these crystals. Poling techniques used to overcome the high conductivity include using crystals with lower conductivity (particularly RTA), ion exchange to block the conductivity at the surface, and poling at low temperature to reduce the ionic conductivity (Karlsson and Laurell 1997, Karlsson *et al.* 1996, Rosenman *et al.* 1998).

Other ferroelectrics have also been suggested for periodic poling, but they are far less common than any of the LN or KTP crystal family. Some of the crystals that have been used in research are SBN, $BaTiO_3$, $KNbO_3$, and $MgBaF_4$. Except for SBN, no results on periodic crystals suitable for device use have been published for these materials. The search for other ferroelectrics suitable for periodic poling is motivated by a desire for materials operating further in the UV and IR than is possible given the transparency range of crystals in the LN or KTP families. LBO and BBO, although important commercial crystals for visible and UV generation, are not ferroelectric.

Certain semiconductor materials have large nonlinearity and good thermal and damage properties, but they lack birefringence for phasematching and they are not ferroelectric for periodic poling. Such materials are particularly useful because they have infrared transmission covering the important atmospheric windows (3–5μm and 8–12μm) and wavelengths with important spectroscopic activity where it is difficult to find suitable laser sources and nonlinear materials. For this IR range, the coherence lengths are typically $\sim 100\mu$m, so that it is possible to employ the method of stacking plates of rotated crystals to achieve quasi-phasematching. The technique of diffusion bonding, in which the stack is fused with heat under pressure to form a monolithic crystal, has been used to make this approach more practical. Diffusion-bonded GaAs stacks with up to 45 layers have been built, and SHG of CO_2 lasers at $\sim 5\mu$m and difference frequency generation out to as long as 16μm have been demonstrated (Lallier *et al.* 1998, Zheng *et al.* 1998). Diffusion-bonded ZnSe and CdTe have also been investigated for this use. The main disadvantages of this approach are the labour-intensive fabrication method, the residual loss per interface which limits the number of plates, and the thickness variation in the plates which causes period errors that limit the useful interaction length of a stack. Fabricating periodically patterned semiconductor crystals using lithographic patterning overcomes the problem with period error. A method of patterning wafer surfaces with a twinned crystal orientation and then using this as a template for subsequent regrowth has been demonstrated, and waveguide devices have been fabricated (Yoo *et al.* 1995). This approach may be especially important for telecommunication devices at 1.3–1.5μm, particularly wavelength division multiplexing (WDM) systems in combination with diode lasers and other integrated optical components.

3.6   Summary of poling methods

In summary, the electric-field poling technique has proven to be a practical process for enabling the fabrication of materials for QPM devices, and as a result periodically poled ferroelectrics are now commercially available. The two most common of these are PPLN and periodically poled KTP (PPKTP). PPLN for IR interactions ($\sim30\mu$m grating periods) can be obtained in lengths of up to 80mm and thickness up to 1mm. For visible interactions ($\sim 6\mu$m periods), there have been impressive laboratory demonstrations using PPLN, but these crystals are less available commercially. PPKTP can be made in higher aspect ratio (*i.e.* ratio of crystal thickness to domain period) than PPLN, which means shorter periods for visible interactions and thicker crystals for IR interactions. However, crystal lengths are limited by available substrate dimensions. The variants periodically poled RTA and KTA have also been demonstrated, but they have far more limited availability due to the relative immaturity of the substrates. PPLT is not commercially available, but it could become important due to its deeper UV transparency. The usefulness of these materials for devices will be described in the next section.

4   Devices demonstrated

With the increasing availability of periodically poled ferroelectrics, many QPM devices are being reported. Some demonstrations are QPM implementations of standard device types used in nonlinear frequency conversion based on conventional crystals, with perhaps some improvement due to the high nonlinearity or extended tuning range with no walk-off. Other demonstrations are of devices that are novel in that they are uniquely enabled by the flexibility of QPM design. In this section, some of the key demonstrations will be described, including prospects for commercial devices. As a prelude to the devices themselves, this section will start with a discussion of the advantages and disadvantages of quasi-phasematching for frequency conversion applications and describe aspects of QPM materials that are especially relevant for device performance.

4.1   Advantages of quasi-phasematched materials

Because of the time and cost required to develop a new nonlinear material, an important advantage of quasi-phasematching is that it extends the operating range of existing materials. It is fortunate that the current periodically poled ferroelectrics have a transmission range (0.35–5μm) that covers many important frequency conversion applications. In addition, the high nonlinear drive possible with QPM material is a good match for frequency conversion of commercial lasers, particularly diodes and diode-pumped solid-state lasers. Another significant advantage of quasi-phasematching with periodically poled ferroelectrics is that the electric-field poling technique involves fabricating crystals on mature substrate wafers using lithography and planar processing methods that have been well developed for the microelectronics industry. In some sense, the vision for quasi-phasematching is that a material like PPLN might serve as an analogue to silicon in microelectronics, that is, a highly developed substrate, grown in large size and volume, that can be used for fabricating a multitude of different devices.

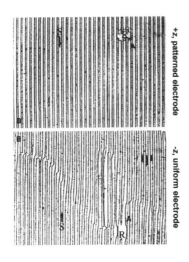

+z, patterned electrode

-z, uniform electrode

PPLN crystal
0.5 mm thick
15 μm period

Shown in orthographic projection with corresponding points "A" and "B" noted.

"S" marks regions with the domain orientation of the original substrate; "R" marks domain-reversed regions.

"A" marks a lithography defect in the electrode on the +z surface.

Figure 24. *Typical PPLN crystal showing crystal defects that degrade the pattern fidelity on the bottom surface relative to the top surface. However these defects do not introduce long-range period errors so they do not seriously degrade conversion efficiency.*

Substrate availability is an important factor for successful commercial use. In this regard, PPLN is by far the most advanced, since it has benefited from the use of LN in surface acoustic wave modulators which are part of consumer electronics products (Bordui and Fejer 1993, Räuber 1978). Acoustic-grade LN is grown literally in quantities of tons per year. Optical-grade LN (i.e. no strain and low index variation) has also been aided by the reasonably appreciable integrated optics market which consumes more than 30,000 wafers per year. An important advantage of LN for taking advantage of larger markets is that it can be grown in high quality relatively quickly and easily using Czochralski growth from a melt; an optical-grade boule 100mm diameter by 120mm long grows in 1–2 days. KTP is a widely used nonlinear crystal in the context of the laser industry, but its production is far below that of LN. It is also comparatively difficult to grow, requiring solution growth (or the even more difficult hydrothermal growth) and taking 3–6 weeks to produce crystals ~ 20mm long. Because of the layer-by-layer growth in the solution, the boules have central defects around the seed and the substrates for poling are cut across the growth layers and hence are subject to compositional variations. Even LN, despite its maturity level, appears to contain crystal defects that affect the uniformity of periodically poled crystals, as shown in Figure 24. Nonetheless, because defects like those shown do not introduce errors in the long-range period, crystals like this perform in frequency conversion experiments with near theoretical conversion efficiency. However compositional variation can cause significant efficiency degradation comparable to period errors.

An important advantage of QPM materials is that they have large effective non-linear coefficients d_{eff} even after accounting for the $2/\pi m$ reduction factor for quasi-phasematching (see Equation 7). In the case of periodically poled ferroelectrics, this is due to the fact that the nonlinear susceptibility is highest for interactions with all waves polarised in the extraordinary direction. This geometry is only accessible when

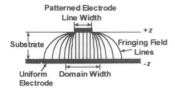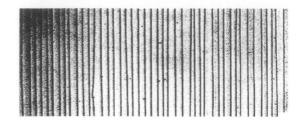

Figure 25. *Typical -z face of a PPLN crystal 0.5mm thick, 30μm period. The +z face has much better fidelity to the electrode pattern. Assuming 10% deviation in the average duty cycle and 10% random variation about the mean, the sample shown here would have 86% of the nonlinear coefficient for first-order QPM compared to a crystal having ideal 50/50 duty cycle.*

the requirement for achieving phasematching through birefringence is lifted (e.g. for LN $d_{33} = 25$pm/V, $d_{31} = 4.6$pm/V; for KTP $d_{33} = 15$pm/V, $d_{31} = 3.7$pm/V). Since the non-linear coefficient affects the frequency converted power by its square (see Eq. 8), a higher nonlinear coefficient has a significant impact. However, all properties (phasematching, transmission, nonlinearity, homogeneity, damage, mechanical stability, lifetime, availability, etc.) must be simultaneously satisfied, so that in practice a high nonlinear coefficient does not necessarily make the material useful.

4.2   Major issues with quasi-phasematched materials

The high nonlinear coefficient of periodically poled material is obtained when the duty cycle of the domain grating is optimised. The duty cycle deviates from the defined electrode pattern because of fringe fields around the grating lines and the interaction of these fields with the substrate (e.g. inhomogeneity in the coercive field). For the case of normally distributed errors, the overall efficiency reduction due to both average and random duty cycle errors is the product of factors from Table 1:

$$\eta \propto \sin^2(\pi D)\left[1 - \frac{\pi^2}{2}\left(\frac{\sigma_\ell}{\ell_c}\right)^2\right]$$

As shown in Figure 25, a typical periodically poled crystal might have 10% deviation of the average duty cycle and 10% random variation, so the overall first-order efficiency is only reduced to 86% of its value for the case of ideal 50/50 duty cycle. This level of performance is a realistic expectation for commercially produced crystals.

Besides conversion efficiency, there are other factors which might indicate the need for duty cycle control. One issue is that there are coincidences where the same grating period can phasematch multiple processes at the same temperature, either other wavelengths or higher-order interactions. It is possible to suppress all even-order interactions by precisely controlling the duty cycle to be exactly 50/50. The duty cycle also effects the magnitude of photorefractive damage, as shown by the analysis in Taya *et al.* (1996). Compared to a single-domain crystal, the alternating domains in a periodically poled crystal lead

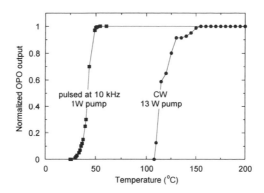

Figure 26. *In a 1064nm pumped PPLN OPO photorefractive damage is eliminated by heating. The temperature required depends on the thermal loading and the sensitivity of the device to the losses introduced by photorefraction.*

to reduction of the space-charge fields that couple to index perturbations which are the source of photorefractive phenomena. A perfect 50/50 duty cycle results in the greatest reduction of photorefractivity since the periodic space-charge field has no DC component in its Fourier series description. However, to obtain a reduction of photorefractivity within 10% of the optimum requires controlling the duty cycle to better than ±1% because even a small deviation from 50/50 duty cycle introduces a large DC component. Duty cycle control is especially important for PPLN, since LN exhibits a large photorefractive effect. It is difficult to maintain tight tolerances on duty cycle control with electric-field poling; nevertheless, reduction of photorefractive damage in PPLN has been experimentally observed for visible interactions. For IR interactions, the reduction in photorefraction is less effective since, although the periods are longer and hence domain wall position is easier to control, the magnitude of the reduction even with ideal 50/50 duty cycle is proportional to the square of the ratio of the grating period to the beam radius. For same size beams, the longer grating periods for IR interactions (e.g. 30μm for an IR OPO in PPLN, vs. 6μm for 532nm SHG) are over an order of magnitude less effective at reducing photorefraction. Thus even though the photorefractivity is less strongly driven by longer wavelength illumination, IR PPLN crystals must be operated at elevated temperature to eliminate photorefraction, as shown in Figure 26 for the case of a 1064nm pumped OPO. Another contributor to photorefractive susceptibility in IR devices is visible light generated by extraneous processes such as SHG or sum frequency generation (SFG) (e.g. 1064nm pump and 1550nm signal produce 630nm). The power generated by these extraneous interactions may be enhanced over what would be obtained in a single-domain crystal because the process may be coincidentally phasematched on a higher QPM order or the domain sections may act as a randomly phased stack giving an output that is the incoherent superposition of the contribution of each section (i.e. the output is larger by the factor of the number of domain sections, which may be over 1000 times).

Photorefractivity is the type of material problem that is hard to anticipate, but can severely limit the material applicability. In the case of PPLN, MgO:LN is an alternative that has reduced photorefraction at room temperature. The periodically poled version of

MgO:LN has been demonstrated, but fabrication is difficult because the domains spread laterally beyond the electrode pattern even more so than in the case of standard PPLN. Stoichiometric LN has been suggested because of reduced absorption and photorefraction, but it is not available due to difficult growth. The KTP family of crystals also shows photorefraction, although less than LN or LT. However, KTP crystals are subject to photo-induced absorption (gray tracking), and they may have inhomogeneities in optical and poling properties due to the substrate nonuniformity.

Thermal effects cause significant problems that are usually evident in nonlinear crystals when scaling to high average powers. Thermal effects arise from heating of the material due to absorption which couples to refractive index perturbations through the thermo-optic coefficient dn/dT, leading to thermal lensing and dephasing. Thermal effects have been reported in both PPLN and PPKTP (Arie *et al.* 1998, Miller *et al.* 1997). While the linear absorption and thermo-optic coefficient values are comparable for these two materials, in high-power visible light generation, nonlinear absorption may dominate. Green-radiation-induced infrared absorption (GRIIRA) (i.e. increased absorption at 1064nm due to the presence of SHG at 532nm) leads to a strong thermal lens in PPLN which has limited SHG to <2W. Note that GRIIRA can be bleached out under some conditions (Batchko *et al.* 1998). There is also research underway seeking to lower the absorption of the substrates. There are existence proofs for high-power visible generation in KTP, although GRIIRA has also been observed (Krupke, this book). The measurement of GRIIRA in KTP is difficult due to complicated temporal evolution of the effects.

4.3   Quasi-phasematched materials in conversion devices

In spite of the issues related to thermal effects, absorption, photorefraction etc., the periodically poled ferroelectrics have been of fundamental importance in making possible the demonstration of a variety of QPM devices. While many of these devices are straightforward QPM versions of frequency conversion devices that are common with conventional crystals, they often have better performance in some regard due to the high nonlinear drive, no walk-off, or extended noncritical phase matching range of QPM materials. Examples of such devices include SHG, difference-frequency generation (DFG), and OPOs. In some devices, QPM has made a decisive difference in practicality; a particularly noteworthy instance is the cw singly resonant OPO (SRO).

QPM SHG in waveguide devices has provided much of the impetus for work on periodically poled ferroelectrics because of the high conversion efficiency possible with tightly confined fields which propagate without diffraction. SHG of blue light from diode lasers was a strong motivating factor for much of this work. Most effort on QPM from 1980 to 1990 was focused on waveguide devices because the fabrication methods of the time produced periodically poled surface layers and the power handling capability of waveguides was a good match for conversion of single-mode diode laser sources. However many of these fabrication methods produced domains with non-ideal shape (e.g. triangular cross sections) which gave poor overlap of the nonlinearity with the waveguide mode and limited typical efficiencies to ~10%/Wcm^2 (Webjörn *et al.* 1989). Electric-field poling can produce domains with the near-ideal rectangular shape, and conversion efficiency of 600%/Wcm^2 in these more optimal devices has been realised (Yamada *et al.* 1993).

QPM SHG in bulk devices was first demonstrated with periodically poled ferroelectrics using material produced by growth modulation methods. The overall efficiency of these crystals is limited because of the short interaction length due to period errors. A demonstration of blue light generation using first-order SHG, with a 1.7μm-period crystal fabricated by Czochralski growth modulation of doped LN, had an effective crystal length limited to 200 domains due to period error (Feng *et al.* 1989). SHG in a resonant cavity to enhance the fundamental power improves the overall efficiency even with short crystals. Using a PPLN crystal fabricated by the laser-heated pedestal method (3.7μm period, 1.24mm long), 1.4W of 523nm was generated for 4.25W of 1064nm input (65W circulating) (Jundt *et al.* 1991). The electric-field poling method produced crystals with long interaction lengths allowing efficient single-pass cw SHG in simple bulk devices. The ideal efficiency for a confocally focused bulk interaction in first-order PPLN is 4%/Wcm, and cw SHG of 532nm producing 1.8W with 42% efficiency in a single pass was demonstrated with a 50mm long electric-field-poled PPLN crystal (Miller *et al.* 1997). Blue light SHG of 473nm produced 450mW average power with 40% single-pass efficiency using a 15mm long PPLN crystal and a pulsed laser (Ross *et al.* 1998).

QPM DFG devices are motivated by IR generation for spectroscopy and telecommunications. DFG in a PPLN waveguide has potential application in WDM systems as a frequency shifter. In one advanced concept, the periodically poled frequency converter section is combined with a tapered waveguide input for efficient launching into the lowest-order mode and a waveguide coupler to mix the two beams in the DFG section, all fabricated as a single integrated optical component on an LN substrate (Chou *et al.* 1998). Bulk DFG in PPLN by mixing a pump laser with a tunable diode laser is a promising technique for spectroscopy and gas sensing in the 1.5–5μm range (Petrov *et al.* 1998).

QPM OPOs are a topic of vigorous research activity. The first demonstration of a QPM OPO was in 1994, but now just 4 years later there are QPM OPOs in all regimes from cw to fs ((Myers *et al.* 1995b); see also Dunn and Ebrahimzadeh in this book). PPLN and PPKTP for OPOs are particularly timely arrivals because of their good match for being pumped with cw-diode-pumped Q-switched lasers which are undergoing extensive commercial development (Myers and Bosenberg 1997). These lasers have relatively low peak power because of the cw diode pumping that limits energy storage, especially at high repetition rate (>10kHz) where the pulse lengths are longer. PPLN OPOs pumped by this type of laser have been demonstrated with threshold pulse energies as low as 6μJ. The performance of this PPLN OPO is most appreciated when tuning is considered, since the low threshold can be maintained even over a broad tuning range. Compared to previous uses of LN for mid-IR generation, the tuning range of PPLN OPOs extends farther into the IR because of better transmission for the extraordinary polarisation used in a d_{33} interaction, compared to birefringent phasematching in LN where only an ordinary-polarised idler is allowed. Generation of wavelengths out to $\sim$ 5μm has been demonstrated with Q-switched and cw PPLN OPOs, and output beyond 6μm has been reported with PPLN OPOs synchronously pumped by mode-locked lasers (Lefort *et al.* 1998).

The low threshold over a broad tuning range is possible because noncritical phasematching can be obtained for any wavelength at any temperature by adjusting the grating period. In conventional birefringent phasematching, tuning usually introduces walk-off so that low threshold is possible only at a limited operating range near the noncritical point. Another advantage of having low walk-off is that it is also possible to operate at high repetition rates where the pulses of cw-diode-pumped solid-state lasers typically have less energy and longer length (Myers and Bosenberg 1997). The resulting lower peak power makes the OPO threshold climb for given pump spot size, so tight focusing is desirable to keep the intensity high. Focusing can not be increased arbitrarily because eventually diffraction limits the interaction length, so there is a well-known optimum focusing condition to balance the area of the waist with expansion due to diffraction. However, walk-off further reduces the interaction length so that the lowest threshold with optimal focusing can not usually be reached with conventional crystals. In QPM devices designed for propagation collinear with the grating vector, there is no walk-off so the low threshold limit set by diffraction can be reached. Thus with tight focusing, QPM OPOs have been demonstrated to run at repetition rates >30kHz: higher pulse rates should be possible.

The limiting case of low peak to average power is that of cw, and QPM OPOs have demonstrated their most impressive performance in this regime. The thresholds of cw OPOs are generally so high that doubly resonant schemes are typically required to build up the field enough for oscillation with practical laser sources. Particularly compact and efficient would be an OPO pumped directly by a cw diode laser. With PPLN, thresholds below 50mW have been obtained, low enough to be reached with commercial off-the-shelf diode laser pumps (Myers 1995b). In these devices, the OPO signal and idler were resonated, which leads to well-known stability problems in trying to satisfy the double resonance condition. An alternative approach is to resonate the pump and signal field (pump-resonant OPO), which has better stability than the DRO because the pump has a constant wavelength independent of the signal. This approach has the complication that the pump must be a single-frequency laser to allow resonant enhancement, but since a common use for these tunable IR systems is spectroscopy, the single-frequency pump is often desirable anyway. With a stabilised output of <10MHz/min and no mode hops over more than 10 hours, a PPLN pump-resonant OPO has been demonstrated as a practical spectroscopic instrument (Schneider *et al.* 1997). Besides resonating the pump in the OPO cavity, another scheme for enhancing the pump field is to place the OPO inside the laser cavity itself; see Dunn and Ebrahimzadeh, this book.

A particularly simple implementation of a cw OPO involves a resonant cavity only for the signal wave. This singly resonant OPO (SRO) has a high threshold because it does not benefit from field enhancement of either the pump or the idler to increase the gain. Its high threshold prevented experimental demonstration of a cw SRO until 1992, and even then the pump source was a custom-built injection-locked and resonantly doubled Nd:YAG laser and the OPO operated only in a narrow region around the point of noncritical phasematching in the KTP crystal with no significant tuning. With PPLN, the cw SRO has become a straightforward and relatively simple device as shown in Figure 27 (Myers and Bosenberg 1997). It can run with commercial off-the-shelf cw-diode-pumped Nd:YAG lasers, multi-longitudinal mode or single frequency. Even though the threshold increases as the cube of the pump wavelength, the high gain of PPLN allows direct pumping at 1064nm instead of 532nm as was previously the only way possible. Using 50mm long PPLN crystals, typical thresholds are ~ 4W with 1064nm pump and ~ 2% round-trip signal loss. As shown in Figure 28, at 2–2.5 times above threshold, the cw SRO reaches >90% pump depletion, with output power of over 3.5W at 3.25μm idler. At longer idler wavelengths, the power drops off mostly because of absorption as the wavelength approaches the IR edge of LN, but output has been obtained well into the 4–5μm region. The cw SRO can run in a single longitudinal mode even with a multi-mode pump laser because the gain bandwidth is sufficiently large that all pump laser modes can pump a

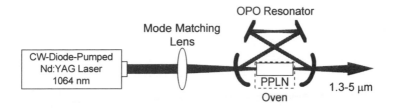

Figure 27. *A QPM CW SRO using PPLN is simple to implement with commercial off-the-shelf components. Both linear and ring cavities have been demonstrated, but the ring resonator is generally preferred for stability reasons.*

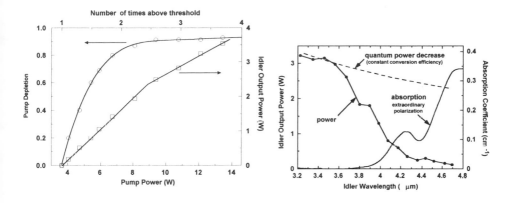

Figure 28. *Pump depletion, idler output power, and tuning of a 1μm pumped PPLN cw SRO.*

single mode of the OPO signal while the idler wavelength adjusts to conserve energy. Typical linewidth of the signal with a multi-mode pump (2.2GHz linewidth with 10 axial modes) is 0.02cm^{-1} in a single axial mode. The device can run on a single axial mode for several hours, in which case the amplitude stability is $<\pm 1\%$. For longer observation periods when mode hops may occur, the amplitude stability is $\pm 5\%$. An intracavity etalon can be inserted for additional linewidth control and tuning. The remarkable performance of the cw SRO based on PPLN has led to commercial development of the system shown in Figure 29 with several systems delivered so far.

4.4 Novel schemes enabled by QPM

In addition to the conventional frequency conversion devices described above, there are a number of novel devices that are only made possible due to the engineerable properties of quasi-phasematching. One category of such unique designs involves patterning different gratings in different parts of the crystal, so that the phasematching properties depend on which part of the crystal is placed in the pump beam. A particularly useful example of

Figure 29. *Example of a PPLN cw SRO. The picture on the left shows the pump laser (Lightwave Model 220) and OPO platform, in this case with an acousto-optic modulator in the pump beam train. The picture on the right shows the OPO platform with the cover removed. The performance of this device is similar to that shown in Figure 28. Amplitude stability < 1% rms has been observed over hours of operation.*

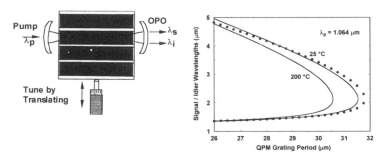

Figure 30. *A multi-grating QPM OPO uses a PPLN crystal having different sections with different periods. Simply translating the crystal through the pump beam provides wide tunability while maintaining noncritical phasematching .*

this is a multi-grating crystal used to make a widely tunable OPO (Myers and Bosenberg 1997). The OPO is inherently a widely tunable device, but often with conventional crystals tuning is limited because the interaction falls outside of the phasematching acceptance bandwidth. By tuning the grating period, the entire tuning range of the OPO can be accessed while still maintaining the noncritical interaction. A grating-tuned OPO can be implemented by fabricating different grating periods placed side-by-side on a single crystal, as shown in Figure 30. With this design, PPLN OPOs have been demonstrated with tuning covering the IR transmission range. Crystals of this multi-grating design are now available as commercial products. Continuous tuning coverage can be obtained by temperature tuning in each discrete grating section, or by using a fanned-out grating design.

Another category of novel devices involves positioning different grating patterns se-

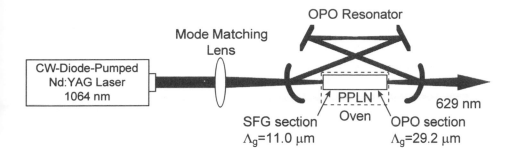

Figure 31. *CW SRO with intracavity SFG. The cascaded process is implemented in a single monolithic PPLN crystal containing both SFG and OPO grating sections. The OPO resonator mirrors are all high reflectors at 1540nm to provide a high circulating field to mix with the pump, and output coupling is provided for the 620nm sum frequency. 2.5W cw of red light has been demonstrated with this approach.*

rially, so that multiple processes can be cascaded in a single monolithic crystal. The electric-field patterning process lends itself readily to fabricating these sorts of devices because the layout is straightforward on a lithographic mask. The major issue is optimising the particular design for both sections to operate simultaneously at a single temperature. Errors in the calculated periods from the Sellmeier equations may cause the two cascaded sections to have different peak phasematching temperatures but this can be corrected by experimentally measuring the actual phasematching temperatures and then adjusting the grating periods in the mask design until the phasematching peaks match to within the tolerance needed to reach the desired efficiency.

Cascaded processes that occur within a resonant cavity have recently been the subject of theoretical and experimental investigations, and their single-crystal implementation in a periodically poled ferroelectric is an attractive approach. A cascaded process which has been shown to be beneficial is an OPO combined with intracavity DFG (Fukumoto *et al.* 1998). If the idler (here meaning the longer of the two wavelengths generated in the parametric oscillator) is the desired output of the device, then the resonating signal can be further down-converted to produce more idler output. Thus in fact, it is possible to generate two idler photons from each converted pump photon, rather than one as would be the case in a conventional OPO. There have been demonstrations of this in PPLN OPOs, with as much as 121% photon conversion efficiency going from 1064nm pump to 3940nm idler. Another benefit of this approach is that the reduction of the signal prevents the back-conversion of the signal back into the pump which typically causes the conversion efficiency to saturate, especially in pulsed OPOs. Thus the combined OPO+DFG has higher energy extraction at higher efficiency than a conventional OPO by itself.

If shorter wavelengths are desired, an OPO cascaded with SFG can be an efficient route. In a demonstration of this, the cw SRO discussed in earlier paragraphs was modified to include an SFG section in the PPLN crystal for sum-frequency mixing of the OPO signal with the residual pump, as shown in Figure 31 (Bosenberg *et al.* 1998). For a 1064nm pump and a 1540nm OPO signal, the sum frequency is 629nm, and over 2.5W cw was produced. The red output was tunable over a limited extent by temperature tuning, since fortunately the phasematching of both grating sections tunes in the same direction with temperature change. Further tuning is possible by using a crystal with multi-grating or fanned-grating cascaded sections.

The flexible patterning possible with periodically poled ferroelectrics allows a range of novel devices that take advantage of the ability to engineer the phasematching properties in a QPM crystal. It is possible to tailor the nonlinear conversion in the frequency, spatial, and temporal domains. Tailoring in the frequency domain allows shaping the phasematching bandwidths, which can be done using Fourier analysis to design grating patterns to obtain the desired spectral shape (Fejer *et al.* 1992b). An important application of this is to broaden the acceptance bandwidth of the nonlinear crystal while maintaining a constant efficiency over as much of the bandwidth as possible, in order to match the phasematching peak of the nonlinear crystal to the laser wavelength which has several nanometres uncertainty in typical commercial diode lasers. The nonlinear conversion can be shaped in the spatial domain by adjusting gain in different parts of the crystal corresponding to different parts of the pump beam. An example of this is to control the non-uniform conversion of a Gaussian beam by reducing the amplitude of the conversion in the centre of the beam, where the high field drives the process efficiently, relative to the wings where the conversion is lower. A device of this type has been built which produces a flat-top SHG beam from a Gaussian input beam (Imeshev *et al.* 1998b). This type of spatial tailoring may also be useful for reducing back-conversion which limits depletion and for controlling gain-induced diffraction and transverse modes in high-energy OPOs. Spatial tailoring can be used to control the amplitude of the beam by adjusting the grating length or the domain duty cycle, and the phase by using curved grating designs. With the standard electric-field poling technique, spatial tailoring is restricted to two dimensions by the constraint of planar processing, but it can be extended to three dimensions by stacking together plates with different patterns. It has been demonstrated that stacks of PPLN plates can be diffusion bonded together to form a single monolithic crystal, and a QPM OPO was built using such a crystal that consisted of three 1mm thick PPLN plates (Missey *et al.* 1997). Finally, tailoring in the temporal domain is also possible by combining the effects of group velocity walk-off with the placement of gratings sections in the crystal (Imeshev *et al.* 1998a). PPLN and PPKTP crystals of this type have been used to produce compressed SHG pulses by chirping the grating period to match the chirp of the fundamental pulse so that spectral components with slower group velocity are converted last and stack up with the faster spectral components which were converted in the front part of the crystal. By repeating this pattern in the crystal, a train of pulses can be generated from a single input pulse .

4.5   Commercial QPM devices

Commercial QPM devices using periodically poled materials have been limited up to the current time, but there is continuing development underway. The interest in blue sources based on SHG of diode lasers led to several companies showing technology demonstrations of diodes doubled in PPLN and PPKTP waveguides. However, these demonstrations have not led to actual commercial products due to the difficulty of developing products based on waveguide frequency converters and the advances in semiconductor lasers for direct blue generation. The first QPM product is a PPLN IR OPO pumped by an acousto-optically Q-switched 1μm solid-state laser (Aculight Corporation). CW SROs have been delivered on government and commercial contracts, and are expected soon as catalogue products. Intracavity OPOs and pump-resonant OPOs may offer opportunities for lower power tunable instruments for spectroscopic and scientific markets.

5   Conclusions

In conclusion, QPM devices have the advantages of higher nonlinear drive and flexible designs, but the materials suffer problems such as thermal lensing, absorption, and photorefractivity that have delayed their incorporation into commercial products. While periodic poling greatly enhances the nonlinear properties of the materials, the underlying problems in their linear properties are all the more apparent. Because of these material issues, high-power visible generation is difficult, but quasi-phasematching is a good choice for low-peak-power systems, which is generally a good match for frequency conversion with diode-pumped solid-state and fibre lasers. The most likely near-term commercial devices are OPOs for IR generation in the 1–10W range, pump-resonant cw OPOs for spectroscopy, and low-power visible sources. Material research is needed to improve absorption and thermal properties or to manage their effects; ongoing work is promising. Quasi-phasematching is expected to find more commercial utility as the materials mature and application niches develop that take advantage of its special attributes.

References

Arie A, Rosenman G, Korenfeld A, Skliar A, Oron M, Katz M and Eger D, 1998, *Opt. Lett.* **23**, 28.

Armstrong J A, Bloembergen N, Ducuing J and Pershan P S, 1962, *Phys. Rev.* **127**, 1918.

Batchko R G, Miller G D, Alexandrovski A, Fejer M M and Byer R L, 1998, in *Conference on Lasers and Electro-Optics*, 1998 OSA Technical Digest Series 6, (Optical Society of America, Washington, D.C.), 75.

Bordui P F and Fejer M M, 1993, *Annu. Rev. Mater. Sci.* **23**, 321.

Bosenberg W R, Alexander J I, Myers L E, and Wallace R W, 1998, *Opt. Lett.* **23**, 207.

Burns W K, McElhanon W and Goldberg L, 1994, *IEEE Photon. Technol. Lett.* **6**, 252.

Byer R L, in *Nonlinear Optics*, P G Harper and B S Wherrett, ed. (Academic, San Francisco, 1977), p 47.

Byer R L, 1994, *Nonlinear Optics* **7**, 235.

Camlibel I, 1969, *J. Appl. Phys.* **40**, 1690.

Chen Q and Risk W P, 1994, *Electron. Lett.* **30**, 1516.

Chou M H, Hauden J, Arbore M A and Fejer M M, 1998, *Opt. Lett.* **23**, 1004.

Edwards G J and Lawrence M, 1984, *Opt. Quantum Electron.* **16**, 373.

Eger D, Oron M, Katz M and Zussman A, 1994, *Appl. Phys. Lett.* **64**, 3208.

Fejer M M, 1992a, in *Guided Wave Nonlinear Optics*, (Kluwer Academic Publishers, Netherlands), p133.

Fejer M M, Magel G A, Jundt D H and Byer R L, 1992b, *IEEE J. Quantum Electron.* **28**, 2631.

Feng D, Ming N B, Hong J F and Wang W S, 1989, *Ferroelectrics* **91**, 9.

Fukumoto J, Komine H, W. H. Long J, and R. K. Meyer J, 1998, in *Conference on Lasers and Electro-Optics*, San Francisco CA, 6, (Optical Society of America, Washington, DC), paper CPD5.

Harada A and Nihei Y, 1996, *Appl. Phys. Lett.* **69**, 2629.
Imeshev G, Galvanauskas A, Harter D, Arbore M A, Proctor M and Fejer M M, 1998a, *Opt. Lett.* **23**, 864.
Imeshev G, Proctor M and Fejer M M, 1998b, *Opt. Lett.* **23**, 673.
Jundt D H, Magel G A, Fejer M M and Byer R L, 1991, *Appl. Phys. Lett.* **59**, 2657.
Karlsson H, Laurell F, Henricksson P and Arvidsson G, 1996, *Electron. Lett.* **32**, 556.
Karlsson H and Laurell F, 1997, *Appl. Phys. Lett.* **71**, 3474.
Kitamura K, Furukawa Y, Ji Y, Zgonik M, Medrano C and Montemezzani G, 1997, in *Conference on Lasers and Electro-Optics*, 1997 OSA Technical Digest Series 11, (Optical Society of America, Washington, DC), 121.
Kittel C. *Introduction to Solid State Physics*, (Wiley, New York, 1986).
Kurimura S, Shimoya I and Uesu Y, 1996, *Jap. J. Appl. Phys.* **35**, L31.
Lallier E, Brevignon M, and Lehoux J, 1998, *Opt. Lett.* **23**, 1511.
Laurell F, Roelofs M G, Bindloss W, Hsiung H, Suna A and Bierlein J D, 1992, *J. Appl. Phys.* **71**, 4664.
Lefort L, Puech K and Hanna D C, 1998, *Appl. Phys. Lett.***73**, 1610.
Lines M E and Glass A M. *Principles and Applications of Ferroelectrics and Related Materials*, (Clarendon Press, Oxford, 1977).
Miller G D, Batchko R G, Tulloch W M, Weise D R, Fejer M M and Byer R L, 1997, *Opt. Lett.* **22**, 1834.
Missey M J, Dominic V, Myers L E and Eckardt R C, 1997, *Opt. Lett.* **23**, 664.
Mizuuchi K and Yamamoto K, 1993, *Electron. Lett.* **29**, 2064.
Mizuuchi K, Yamamoto K and Kato M, 1997, *Appl. Phys. Lett.* **70**, 1201.
Myers L E, 1995a, PhD Thesis, Stanford University.
Myers L E, Eckardt R C, Fejer M M, Byer R L, Bosenberg W R and Pierce J W, 1995b, *J. Opt. Soc. Am. B* **12**, 2102.
Myers L E and Bosenberg W R, 1997, *IEEE J. Quantum Electron.* **33**, 1663.
Petrov K P, Curl R F and Tittel F K, 1998, *Appl. Phys. B***66**, 531.
Prokhorov A M and Kuz'minov Y S. 1990, *Physics and Chemistry of Crystalline Lithium Niobate*, (Adam Hilger, Bristol.
Räuber A, 1978, in *Current Topics in Materials Science*, E. Kaldis, ed. (North-Holland Publishing Co., Amsterdam), p481.
Rosenman G, Skliar A and Katz M, 1998, *Appl. Phys. Lett.* **73**, 3650.
Ross G W, Pollnau M, Smith P G R, Clarkson W A, Britton P E and Hanna D C, 1998, *Opt. Lett.* **23**, 171.
Schneider K, Kramper P, Schiller S and Mlynek J, 1997, *Opt. Lett.* **22**, 1293.
Taya M, Bashaw M C and Fejer M M, 1996, *Opt. Lett.* **21**, 857.
Webjörn J, Laurell F and Arvidsson G, 1989, *IEEE Photon. Technol. Lett.* **1**, 316.
Webjörn J, Pruneri V, Russell P S J, Barr J R M and Hanna D C, 1994, *Electron. Lett.* **30**, 894.
Yamada M, Nada N, Saitoh M and Watanabe K, 1993, *Appl. Phys. Lett.* **62**, 435.
Yoo S J B, Bhat R, Caneau C and Koza M A, 1995, *Appl. Phys. Lett.* **66**, 3410.
Zernicke F and Midwinter J E, 1973, *Applied Nonlinear Optics*, (Wiley, New York, 1973).
Zheng D, Gordon L A, Wu Y S, Feigelson R S, Fejer M M, Byer R L and Vodopyanov K L, 1998, *Opt. Lett.* **23**, 1010.

Medical lasers: fundamentals and applications

Rudolf M Verdaasdonk

University Hospital, Utrecht, The Netherlands

1   Introduction

Lasers systems have been applied in medicine since they first became available. In many early cases, the lasers were still laboratory models. At that point, the 'magic' surrounding lasers was more the driving force for curing than was a scientific understanding of the interaction of laser light with tissue. However, in time, people started to appreciate the interaction more as the results did not turn out to be as magic as expected. More basic science was conducted and strategies and laser delivery devices were developed. With the availability of new laser systems with alternative wavelengths, new applications have been explored. Also, by introducing computer control, procedures could be performed more precisely, in a more controlled manner and faster. Nowadays, lasers have proven to be the ideal source for delivering energy efficiently through very small instruments, especially in the area of minimal invasive techniques or keyhole surgery .

During the developments of these techniques, the sales of some lasers were influenced more by salesmen than clinical demonstrations. Sometimes new laser technologies were pushed on the market without, at the same time, providing adequate knowledge for optimal treatment. Consequently, the physicians rejected promising laser procedures since the technique was too complex and the results were disappointing at that stage.

Therefore it is essential to define a clear goal of what is expected from any treatment and outline a study to bring the treatment to a successful conclusion.

- The **goal** in laser medicine is the selective destruction of tissue that is suspected of malignancy, interfering with physiologic processes or of interfering with social expectancy, while at the same time causing minimal damage to surrounding 'healthy' tissues.

- The **outline** for the development of a new clinical laser application can be separated into phases: a) determination of the clinical goal effect, b) choosing the laser energy source, c) choosing the delivery device, d) choosing the tissue effect, e) understanding the mechanism of action, f) optimisation of the technique.

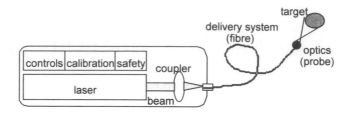

Figure 1. *Schematic of a medical laser system*

In this chapter, we will discuss the path from basic research to clinical application of lasers with this goal and outline in mind.

2   The laser system

The unique features of lasers for medical applications

The motives for using a laser as a replacement for other medical instrumentation or as a unique tool are based on three of the basic characteristics of laser light:

- The monochromatic nature of the light can be used to target specific chromophores in tissue thus enabling either selectivity or homogeneous distribution in a large volume.

- Laser light is produced in a small and minimally diverging beam. This makes it easy to transport the energy through small optical systems, such as miniature mirrors, and apply the energy in a non-contact fashion. Furthermore, it enables efficient coupling into small optical fibres providing transportation anywhere in the human body.

- Average powers and peak powers can be produced at levels sufficient to increase tissue temperatures to hundreds of degrees celsius in very short times.

The medical laser system

In principle there is no difference between an industrial/scientific laser and a medical laser. Usually, a medical laser system is a combination of a laser and a device to deliver to light to the patient (Figure 1). The medical laser has special provisions for safety and reliability compared to 'normal' lasers. From the regulatory point of view, a medical laser has to meet various safety requirements to prevent any beam emerging in an uncontrolled way. This requires safety interlocks to prevent the laser from emitting power as long as particular conditions are not fulfilled. Also, special attention is paid to the electrical circuit to prevent any leakage currents and electromagnetic emissions that could influence other delicate instruments inside an operating theatre and make contact with the patient.

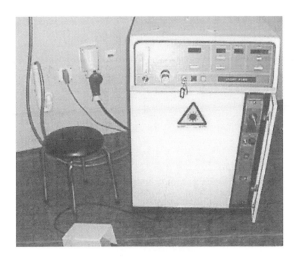

Figure 2. *'Old' generation 100W Nd:YAG laser with three phase current and external water cooling*

From a commercial point of view, the system has to look attractive and has to be easy to use. So the laser is embedded in a handsome box with displays helping through an automated start-up procedure which sets up the laser parameters. With a minimum of training, nurses and surgeons should be able to handle the laser safely. Preferably, the laser system should be transportable and small. However, most of the early medical laser systems did not meet the latter conditions. For example, the first generation of medical Nd:YAG lasers were clumsy, large systems running on three phase power and external water cooling (Figure 2).

Technological developments have improved on this greatly, especially with the introduction of diode lasers either as the primary source or as a pump source.

With the exception of diode and carbon dioxide lasers, most lasers are rather inefficient and a significant amount of energy is wasted as heat. As external water-cooling is not practical in an operating theatre, forced air-cooling is often used. Such systems tend to be noisy. Also, the heat dumped into the operating theatre rapidly increases the ambient temperature by several degrees centigrade, challenging the air-conditioning and blowing air into the sterile field of the operation. These are some issues which must be kept in mind when designing a medical laser system.

The current generation of lasers are becoming more user-friendly by integration of PC technology with coloured touch screens and help menus. One example of this development is a small diode laser system operating at 810nm and producing 60W of power (Figure 3).

Another distinctive feature of medical laser systems is the beam delivery device, either an articulated arm with optics or a fibre coupling system. These parts will be discussed in a later section.

The characteristics of medical laser systems which determine the resultant effect in tissue (Figure 4) are: a) penetration depth of the laser light into tissue (colour/wavelength),

Figure 3. *A 60 Watt diode laser system: an example of the latest generation of medical lasers*

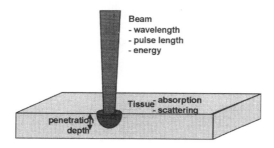

Figure 4. *Basic interactions of a laser beam with tissue*

b) pulse energy, c) pulse length (continuous wave (cw), chopped (ms), free running (μs), Q-switched (ns) or modelocked (ps) pulses) and d) pulse repetition rate (Hz to kHz).

The following section will describe the interaction of light with tissue. This knowledge then allows the physician to choose the appropriate laser parameters for a specific treatment. The choice of which laser to use for a particular application depends on the combination of laser system and delivery device in relation to tissue characteristics and biological response.

3   Light interaction with tissue:

3.1   Characteristics of biological tissue

From a physics point of view, biological tissue may be considered as a matrix of water pockets with coloured particles floating within, divided by membranes (cells). In-between these water pockets, there are smaller and larger channels through which liquid with coloured particles flow (vessels with blood). Tissue is comprised of about 80% water. The solid parts consist mostly of complex carbon chain molecules and some minerals. With this simple model in mind, it is possible to identify the physical properties of biological tissue.

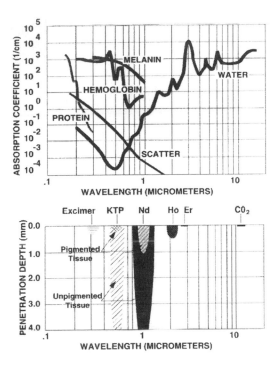

Figure 5. *Absorption spectrum (top) and penetration depth (bottom) in tissue depending on wavelength of laser (courtesy Coherent Medical)*

Optical properties

When a slab of tissue is irradiated with a beam of light the following interactions can be distinguished (Figure 4):

- Reflection: light coming back from the tissue

- Transmission: light leaving the tissue again on the distal side

- Scattering: spreading of light through the tissue

- Absorption: light energy is absorbed by tissue contents and converted to heat.

The extent of these interactions depends on wavelength of the laser light in relation to the optical properties (scattering and absorption) of the tissue. While the matrix and cell membranes are mainly responsible for the scattering of light, coloured pigments or chromophores are mainly responsible for the absorption of light. Figure 5 shows the typical absorption spectrum of tissue. In the visible range, the haemoglobin in the red blood cells and the melanin pigments elsewhere govern the absorption spectrum . However, proteins dominate the UV absorption spectrum and water in the cells dominates the IR spectrum.

The combination of scattering and absorption will determine which area in the tissue will be heated and affected (Figure 4). The depth into the tissue where the laser beam intensity has decreased to 13% (e^{-2} level) is defined as the penetration depth. The typical penetration depth for common medical lasers is shown in the lower part of Figure 5. For lasers in the visible range, there is a large difference in penetration depth with and without the presence of pigments in the tissue.

Thermal properties

Since most of tissue consists of water, the thermal properties of tissues resemble those of water. However while most water is bound up in cells and thus unable to move convectively, the water in the blood stream can transport heat around the body. Such heat transport by the blood is important in normal living and will be of importance during laser treatment (Chato 1985).

Mechanical properties

The cell membrane is composed of proteins and provides the structure and strength of the soft tissues. Thicker protein layers containing less water are stronger and can also be elastic (ligaments, cartilage). When combined with minerals like calcium tissue becomes hard and strong (bone).

3.2   Effects on tissue

During the interaction of light with tissue both desired and undesired effects are obtained. In the optimal treatment, the undesired effects are minimal. The effects are induced by thermal, mechanical or chemical effects, or by a combination of these.

During tissue irradiation, where the absorbed light is usually converted into heat, various dynamic processes take place (McKenzie 1990). Besides physical processes, physiological processes will counteract the inflicted injury and tissue properties will change.

Due to the presence of a temperature gradient, heat will be conducted from the position of absorption to surrounding tissue. Blood perfusion will effectively transport heat away from the irradiated area. The human body may even respond by dilation of the blood vessels to improve cooling which will increase the red tissue colour. If the cooling is not sufficient, the temperature increases to a level at which the blood vessels coagulate and perfusion (and thus cooling) is blocked. The tissue turns to a pale colour. With continuing temperature increase, the tissue dehydrates and the water content will start to vaporise at 100°C. If the water is still enclosed in the cell membranes the pressure will build up until water vapour escapes and at the moment the membrane breaks there is a 'pop' sound. This explosion is referred to as 'the popcorn effect' and might induce mechanical damage due to rupture formation (Verdaasdonk *et al.* 1990). The tissue temperature is clamped at 100°C until all the water is vaporised. From that moment, the temperature rises rapidly to 200–300°C, at which stage the tissue turns brown/black, as it dissociates into carbon and gases. The carbon particles will effectively absorb any light and induce vaporisation of residual tissue components. Bright white light flares may be observed due

tissue effect	temperature range °C	mechanism
hyperthermia	40–50	thermal
necrosis	40–60	thermal / chemical
melting	50–80	thermal
coagulation	70–80	thermal
dehydration	90–100	thermal
boiling	100–130	thermal
carbonisation	200–300	thermal
vaporisation	300– >1000	thermal
pressure re-modelling	–	mechanical
explosive vaporisation	100–130	mechanical
rupture formation	–	mechanical
fragmentation	–	mechanical
plasma formation	1000–2000	thermal / chemical

Table 1. *Overview of the effects of temperature on tissue*

to instant vaporisation of carbon associated with plasma formation (1000–2000°C). The result is a crater in the tissue surrounded by a thermal damage zone.

In Table 1 an overview is given of the range of tissue effects that can be expected dependent on temperature induced. The general scenario as described above illustrates the (complex) dynamic changes that occur during laser-tissue interaction and which will depend on various laser parameters.

3.3 Predicting tissue effects

The effect in tissue can roughly be predicted by considering the amount of energy absorbed in a particular volume of tissue per unit of time (Figure 6). This simplistic relation will be examined in detail below.

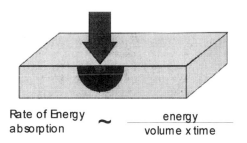

$$\text{Rate of Energy absorption} \sim \frac{\text{energy}}{\text{volume} \times \text{time}}$$

Figure 6. *Simplistic relation to predict tissue effects, describing the rate of energy absorption per unit volume of tissue.*

Amount of energy

The first part of the equation is the amount of energy coming from the light source. Assuming that basically all tissue effects are induced by a temperature increase, a quantity can be derived for the energy needed to raise the temperature of one unit volume of tissue to complete vaporisation at 300°C. In this simplified model we assume that tissue consists of 75% water, that no losses occur during thermal energy transfer to the tissue, that there are no thermal energy losses due to conduction and convection, that the solid tissue component vaporises at 300°C and that we start at a body temperature of 37°C.

Calculation for 1mm^3	~1 milligram of tissue
heating from 37 to 100°C :	0.236J
vaporisation of 75% water at 100°C	1.490J
heating the remaining 25% of tissue from 100 to 300°C	0.198J
vaporisation of 25% of tissue	0.003J
	total 1.927J

So under ideal conditions about 2J energy is necessary to ablate or vaporise 1mm^3 or 1mg of tissue. Just to 'kill' tissue by coagulation only 10–20% of this amount (0.1–0.2J) could be sufficient. For larger tissue volumes it is easier to make the estimation using 2000J for 1 gramme or 1cm^3 of tissue. These quantities are helpful in estimating effects in tissue.

Volume of tissue absorbing the energy

The surface of irradiation (spot size) and the penetration depth of the laser wavelength into the tissue determine the volume (Figure 4). The penetration depth has already been discussed in relation to the optical properties of tissue (Figure 5). In combination with the scattering properties, the energy is dispersed through the tissue. For deeply penetrating and highly scattered wavelengths, the energy is spread in a large volume resulting in a gradual temperature increase in the whole area. In contrast, light of a wavelength that is highly absorbed in a very thin layer at the surface could very rapidly raise the temperature of that small volume to vaporisation temperatures.

The spot size at the surface is of importance. If a particular amount of energy is administered over a large surface area, the amount of energy per unit volume is limited. On the other hand, if a deeply penetrating wavelength is focussed to a small spot at the surface, the energy density can be high enough to induce vaporisation temperatures. Note also that the energy in a spot of a laser beam is usually not uniformly distributed.

Time of tissue irradiation

The last major factor in the effect of tissue is the time of irradiation. At this point a distinction is made between continuous wave lasers and pulsed laser systems.

CW exposure

The time range for continuous wave exposure is roughly between one tenth to tens of seconds. The temperature in a particular volume of tissue will rise as long as irradiation continues. Depending on the temperature levels reached, different tissue effects can be observed. Using a cw laser the irradiation time is in the range of seconds and heat will be conducted from the irradiated area to the surrounding tissue. This will limit the maximum temperature in the target area and induce thermal effects in the surrounding tissues. This mechanism can be used to create a "bloodless scalpel" in laser surgery by cutting (vaporising) through tissue while coagulating the adjacent small blood vessels to prevent bleeding. The typical ablation crater of a cw laser in tissue has a black ring surrounded by a lighter colour. However, heat conduction can also result in undesirable damage to surrounding tissue. This can be controlled by limiting the exposure time or by using breaks between multiple exposures.

In the basic relation (Figure 6) heat conduction is not taken into account and must be considered as a loss. This means that much more energy is needed to vaporise a given volume of tissue if a cw laser is used than if a short pulse laser is used.

To limit heat conduction effects, the exposure time should be as short as possible while delivering the same amount of energy. Using high power continuous wave lasers and keeping the spot size small, tissue effects can be obtained with exposure times between 10 and 100ms. This way small targets such as blood vessels can be coagulated with minimal damage to the surrounding tissue. This strategy is applied for port wine stain treatment. For shorter pulse lengths, tissue effects can only be achieved using the higher peak powers produced by pulsed laser systems.

Short pulse exposures

Pulsed laser systems generate laser pulses with lengths in the range of sub milliseconds down to picoseconds but with a peak power that can be much greater than that produced using cw lasers. In this short time, a considerable amount of energy can be delivered to the tissue, sufficient to increase the temperature far beyond 100°C. In such a short time, the water content in the tissue volume (determined by the spot size and the penetration depth), is instantly turned into vapour. This vapour is created at a high pressure and starts expanding to 1600 times its original liquid volume, in tens to hundreds of microseconds. This way explosive vapour bubbles are formed with a potentially large mechanical force (Figure 7).

At a tissue surface in air, the vapour will escape from the surface into the air with some momentum towards the tissue. This might induce mechanical damage in the tissue. If the vapour is formed inside a small channel deeper in the tissue, the vapour will also push the wall of the channel aside and leave mechanical damage while escaping to the surface. While expanding, the vapour will cool down and turn back into moisture releasing the energy of vaporisation, which is distributed in the environment.

The vapour, explosively expanding from the surface, might take debris and liquid particle with it. This debris consists of vital cells and aerosols. These might contain infectious parts like viruses. Therefore it is important for surgeons and nurses to have adequate plume suction nearby and additionally wear protective gear to prevent inhalation.

If the explosive vapour formation takes place in a water environment (e.g. during an endoscopic procedure), the pressure from the surrounding tissue is equal to the liquid on top of the tissue surface. Consequently, the vapour bubble will expand symmetrically in

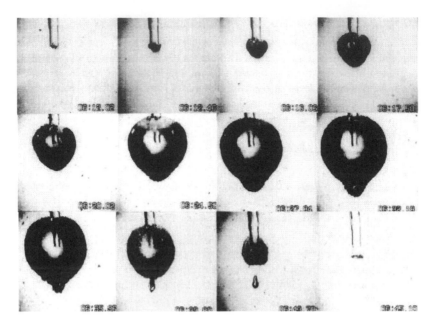

Figure 7. *Sequence showing an expanding and then imploding vapour bubble in water induced by a holmium laser pulse. The frames are 100μs apart covering 1ms in all (these high speed images were captured using the technique described in Section 3.5).*

all directions including into the tissue. The bubble will expand along the path of lowest resistance, creating fissures between tissue layers. The bubble, which can be observed in liquid environment, is also present underneath the tissue surface. Considerable mechanical damage can be induced into the tissue far from the position of the irradiation. The typical maximum size of these vapour bubbles is in the range of 1mm up to 10mm depending on the energy of the laser pulse. The deeper the penetration depth of the laser, the higher is the vaporisation threshold. However, the moment the threshold is reached, a large volume of water is vaporised resulting in a big vapour bubble and potentially more mechanical damage. The mechanical effects inflicted on soft tissue by explosive vapour bubbles can be considered damage. However, it provides the basic ablation mechanism of pulsed laser systems (van Leeuwen *et al.* 1991). The forceful explosions can be applied effectively against hard tissues such as the stones that can occur as abnormalities in the human body (see Section 5.7, lithotripsy).

During expansion, the vapour will cool down and eventually turn back into water, releasing its latent heat of vaporisation. In contrast to vapour expanding in an air environment, the vapour is still confined near the site of irradiation and will heat the tissue in contact with the vapour. At a water/tissue interface there will be a forceful turbulence providing effective cooling of the tissue surface. Within tissue, the heat released after the collapse of the vapour bubble will heat up the surrounding tissue. This thermal energy is rapidly conducted into the environment and within a second, ambient temperatures are restored. If however the next laser pulse is released within 0.5 seconds, the cooling process will not have ended and some heating effects from both pulses will be present. At higher

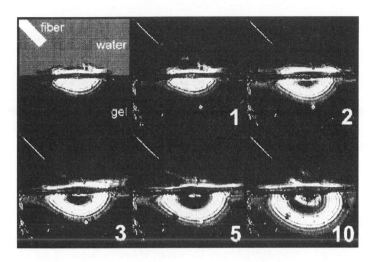

Figure 8. *Heat built-up and conduction in simulated tissue during irradiation with 1J pulses (number shown) at 5Hz from a holmium laser (these pictures were obtained using the thermal imaging techniques described in Section 3.5).*

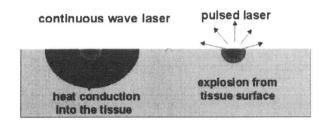

Figure 9. *Comparison of the effects of cw and pulsed lasers on tissue.*

pulse repetition rates, there will be substantial heat conduction to the surrounding tissue resulting in thermal damage (Figure 8). The thermal dynamical effects during tissue irradiation as presented in Figure 8 are studied using the imaging techniques described in Section 3.5 (Verdaasdonk 1995).

Figure 9 shows the main difference between tissue effects induced by cw and pulsed lasers. In contrast to the ablation crater formed by a cw laser, the crater formed with a pulsed laser does not show any charring. Since most of the pulsed laser energy is consumed by the vapour formation, the temperature does not rise far above 100 degrees centigrade, so the tissue is not dehydrated and carbonised. However, if one pulse is followed sufficiently soon after by a series of pulses, the surrounding tissue can dehydrate and the temperature can rise above 200 to 300 degrees, resulting in carbonisation.

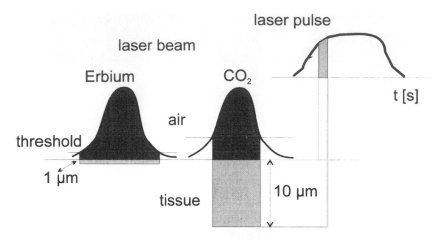

Figure 10. *Schematic representation of the penetration in tissue of a CO₂ and Erbium laser beam with a typical Gaussian distribution. This layer of tissue is instantly vaporised by the first part of the laser pulse. In time a deep channel will be formed by the total laser pulse.*

Effective penetration depth

When laser pulses of microsecond duration are used for tissue ablation, the definition of penetration depth has to be reconsidered. Laser light, at a wavelength which is highly absorbed by tissue, is expected to have a very shallow penetration depth. For example, the CO_2 and Erbium laser have expected penetration depth in order of micrometers (Figure 5 and Table 2, page 222). However, this may not be true in practice and a surgeon expecting only a superficial lesion could be making a serious misjudgement.

The energy absorbed during the first microseconds of the laser pulse can be sufficient to vaporise the top layer of the tissue, creating a hole in the tissue. In the remainder of the pulse the beam enters the hole continuing to vaporise the tissue at the bottom of the crater for as long as the pulse continues (Figure 10). This way the beam might penetrate centimetres into the tissue with the risk of damaging vital structures. Out of view of the surgeon, arteries deep in the body could be perforated, resulting in major bleeding. This phenomenon can be visualised by irradiating a transparent tissue model with a focussed Erbium laser beam. During a $300\mu s$ long pulse, a 15mm long vapour channel is formed in the gel (Figure 11). For obvious reasons, this phenomenon has been called 'the Moses effect'. Surgeons have to be aware of this effect when using focussed beams as, for example, in laparoscopic procedures.

This 'effective' penetration depth can be far larger than the penetration depth as presented in Figure 5 and Table 2, page 222. The effective penetration depth depends on the energy per pulse and is a resultant of the basic relation illustrated in Figure 6.

Figure 11. *High speed image sequence of a deep channel formed during tissue irradiation with a 300µs focussed erbium laser beam.*

3.4   Biological response

The previous paragraph described, in general, the acute tissue effects that can be expected after exposure to laser light. The extent of the effect is a result of the laser parameters, the delivery system and the type of biological tissue targeted.

The acute tissue effect is, however, not the end result or goal. The human body will respond to the induced damage with a biological response in a time span from hours to months after treatment. After the laser has removed tissue, the human body will start to repair the damage largely independent of the intentions of the surgeon, and without any distinction between the damage being caused on purpose or as an undesirable side effect.

In most cases the surgeon removes or kills tissue with the intention that it should stay away (if it was causing an obstruction) or that healthy or 'normal' tissue will grow back instead. However, the original defect can grow back if not all of the abnormal 'malignant' cells were killed during the treatment. This can result in the recurrence of the defect (e.g. a tumour). The repair tissue will usually consist of fibrous tissue also known as scar tissue. This tissue can start growing back but may not stop at the point of regular shape or size. In the case of an obstruction that has been removed, the repair tissue itself will become an obstruction. If the lesion was on the skin, the scar tissue might have a different texture and content of pigment and in that way form a contrast to the surrounding tissue.

In a large a number of applications, however, it is the biological response that makes the use of lasers unique compared with other surgical instruments. Due to the controlled way of energy delivery and minimal collateral tissue damage, the biological response can be exactly as intended and expected. In many cases there exists no alternative to the laser treatment.

3.5   Imaging techniques to study laser-tissue interaction

Various techniques can be applied to study the interaction of medical lasers with biological tissues. Since the major effect is thermal, temperature measurements are performed commonly, using either thermocamera techniques or thermocouples. Both these methods have limitations as to resolution or to interference with the medium of interest (Torres *et al.* 1990). Besides the thermal effects, mechanical effects occur within very short time frames, especially when pulsed lasers are involved. Therefore a multipurpose setup was developed to study thermal and mechanical effects of continuous wave (CW) and pulsed lasers during interaction with tissues with a high resolution in both temporal and spatial

regimes. In this setup, which is based on Schlieren techniques , very small changes in
optical density of the media can be observed. Such changes are induced by temperature gradients or local stresses and are colour-coded resulting in an enormous contrast
enhancement (Howes 1984). High-speed images can be obtained by using exposure times
down to nanoseconds. This method of real-time visualisation of the laser-tissue interaction contributes to a better understanding of the mechanism of action (Verdaasdonk
1995).

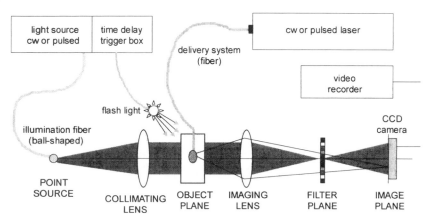

Figure 12. *Schlieren setup to study laser-tissue interaction.*

Colour Schlieren techniques

In Figure 12 the optical setup, which is based on Schlieren techniques, is illustrated. A
continuous or pulsed white light source is coupled into a fibre with a ball-shaped end. The
light emitted from the fibre is focused due to the spherical fibre end and then diverges.
The focal point of a collimating lens coincides with the focus of the fibre. The diameter
of the lens is matched with the divergence of the beam so that all the light is collimated.
A rectangular tank filled with water is positioned between the collimator and imaging
lens, the 'object' plane. The walls are perpendicular to the parallel beam to prevent any
optical distortion due to refraction. Within this tank conditions are created to study laser
tissue interaction. The imaging lens will focus the parallel beam in its focal point on the
optical axis. However, rays can be deflected due to variations in the refraction index or
irregularities in the medium in the object plane induced by local stresses or temperature
gradients. These rays will cross the focal plane at a particular distance d from the optical
axis (Figure 13). The non-deflected rays will be focused on the optical axis. By inserting
a mask or a filter in the focal plane of the imaging lens, it is possible to block out the
non deflected rays, preventing them reaching the image plane or to earmark rays crossing
the plane at certain positions. This process of modifying the object information in the
focal or filter plane is known as spatial filtering. By blocking the rays crossing the optical
axis, only refracted and diffracted rays will pass the filter plane and form an image at
the image plane. This results in an enormous contrast enhancement of the image of the
perturbations of refractive index, as it is only the rays that have been perturbed that will

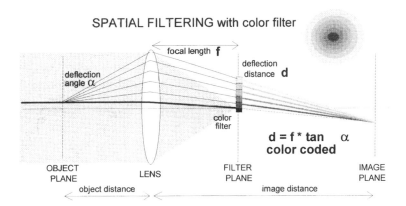

Figure 13. *Schematic of spatial filtering using a block filter. Surrounding the block filter by coloured rings (as shown in Figure 12) permits colour coding of the deflected beams that reach the image plane.*

contribute to the final image (Hecht 1987). Depending on the deflection angle α and the focal length f of the transform lens, a ray will pass the filter plane at a deflection distance d from the optical axis (Figure 13) given by

$$d = f \tan \alpha \tag{1}$$

The information on the degree of deflection can be preserved by colour coding the rays coming through the filter plane by using a colour filter. Referring to Figure 13, the simple block filter is replaced by a new filter which consists of concentric rings of discrete colour bands separated by small black rings. The centre of the filter is a black dot which blocks the background light. Adjacent to the black dot, going away from the centre, the colours shift gradually from blue to red.

Rays passing the filter plane will be colour coded depending on the deflection distance d (and hence deflection angle α) and will be reconstructed to an image at the image plane. The generated colour image will show in a position dependent manner, the degree of deflection in the object plane. From each colour, the deflection angle α can be determined. The angle α is related to the variation in the refractive index of the medium in the object plane. The colour image can be interpreted as a thermal image when the relation between refractive index and temperature gradient is known. The black rings in the filter will result in black lines in the image separating the discrete colours giving an impression of 'isotherms' (Figure 8).

The position of the imaging lens, the filter and the CCD camera are chosen depending on the magnification desired according to the lens formula. The diameter of the filter determines the dynamical range of temperatures that can be visualised. Using an x-y microtranslator, the filter can be optimally aligned on the optical axis. The CCD camera is positioned in the image plane. Additional filters can be used to filter out scattered light from the primary laser wavelength. To obtain microsecond resolution, a video camera with high speed mode can be used.

High speed Schlieren setup

The Schlieren setup can be used directly for high speed photography by using a video camera with a high-speed shutter, for example, going down to 100μs. However, the light from a cw white light source will not be sufficient. Instead, a pulsed white light source can be used if enough light can be coupled into the illumination fibre. Using normal video exposure times (20ms), the flash lamp can 'freeze' images with illumination times in the microsecond range (Verdaasdonk 1995). This time range is sufficient to capture thermal phenomena using the colour-coded technique. However, for mechanical phenomena shorter illumination times are needed. To image micro explosions and shockwaves that might be induced by pulsed lasers in tissue, exposure times down to nanoseconds are needed. The only light source that can provide light flashes in the nanosecond region, with sufficient light transported through the illumination fibre, would be a pulsed laser in the visible range. Since the light source is monochromatic it is not possible to colour code the degree of deflections of the rays in the image. Instead quantitative data can be obtained using specially designed spatial filters or by using a multiple wavelength laser, a broadband dye laser or laser induced fluorescence light.

By controlling the time between the laser pulse interacting with tissue and the light flash for exposure, it is possible to study the dynamics of the interaction with a temporal resolution of microseconds and even down to nanoseconds. A trigger box drives the light flash for exposure after a pre-set time delay from the start of the laser pulse interacting with tissue (Figure 12). By increasing the delay time, a sequence of images of individual laser pulses can be obtained showing the dynamics of the interaction. Using a laser source with a high repetition rate as the flash light, it is even possible to make multiple exposures within one image. Figure 14 shows such an image obtained with a copper vapour laser which emits 10ns pulses of green/yellow light with a repetition frequency of 10kHz. During the 1ms exposure time, 10 flashes, 100μs apart, show the expansion of a cavity in a water surface.

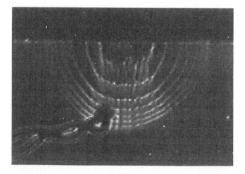

Figure 14. *Multiple exposure of an expanding cavity at a water surface during interaction with a pulsed laser*

High speed conventional photography setup

In addition to Schlieren imaging, a separate flash-light, directly illuminating the target, can be used for conventional high-speed imaging (Figure 12). The same triggering and time delay technique can be applied to capture images during interaction of laser pulses with tissue using a white light flash of several microseconds. Ordinary electronic flash lamps for photography might be used for this purpose. The arc lamps in stroboscopes are better as they provide $1\mu s$ light flashes with a repetition rate up to several hundred Hertz.

These two techniques can be combined as shown in Figure 12. This way it is possible to capture the thermal and explosive mechanical effects in one image. Figure 31 shows an explosive vapour bubble at the surface of hot tissue layer. The bubble going upwards is captured by the conventional imaging and the thermal gradient in the tissue is observed by the Schlieren technique.

Simulation of laser-tissue interaction

The Schlieren technique can not be applied during 'real' laser-tissue interaction in (living) biological tissues as it requires the tissue to be transparent. Therefore a transparent "model" tissue is used to simulate the interaction. A transparent polyacrylamide gel is used which is assumed to have similar thermal properties to biological tissues. It might also be a reasonable approximation to the mechanical properties of soft tissues. Laser-tissue interaction is simulated in the object plane of the Schlieren setup (Figure 12) by using one of the arrangements shown in Figure 15. A medium or layer of tissue is exposed directly to a laser beam or indirectly through an optical fibre. The medium itself or a slab of tissue will absorb the laser light. Consequently, heat is generated and will diffuse through the medium or transparent model tissue.

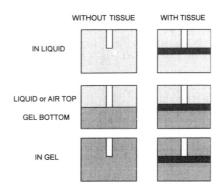

Figure 15. *Conditions to simulate the interaction of a focussed, or fibre delivered, laser beam with artificial (gel) and/or real tissue in a Schlieren setup*

The wavelength of the laser studied might be absorbed directly by the medium or by the gel. It is also possible to dissolve an absorbing dye in the medium as long as the absorber does not influence the transmission of visible light. These conditions are depicted in the first column in Figure 15. To simulate the in vivo situation as closely as possible, a slab of the original tissue is used for the absorption and scattering events (Figure 15, the second column). In tissue or gel, the heat transfer can only take place by heat conduction. In the case of a liquid environment, convection is also involved and this is a very effective way to transfer heat. An air environment, on the other hand, will act as a near-perfect insulator (Figure 15, second row).

Calibration of the Schlieren setup

To interpret the colour image as a temperature image, the relation between colour and temperature must be calibrated. The relation between the deflection distance and the temperature (Figure 13 with rings) depends on the angle of deflection α, the focal length of imaging lens, the object distance, the image distance, and the symmetry of the refractive index distribution. The symmetry of the refractive index distribution is of major importance. This calibration can only be performed assuming a particular symmetry. The temperature gradient can be approximated by either a uni-directional or an axially symmetrical distribution (Verdaasdonk 1995).

Typical tissue effects of common medical laser systems

After this general overview of the way that laser parameters induce particular tissue effects, it is possible to present a table showing the characteristics of common medical laser systems and their typical tissue effects (Table 2, page 222). The application of different lasers for particular clinical treatments will be described in Section 5.

4 Laser light delivery systems

An important part of a medical laser is the delivery system for controlled transport of light between laser source and target. This system can also be used to transport light back from target to detector for diagnostic purposes (Katzir 1993). The design of the delivery system will be determined by the characteristics of the laser system on one hand, and by the tissue application on the other hand. In most laser systems, the laser light is transported to the target using fibre optics. However, for far infrared wavelengths and for high peak power pulsed lasers transmission through fibre optics is not possible due to high transmission losses or damage to the fibre. Instead, the beam is transported through an articulated arm consisting of hollow tubes with reflectors in the joints.

4.1 Articulated arm

When wavelength or peak power does not permit transmission through optical fibres, the laser beam can be transmitted using reflecting surfaces such as mirrors or prisms (Figures 16 and 17). Since the spatial properties of the laser beam are preserved using articulated arms, this may be the delivery system of choice for some applications. This is especially the case when the irradiance distribution in the spot has to be controlled or the beam has to be focused to a small spot.

Figure 16. *Medical CO_2 laser with articulated arm*

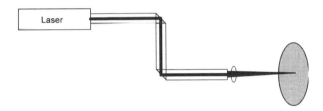

Figure 17. *Schematic of an articulated arm with focussing optics*

By choosing the proper materials, reflectors can be made for most wavelengths and for very high peak powers. An articulated arm usually consists of six to eight mirrors mounted on rotating holders to provide steering in any direction. The holders are connected to each other by a set of rigid tubes. If properly aligned, the beam will exit the final arm at the same position and angle independent of the position of the freely movable tubes. The alignment of the arm is very critical. In earlier experience with medical CO_2 lasers, the arms were easily misaligned, especially after transportation. However, the construction of the recent generation of articulated arms has proven to be more robust and reliable. The articulated arm is standard on all CO_2 laser systems, although it might be replaced by fibre optics in the near future (Gannot *et al.* 1994, Snaijer *et al.* 1998). However, it will remain the means of transportation for the delivery of the high peak power laser pulses from Q-switched laser systems for example. In principle, the original beam characteristics do not change during transportation through an articulated arm. However, the beam diameter increases depending on the beam divergence and length of the articulated arm. So the beam coming out of the articulated arm has a diameter of several millimetres and usually has a Gaussian distribution.

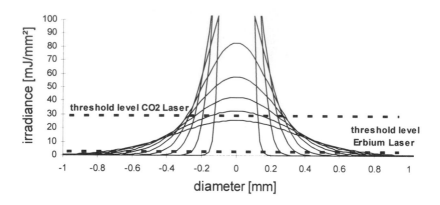

Figure 18. *Gaussian intensity distributions at various distances from the waist of a focussed beam. The dotted lines show the part of the beam where the power density is above the threshold level for tissue ablation for the Erbium and the CO_2 laser.*

Focussing optics

To obtain ablation effects in tissue and to have precision and control, an optical system is present at the end of the articulated arm to focus the beam onto the target tissue. The optical components must be compatible with the wavelength of the beam. In the case of CO_2 lasers, lenses of fluorozirconate glass are used. Depending on the application various focal lengths are employed, and spot sizes as small as tens of micrometres can be obtained.

The influence of the Gaussian distribution can be best illustrated with a laser having a very superficial penetration depth, such as the CO_2 laser. For larger spot sizes, the energy in the centre of the beam is sufficient to vaporise the tissue, while toward the rim of the spot the tissue is heated just to coagulation temperatures. Within the spot it is possible to distinguish a part that is above the ablation threshold of tissue (Figure 18). By focussing the beam to smaller spot sizes the energy density increases and a larger fraction of the area within the spot exceeds the ablation threshold. For a focussed beam, the spot size is related to the distance between the focussing lens and tissue surface. Just by varying the distance to the tissue, the energy density within the spot can be arranged to be above or below the ablation threshold (Verdaasdonk *et al.* 1995).

In Figure 19 the area of ablation within a focussed CO_2 laser beam is defined, tissue within this area will be ablated. This way the surgeon has a flexible choice to create the desired effect in tissue. On the other hand, it also makes the application less controlled since a small variation in distance can make the difference between coagulation and vaporisation of tissue. Thus for treatment, the focussing optics are in hand pieces with spacers to keep the distance between the lens and the tissue constant.

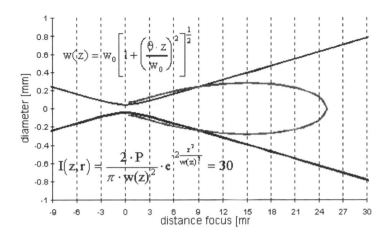

Figure 19. *Outline of a focussed Gaussian beam ($1/e^2$ contour) as defined by the upper formula where the diameter w depends on the distance z from the focal point. The 'ellipse' represents the area within the beam where the power density I is above the ablation threshold for tissue for the CO_2 laser.*

4.2 Optical fibres

If the wavelength and intensity permit, optical fibres are usually the preferred method for transporting laser light. The choice of material and construction of a fibre is determined mainly by the wavelength of the laser. In the range from 300–2500nm, where most lasers operate, silica fibres are commonly used. The most interesting wavelengths outside this range are in the infrared (IR), 2.94μm of the Er:YAG laser and 10.6μm of the CO_2 laser and in the ultraviolet (UV), the 193nm and 248nm of the ArF and KrF excimer lasers, respectively.

Silica fibres for visible and near-IR light

Silica fibres are highly transparent to wavelengths in the visible spectrum where losses due to absorption are in the order of tenths of a percent per meter. Absorption increases for UV and IR wavelengths. The absorption characteristics of the fibre are dominated by the concentration of hydroxyl radicals (OH^-). Hydroxyls form in the silica as a result of free water present during the manufacturing process of the fibre. Low-OH concentration fibres have superior transmission in the near IR (up to 2400nm, where absorption by silica itself takes over). These low-OH fibres are especially suitable for transmission of wavelengths of diode lasers around 950nm, the 1.34μm wavelength of the Nd:YAG laser and wavelengths around 2μm of Holmium lasers where the absorption of water peaks. High-OH concentration fibres have less loss in the ultraviolet and are relatively easier to produce. For transmission of UV wavelengths down to 300nm most silica fibres can still be used. However, for shorter wavelengths, the material of the fibre cladding influences the transmission. Plastic clad fibres are sensitive to UV light. Due to destructive effects of the UV light on plastic, the transmission will drop rapidly with time. Some hard polymer claddings are more resistant to UV light but the silica clad fibres are superior. Although silica/silica (or hard clad) fibres have a better performance in the UV, they need an additional soft coating to improve their mechanical properties. Such fibres will transmit down to 200nm (Magera and McCann 1992).

Fibres for IR wavelengths

For the IR region beyond 2500nm, materials other than silica are being used. These IR fibres can be classified into three categories: IR glasses, crystalline fibres and hollow waveguides (Merberg 1993).

IR glasses

For the region between 0.5 and 4.5μm, fibres made of fluorozirconate and fluoroaluminate glass can be used . These fibres are commercially available in diameters from 100–400μm. They have a high damage threshold but can not be used above 150°C due to their low melting point. The refractive index is similar to silica ($\sim$ 1.5) and the transmission over several metres is above 95%. This makes these fibres suitable for applications with the 2.9μm Erbium:YAG laser (Helfer et al. 1994). This laser is being investigated for ablation of hard tissues such as bone and teeth. For longer wavelengths (4 to 11μm) chalcogenide glasses become of interest. Fibres of this material are available in 150 to 500μm diameter. However, they are toxic and their mechanical properties are inferior to silica fibres. The combination of the Fresnel reflection due to the high refractive index (2.8) and relatively high absorption, results in losses of several tens of percent per metre (Harrington 1992).

Crystalline fibres

Better candidates for the mid-IR range can be found among the crystalline fibres. Sapphire can be grown to a single crystal with a diameter of 200 - 400μm, which is strong, hard and flexible (Barnes et al. 1995). Sapphire, however, has a high index of refraction (1.75), which produces rather high reflection losses at each surface. Sapphire is a good choice to transmit Er:YAG light but is less suitable for the CO_2 laser. Silver- and Thallium-halide polycrystalline alloys (e.g. KRS-13), in contrast, can successfully transmit even high power CO_2 light. From this material good quality fibres are manufactured with transmission losses of a few dB/m and which are insoluble in water, non-toxic and fairly flexible.

Hollow waveguides

Hollow waveguides consist of flexible tubes with a 'core' of air. They transmit light in the whole IR range with high efficiency (Matsuura et al. 1995). One type incorporates metallic, plastic or glass tubing which is coated on the inside with a metallic or dielectric film with a refractive index $n > 1$ ('leaky' guides). Another type comprises waveguides with a dielectric coating of $n < 1$ for 10.6μm on the inside of hollow glass (Abel et al. 1994) or crystalline tubes, resulting in an attenuated total reflectance on the inside of the

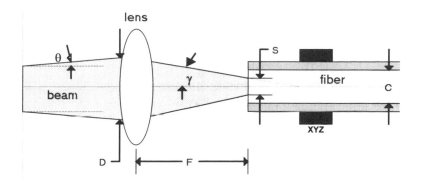

Figure 20. *Schematic of a fibre coupler*

tube ('ATR' guides). They are manufactured in lengths of just over 1m with an inner diameter of 250μm . The losses are 0.4–7.0dB/m depending on core size. The losses due to bending are inversely proportional to the core radius. Power transmissions of over 100W have been achieved. The damage threshold for high power densities is comparable with solid core fibres.

4.3   Fibre coupling

Medical laser systems must have a means of reliably coupling light into removable optical fibres. A large range of fibre sizes are embedded in dedicated connectors which ensure self-centering when connecting to the laser system.

For coupling to the fibre, the beam has to be focused down to a spot of a few hundred μm depending on the diameter of the fibre. The material of the focusing lens has to be transparent for the wavelength of the laser light. The lenses are usually made of material similar to the fibre itself and the surfaces are coated to reduce losses caused by Fresnel reflections.

The efficiency of coupling light into the delivery system depends on the diameter of the laser beam D in combination with the focal length F of the lens and the divergence θ of the light (see Figure 20). These quantities determine the minimum spot size s to which light can be focused before coupling into the delivery system:

$$s = 1.27\lambda F/D \quad \text{for mono-mode Gaussian beams}$$
$$s = \theta\,F \quad \text{for multi-mode beams.}$$

Typically, for most medical lasers, the diameter of the laser beam is a few millimetres and its divergence is 5–20mrad (1–4 degrees). A spot size near 100μm can easily be achieved allowing coupling into waveguides of comparable size. However, the coupling of light from high power diode lasers, which have recently been introduced in the medical field, is a challenge. Due to the high beam divergence of diode lasers (tens of degrees), it is difficult to obtain an efficient coupling into silica fibres with diameters below 600µm.

Currently, most medical laser systems with wavelengths suitable for fibre delivery, have coupling systems that fit SMA-905 terminated fibres. SMA couplers have now also been accepted as the standard for high power laser delivery. This is in contrast with older medical lasers systems, each of which had its own dedicated coupling system.

When focusing the laser beam to a small spot, the power density will increase dramatically. Especially in combination with a pulsed laser system, the peak power density can easily exceed the damage threshold of the delivery system. This threshold is wavelength dependent, with UV wavelengths having a lower threshold than the IR wavelengths. Intense optical electrical fields induce a dielectric breakdown of either the air near the fibre interface or of the fibre itself. Sometimes the laser system itself can be adapted to avoid such problems. For example, the pulse length of 308nm excimer lasers was stretched from 20 to 120ns to provide coupling into fibre delivery systems for vascular applications.

Transmission losses

Possible energy losses during transportation through fibres may originate from the coupling due to a mismatch of spot and fibre diameter, impurities on the entrance surface and Fresnel reflections. The same factors apply to the exit side of the fibre. The absorption by the fibre material itself can contribute to the losses substantially. While silica fibres have an attenuation of only 0.1dB/km for wavelengths around 1.5µm, attenuation in order of dBs per meter are not unusual for IR fibres. For UV wavelengths fluorescence and Raman scattering effects can decrease transmission efficiency. Further losses can originate from sharp bends in the fibre and from local stresses on the core. The bending of a fibre is primarily limited by its mechanical properties. However, before the core of the fibre breaks, the light will start leaking out of the fibre. Hollow fibres are particularly sensitive to sharp bends (Abel *et al.* 1994). Silica-core fibres have properties superior to other fibres.

4.4   Irradiance distribution of a fibre delivered beam

It is important to understand the irradiance distribution of the beam coming out of the fibre. For energies used for therapeutic applications only large diameter waveguides ($>$ 100µm) are considered and such waveguides will be multi-mode. The spatial distribution of the irradiance at the exit surface of a multi-mode waveguide is equal to the spatial distribution of the refractive index (Mahlke and Gössing 1987). Since the core consists usually of a material with a uniform refractive index (e.g. step index fibres or hollow waveguides), the spatial distribution of the irradiance at the exit surface will also be uniform. Theoretically, the angular distribution of the irradiance at the surface of a multi-mode waveguide will be Gaussian according to:

$$I(\alpha) = I_0 \exp(-2\alpha^2/a^2)$$

where I_0 is the irradiance at the fibre surface on the optical axis, α is the angle with the optical axis and a is the divergence angle of the exit beam.

Taking these characteristics into account, the irradiance distribution of the exiting beam of a multi-mode fibre is uniform at and near the fibre interface. However, it develops

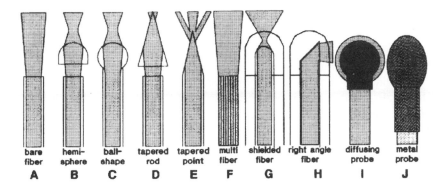

Figure 21. *Overview of modified fibre tips*

into a Gaussian distribution away from the fibre. This behaviour applies for the 'ideal' use of a multimode fibre. The output can vary considerably depending on launch conditions, fibre length, bend configuration and local stress. Due to the special angular and spatial distribution of the irradiance of the output beam, the beam can not be modelled using Gaussian beam optics. Ray-tracing can be used to simulate both Gaussian and irregular spatial distributions in the near and far field of optical fibres. The spatial distribution of a beam coming directly from the laser is usually Gaussian (mono-mode) or a summation of Gaussian profiles (multi-mode) and Gaussian beam optics can be applied (Verdaasdonk 1991). These properties are important for the design and computation of target optics or modifications of the fibre tip.

4.5 Modified fibre tips

The irradiance distribution of a beam leaving the distal end of a fibre ('fibre tip' or probe) can be described as a 'torch'. The beam can, however, be adapted for special applications by (optically) modifying the fibre tip (Verdaasdonk 1995). A selection of modified fibre tips is presented in Figure 21. Typically, the modification of the fibre tip is made from the fibre material itself (fused silica) or parts, made of silica, sapphire or metal, are added (Figure 22). Ball or taper shapes are used to increase the power density at or just in front of the fibre tip. To preserve the optical effect in an aqueous environment an optical shield is added covering the tip. The beam can be directed at a right angle out of the fibre making use of total internal reflection or adding a reflector on the fibre tip. A ball of highly scattering material positioned on the fibre tip creates an isotropic irradiating probe. All the laser energy is converted into heat when a metal cap covering the fibre turns it into a 'hot tip'. A single fibre can be replaced by a bundle of smaller fibres covering the same surface area while increasing the flexibility of the delivery system. The small fibres can also be positioned in various shapes, such as a ring or a bar. Energy can be transmitted preferentially through individual fibres or part of the fibre bundle. The other way round, the system can be used for feedback of, for example, fluorescence light coming from tissue in front of the tip to enable diagnosis (Vo Dinh *et al.* 1995).

Shaping fibre tips

There are several methods to modify the tip of a bare silica fibre. The core of the fibre can be heated above the melting temperature of silica (> 1500°C). Due to the surface tension of the liquid silica tip, a droplet with an almost perfectly spherical shape will form (Ward 1987). If the fibre core is heated locally while the fibre is pulled apart a tapered tip is formed. The angle of the taper will depend on the diameter of the fibre, the melted volume and the pulling force. The very tip of the taper can easily be shaped into a small hemisphere by melting or be polished to a flat surface.

Various methods are used for heating of the silica. A small oxygen-enhanced burner can do the job. It is easy, inexpensive and quick and when required, it can even be performed in the operating theatre (Verdaasdonk *et al.* 1995). CO_2 laser light is effectively absorbed by the silica and can be used to melt the fibre tip or fuse for example, the fibre core to a sphere. Some researchers have employed the laser itself (e.g. Argon or Nd:YAG) to melt the tip by positioning it in the centre of a spherical hole of a stone while transmitting several watts of light through the fibre (Russo *et al.* 1984). The IR radiation from the stone will eventually melt the fibre tip.

The fibre tip can be etched using chemicals. By retracting a fibre slowly at a controlled speed out of a solution of hydrofluoric acid (HF), the diameter of the fibre will become smaller towards the end resulting in a taper shape (Chang *et al.* 1978). However, special precautions should be taken working with the highly toxic HF.

Polishing is the standard technique to condition fibres for optimal transmission. Besides the normal flat surface, the tip can be polished at an angle to accommodate a totally reflecting surface for a side-firing fibre. Rotating the fibre while polishing at a small angle will result in a tapered tip.

Other materials can be fixed to the fibre end to direct the light distribution. An optical shield will provide an air interface even in a water environment. Highly scattering material such as nylon, teflon or ceramic materials of various shapes can be attached to the fibre to diffuse the light from the tip (McKenzie 1988). A scattering material can also be grown on top of a bare fibre tip for example by using a UV-light-curing polymer in which a highly scattering material is dispersed (van Staveren *et al.* 1995). Diffusing probes are used for photodynamic therapy or hyperthermia applications.

Small fibres can be stuck together in the desired shape such as a ring forming a so-called multi-fibre catheter . The fibres should be packed close together to have a minimum of 'dead' space between the fibres (Verdaasdonk *et al.* 1994).

4.6   Properties of fibre delivery systems

The optical and thermal behaviour of modified fibre tips is governed by the optical and thermal properties of the materials of the probes and of the environment in which they are used.

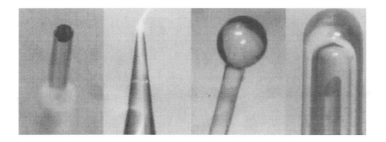

Figure 22. *Modified fibre tips: (left to right) bare fibre, tapered point, ball-shaped fibre and right angled fibre*

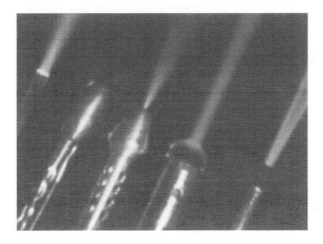

Figure 23. *Beams emerging from various modified fibre tips in water*

Optical properties

The refractive index mismatch between material and environment determines the degree of the optical interaction. A transition with a large index change can strongly affect the direction of a beam, but it can also introduce undesirable reflections and hence power losses (Verdaasdonk *et al.* 1991). A photograph of the beam will give a good first qualitative impression of the beam divergence and possible secondary beams (Figure 23). If the delivery device is used with a non-visible wavelength, the device can be coupled to a laser with a visible wavelength. However, one should take the wavelength dependence of optical components into account. Beams can be visualised using smoke in air or diluted ink in water. This simple method of visualisation can show the dramatic difference of the beam shape in air as compared to in water (Verdaasdonk *et al.* 1991).

Ray-tracing

Theoretical calculations can be very helpful to predict and optimise the optical behaviour of a particular delivery design. The starting conditions of the ray-tracing are important since they should simulate the irradiance distribution as to the spatial and angular distributions of a multi-mode fibre. Due to this unique distribution standard geometrical optics is not accurate enough. The optical behaviour of hemispherical, ball-shaped and tapered fibre tips of silica and sapphire have been discussed previously (Verdaasdonk *et al.* 1991). It was shown that for spherical probes submerged in liquid media or in contact with tissue, the focusing is less pronounced or even absent due to the limited change in the refractive index of the tip and the environment. In tapered fibres, rays are reflected towards the tip but they 'leak' out of the scalpel before reaching the tip. The angle and the position at which a ray refracts out of the scalpel ('leaks') changes in discrete steps depending on the taper angle. For taper angles as small as 5 degrees, a power density increase of over 300 can be achieved. If the scalpel end is hemispherical rather than flat, the maximum increase in power density is reduced by 10–30% due to internal reflection losses.

Most side-firing fibre designs consist of a bare fibre tip that is polished at such an angle that the rays inside the fibre reflect at the angled exit surface because their angle of incidence exceeds the angle of total reflection. Ray-tracing will show that for a fibre with a NA of 0.22 in an air environment, the tip should be polished at an angle $> 40°$ to serve as a total reflector for preventing energy loss in secondary beams that refract out of the fibre tip. The exiting beam will be asymmetric. In the plane perpendicular to the fibre axis, the fibre core will act as a cylindrical lens and focus the beam. In a water environment the total reflection behaviour is lost. An optical shield can then be used to preserve the reflecting silica-air interface. For shielded configurations one has to be aware of secondary beams due to internal reflections.

Power output

For an accurate dosimetry during a laser treatment it not always possible to rely upon the reading of the display of the laser. Although most lasers have a power calibration port on the system, the measurement can not always be performed in a condition which is comparable with the clinical application. For example, the beam of side-firing fibre tips used for the treatment of benign prostatic hyperplasia (see urology section) exits the fibre at an angle near 90 degrees and the tips are used under water with relatively high input powers (ranging from 20 to 80 Watt). To determine the effective output power in laser treatments a special power meter called 'Aquarius' was developed. The fibre is placed inside a water-filled container and aimed at a thermopile detector head through a glass window. This measurement is a good representation of the power that actually reaches the tissue and that causes the therapeutic effects. The power meter also permits power measurements in sterile conditions so the change in power during a clinical treatment can be monitored. The transmission behaviour of various designs of side-firing fibres was evaluated with the 'Aquarius' showing 50% variations in transmission between different types of fibre. The transmission also varied dramatically during clinical application (van Swol *et al.* 1995 and 1996).

Thermal properties

Heat may be produced either by absorption of light by the probe itself or by the tissue in contact with the probe. High heat conduction helps to control excessive local probe temperatures which may protect the probe from being damaged. It is attractive to make probes of sapphire due to its high thermal conductivity combined with a high melting temperature. The performance of the probe is influenced by the wavelength, pulse duration and the geometry of the fibre tip.

Besides delivering laser light to the tissue, the delivery device itself can have a substantial contribution to the laser effect on the tissue. When the tip of the fibre (probe) is used in-contact with the tissue, the physical properties of the probe become particularly important. The physical processes involved in the interaction between fibre delivery systems and the tissue target can be differentiated in terms of optical, thermal and mechanical interactions (Verdaasdonk *et al.* 1990). The physical processes involved depend mainly on the means of application: non-contact versus in-contact with the tissue. There is also a significant difference in effects between the delivery of continuous wave and pulsed laser energy.

5 Applications of medical laser systems

This section describes various medical laser applications. These are either performed routinely or have been developed recently as the state of the art supported by the physics and research discussed in this chapter.

5.1 Dermatology and aesthetic Surgery

Localised superficial skin lesions

Focusing handpieces coupled to CO_2 lasers (Figure 17) are used for precise vaporisation of skin lesions like warts and condilomata (Wheeland 1995). The hand-pieces have relative short focal length lenses (100–150mm). The spot size on the tissue will be determined by the focal length and the distance between the lens and the tissue. Since the power density is proportional to the square of the distance from the focus, the distance is very critical for a controlled tissue effect (Verdaasdonk *et al.* 1995). Typically, over a distance of several centimetres, the spot size ranges from 0.1–3mm, a 900 fold variation in power density! To keep the power density constant, a combination of two lenses is used to create a collimated beam. These collimated hand pieces decrease the diameter of the beam, while the spot size is fixed and relatively independent of the distance to the tissue.

Large superficial skin lesions

Wrinkles of the skin can be smoothed by scanning a CO_2 laser beam across the skin (Weinstein 1994). This application has recently become very popular in aesthetic surgery. Within 45 minutes the whole face of the patient can be treated using the scanning device to remove the epidermis. Although the superficial skin is peeled by ablation, there may

scanning pulsed laser   scanning CW laser

spot 2.3 mm   spot 0.3 mm
500 mJ   20 mJ for 300 mm/s
20 W for 40 Hz   20 W

Figure 24. *Strategies for scanning large surfaces with cw and pulsed CO_2 lasers*

also be a considerable thermal effect. The heating induces shrinkage of collagen strings in the dermis resulting in a stretching of the wrinkles at the surface. It is still unknown what will happen to the thermally damaged collagen in the future.

The scanning systems consist of reflectors on a two axis system, which irradiate large surface areas quickly and in a controlled manner from a distance using a processor-controlled pattern. This pattern can either be preset from a menu of options or programmed for a specialised treatment. The tissue is irradiated more evenly, more accurately and faster than can be achieved manually.

Depending on the type of CO_2 laser there are two strategies. In the case of continuous wave CO_2 lasers, ablative temperatures are obtained instantly, by focussing the beam to a small spot with a very high power density. The beam is moved in a spiral shape at high speed to obtain a superficial ablation and even surface. For pulsed CO_2 laser systems the scanners have a choice of pre-set patterns with adjustment for the size and overlap between individual spots (Figure 24). In combination with the pulse energy and repetition of the pattern the ablation depth and adjacent thermal effects can be controlled. The speed of tissue ablation over a large surface is impressive.

Vascular lesions (port wine stains)

"Port Wine Stains" are caused by an excess of blood vessels near the surface of the skin. These can be preferentially destroyed by using a laser wavelength that is strongly absorbed by the blood. The laser beam is scanned across the surface in a controlled manner. Various scanning systems have been developed for the selective treatment of these vascular lesions. A scanning system 'Scanall' developed by Visiray (Hornsby, Australia), video-captures and digitises the lesion prior to treatment. A computer calculates the optimal treatment and path for scanning of the stain. The scanning speed is adjusted for the proper energy dose per unit area. The yellow beam from a copper vapour laser (578nm) is focused and scanned over the skin of the patient. The exposure time per position is just sufficient to heat and damage blood vessels, with minimal thermal relaxation in the surrounding area to have an optimal preferential effect. A French research group has developed the Hexascan (Mordon *et al.* 1993). The skin is irradiated with consecutive pulses of tens of milliseconds duration. The 1mm diameter spots fill up a hexagonal shape in such way that each following spot is fired as far away as possible from the previous one. Thus heat accumulation can be prevented in one area (Apfelberg and Smoller 1993). The total diameter of the hexagon can be pre-set from 1 to 13mm. The light energy which is usually from an argon laser, is transported through a 400μm fibre to the handpiece in which the scanner is incorporated. The exit end of the fibre is moved around in the object plane of a lens. The image focused as a spot on the surface of the target tissue will scan the surface according the motion of the fibre in the object plane. The exposure time of the mechanical shutter is a minimum of 30ms.

Similar to the Hexascan, a scanner has been developed specially for copper vapour laser applications. The exposure time is controlled both with a mechanical shutter and the scanning speed of the spot along the surface. The effective exposure time is in the range of 1 -100ms. In combination with a large copper vapour laser system producing over 10W of yellow light, therapeutic fluences of $10J/cm^2$ can be obtained within a few milliseconds thus preventing collateral thermal damage. This combination is potentially a good alternative for the flash lamp pulsed dye laser (Ross *et al.* 1993).

Pigmented lesions

Pigmented lesions, for example, tattoos, are removed using Q-switched Nd:YAG or ruby laser pulses delivered through collimated hand pieces with a spot size of several millimetres. The laser energy is preferentially absorbed by the coloured particles, which explode and break into smaller pieces. Subsequently, the small particles are reabsorbed by the body (Zelickson *et al.* 1994). The lesion will bleach in time after several treatments.

5.2 Opthalmology

Inner eye treatments

Opthalmology was in the early sixties, next to dermatology, the first field for the application of lasers. Lasers have been applied successfully since then as their unique characteristics enable treatment of the rear of the eye without any surgery. For example, diabetic patients often have a growth of blood vessels on the retina. To stop this, laser energy can be focussed on the retina through the transparent eye, causing partial coagulation of the offending vessels. Similarly ruptures of dissections of the retina can be sealed. These procedures are performed on an out-clinic basis within very short times.

Another simple procedure is the treatment of secondary glaucoma. Due to an adverse reaction of the human body on lens implants, a diffuse membrane encapsulates the artificial lens blocking clear vision. By precise focussing of the beam from a Q-switched Nd:YAG laser, a micro explosion is induced in the membrane, tearing it apart. The effect is similar to the opening of a curtain. The patient walks out of the doctor's office within 10 minutes with restored vision.

Cornea shaping

By adapting the curvature of the cornea, the eye can be corrected for near and far sight by up to 4–5 dioptres. Pulses of UV laser light can be used to reshape the cornea by selective removal of tissue. The 10ns pulses of the 193nm ArF excimer laser light only

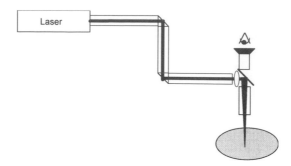

Figure 25. *Schematic of endoscope/microscope applicator*

ablate a few micrometres of tissue at a time (van Saarloos and Constable 1993). The beam is delivered by a mirror system and the spot covers the whole cornea. One method of ablating the cornea to the desired shape is by using a varying diaphragm. By ablating a pattern of rings in the surface of the cornea, the curvature is adjusted. Another method uses a contact lens with the 'negative' shape for the adapted lens. This contact lens is ablated on top of the cornea. At some point, the beam will perforate the thinner parts of the contact lens and start to ablate the cornea locally as well. Thus the depth of the ablated cornea is controlled by the local thickness of the contact lens.

5.3   Keyhole surgery for superficial lesions

Mucosa defects in mouth, larynx and ear

As shown before, the CO_2 laser is effective for superficial ablation of small lesions for example, on mucosa. These spots can be reached by directing the laser beam through natural 'openings' in the human body. To provide clear vision on the area of treatment, the distal end of an articulated arm can be coupled onto an operating microscope. The beam is aligned along the optical path of the operating microscope using a small 90° degree deflector (Figure 25). A joystick is used to guide the beam through the field of view by slightly tilting the angle of the deflector. The beam is focused on the target tissue using optics with a focal length compatible with the optics of the microscope, usually in the range from 200–400mm. Again the spot size (power density) will determine the resulting tissue effect. The view through the microscope enables precise ablation of tissue with visual feedback. The beam can be aimed accurately through narrow openings. Microscope manipulators are commonly used by ENT surgeons (Ossoff *et al.* 1994), for example, to ablate lesions on the larynx, and by gynaecologists, for example, for cervix ablation.

Endoscopic procedures

To enable exposure of tissue inside the body, the articulated arm of a CO_2 laser is coupled onto a rigid endoscope (Figure 25). The laser beam is narrowed by a long-focal-length lens (300–500mm) and aligned axially through a working channel of a rigid endoscope coaxially with the optics for viewing. The laser energy can be used in cavities, which are filled with air as in laparascopic (Lanzafame 1990) and thoracoscopic procedures (Wakabayashi 1991). Sometimes, the fluid in cavities like the knee or bladder is temporarily replaced with CO_2 gas to enable laser exposure. In general surgery, this delivery technique is applied to treat small superficial lesions on intestines, while in gynaecology, endometriosis on the outside of the uterus or bladder is ablated.

Recently, a comparable delivery device coupled to a 800W CO_2 laser was used to drill 1mm diameter channels 10 30mm in depth through the heart muscle (Frazier *et al.* 1995). This method is under investigation by cardiac surgeons as an alternative for bypass surgery. The hypothesis behind this technique, called transmyocardial revascularisation (TMR), is that oxygen-rich blood flows from these channels directly into the heart muscle (Sachinopoulou *et al.* 1996).

5.4   General surgical applications

Resection of (blood rich) tissue using 'dirty' tips

Tapered fibre tips delivering Nd:YAG, diode, argon or frequency-doubled Nd:YAG laser light, are used as an alternative to the surgical knife to cut through blood-rich tissue with minimal bleeding (Gallucci *et al.* 1994). The high power density at the tip of the taper will initiate tissue carbonisation. Some of these carbonised particles will adhere to the tip creating a hot needle by absorbing laser light effectively at the surface of the fibre tip itself. From that moment, the tip can be moved steadily cutting the tissue like a 'hot knife through butter'. Heat conduction coagulates the microvasculature in the direct environment. Surgeons, normally familiar with electro-surgery, have to be trained to hold the laser tip at one position for a few seconds at onset before moving it, to allow the tip to heat up. The laser exposure should be stopped when not in-contact with tissue to prevent damage to the tip. Laser scalpels are preferred to electro-surgical scalpels for resection of muscles, for example, the tongue. Electric currents tend to stimulate the muscle resulting in jagged contractions interfering with the surgery. Also, the electrical currents have a preference for the path of lowest resistance such as blood vessels, which make the tissue effect unpredictable. In the case of a laser scalpel the thermal effect is uniformly distributed around the fibre tip. Laser scalpels have been applied successfully in for example, tumour resection from the tongue and liver with minimal loss of blood.

Surgeons confirm that the fibre tips, which have become 'dirty' due to adherence of carbon particles during application, work more effectively than new 'clean' probes (Barroso *et al.* 1995). These absorption kernels on the probe are helpful for initiating tissue ablation when the wavelength of the laser light is poorly absorbed by the tissue itself. The carbon particles reach ablation temperatures after absorbing only a minimal amount of laser light. The 'hot' spots on the probe surface carbonise tissue in direct contact (Figure 26). From that moment on, the 'black' tissue itself will keep the ablation process going (Verdaasdonk *et al.* 1990).

To make new fibre tips 'work' instantly, manufacturers of fibre delivery systems apply such a coating on purpose in a controlled way during the manufacturing process. The coating and the probe itself should be resistant to high temperatures. White flashes, sometimes seen during tissue ablation, indicate that plasma temperatures (1500 2000°C)

Figure 26. *Tissue effect using a 'dirty' ball tip*

are reached. This explains the degradation of the surface of probes made of silica and even sapphire (Verdaasdonk *et al.* 1991). After tissue ablation the probe surface becomes frosted and pitted. This degradation is beneficial when starting with a 'clean' probe. The adhesion of carbonised tissue to the surface will enhance the performance of the 'dirty' tip. To prevent probe damage, some laser systems have a feedback system monitoring the temperature at the fibre tip and detecting white flashes. However, the white flashes are inherent to the ablation of carbonised tissue and are part of the tissue ablation process.

5.5   Neuro-surgery

'Black' fibre tips for endoscopic treatment of hydrocephalus

A defect in the "fluid-regulation" system of the brain can result in a build up of fluid in the ventricles of the brain increasing the intercranial pressure and resulting in serious neurological defects. This is called hydrocephalus. Normally, this problem can be treated acutely by inserting an external drain into the ventricle system. For the long term there is a fair chance of infection or occlusion of this drain. As an alternative, a 'natural' path for drainage can be made, by puncturing the membrane that forms the bottom of the third ventricle. Due to the presence of delicate and vital structures just underneath this membrane, it is essential to create this hole by precise ablation of tissue without peripheral damage. For this purpose an atraumic ball-shaped fibre was developed with a coating at the surface that absorbs over 90% of the laser energy.

The 'black' tip used with only 1W for half a second, will instantly vaporise a thin layer of tissue in contact with the probe surface leaving a small lateral coagulation zone.

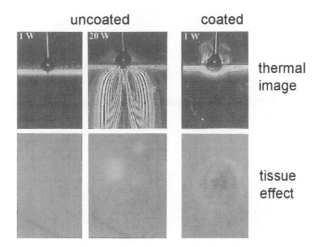

Figure 27. *Comparison of the thermal and tissue effects of coated and uncoated fibre tips*

In contrast, 'clean' fibre tips need high powers (> 10W) for several seconds to achieve ablative effects in tissue and meantime the beam penetrates deeply, heating a large volume of tissue (Figure 27).

These coated fibre tips have been applied successfully to over 100 patients for treatment of hydrocephalus using miniature flexible endoscopes (Vandertop *et al.* 1997). A 400μm fibre with a 800μm 'black' ball tip was advanced through the 1mm working channel of the 2.3mm endoscope to perform the procedures in a minimally invasive manner. The floor of the third ventricle was perforated using multiple exposures of 1–2.5W for 0.5–2.0 seconds from a 810nm diode laser, ablating thin layers of tissues. Small holes were linked to one big opening to form a natural passage for the abundant fluid responsible·for the hydrocephalus as illustrated in Figure 28. The procedure was performed safely without the risk of perforation of the basilary artery that is located just below the floor of the third ventricle. Any cw laser light that can be transported through optical fibres is suitable for this procedure. Since only a few Watts of laser power is needed, compact and efficient diode lasers as shown in Figure 3 are the laser systems of first choice.

Neuro-surgery bypass technique

Patients who suffer from brain ischaemia or a dangerous arterial aneurysm, might be effectively treated by a bypass of the diseased artery. In a similar manner to bypass surgery on the coronary arteries on the heart, a donor vessel is used to bypass an obstruction. For performing cardiac bypass surgery, the receiving artery is temporarily clamped and the pumping function of the heart is taken over by a heart-lung machine. For the neuro-surgery bypass procedure, however, the flow through the receiving artery can not be interrupted because this would directly result in brain damage. Therefore, a special method has been developed to create a high-flow bypass in the brain without occlusion of the recipient artery (Tulleken *et al.* 1993). On the outside wall of the recipient artery, an

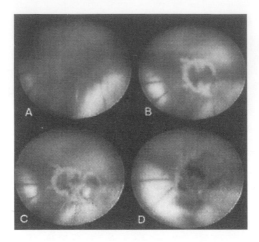

Figure 28. *Endoscopic view of a ventriculostomy using a 'black' fibre tip. A shows the ventricle before treatment; B to D shows the gradual removal of tissue using the heated fibre tip to allow a free passage of fluid.*

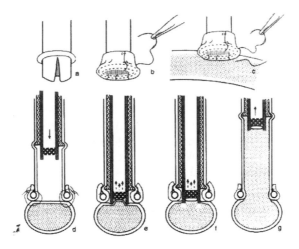

Figure 29. *Schematic of non-occlusive anastomosis technique*

end-to-side anastomosis is prepared without an entrance to the receiving artery (Figure 29). Through a temporary artificial side-branch of the donor artery, a multi-fibre catheter is introduced to punch a hole in the wall of the receiving artery.

This catheter consists of a ring of 170×60µm fibres around a 2mm diameter grid (Figure 30). Through the central lumen of the catheter, the wall of the receiving artery is sucked to the grid. During 5 seconds 200 pulses of 10mJ from a 308nm excimer laser are used to ablate the tissue in front of the ring of fibres. After the catheter, with the circular round wall flap attached to the grid, has been withdrawn, the temporary side-branch is

Figure 30. *Special multifibre ring catheter for perforation of the arterial wall. The fibres are in the outer ring, suction is applied via the inner set of holes.*

closed immediately. In this way the anastomosis is completed without interruption of the blood flow of the receiving artery. This technique has been performed successfully in 100 patients over the last three years. The patients tolerated the procedure well. No new neurological deficits were noted after the operation (Tulleken and Verdaasdonk 1995, 1997). Besides neuro-surgery, this technique has great potentials for cardiac-surgery. Bypass-surgery on the beating heart is under investigation (Borst *et al.* 1996).

5.6   Orthopedic surgery

The holmium laser has become an accepted tool for the treatment of defects of the cartilage in the knee joint during athroscopy (Buchelt *et al.* 1992). The laser pulses are delivered, non-contact, through a low-OH fibre incorporated in a rigid handpiece at a 30 40 degree angle a few millimetres above the tissue surface. The explosive vapour bubble will 'shave off' any tissue strings on top of the injured cartilage (Figure 31 top row). After collapse of the vapour bubble, the heat of condensation will melt and smooth the surface of the cartilage (Figure 31, bottom row). This mechanism is also applied to seal fractures in the cartilage (Vangsness *et al.* 1995). The angle of irradiation is important so that the tissue is not directly exposed to the light, which will normally tear the tissue apart.

5.7   Urology

Lithotripsy (stone breaking)

The Holium laser is also successfully applied for breaking into pieces stones that are interfering with the urinary system. In the aqueous environment, a forceful expanding vapour bubble is formed during the laser pulse that will fracture and chip the stone to small pieces (Figure 32). Even the most rigid stones have been fractured successfully.

BPH treatment

Over 25% of the male population above 65 years will have problems with voiding the bladder due to a benign enlargement of the prostate gland obstructing the urethra (Benign

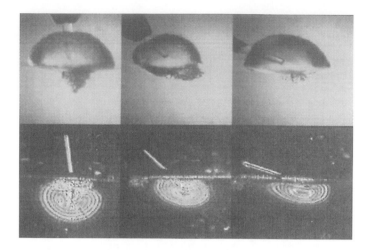

Figure 31. *Combined fast and thermal imaging showing the effect of the angle of irradiation on the thermal and mechanical effects e.g. during cartilage remodelling*

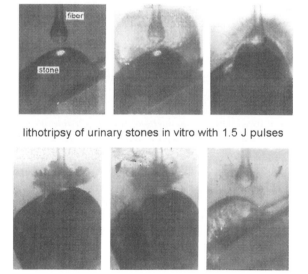

lithotripsy of urinary stones in vitro with 1.5 J pulses

Figure 32. *Fast imaging of exploding vapour bubbles breaking urinary stones to pieces*

Prostatic Hyperplasia, BPH). Normally, during a trans-urethral endoscopic procedure, part of the prostate is resected using a loop wire heated by electrical energy. Drawbacks for this standard procedure are the potentially high blood loss and various complications. As an alternative, the laser treatment of BPH has been popular for the last 3 years. Using high powers of laser light with a deeply penetrating wavelength, such as the cw Nd:YAG laser, the prostate tissue is coagulated and subsequently vaporised to re-open the lumen of the urethra (Gerber 1995). The laser beam, delivered through a fibre introduced through the urethra, has to exit the fibre at a 90 degree angle to irradiate the prostatic tissue. Many types of 'side-firing' laser devices have been introduced for the treatment of BPH using either reflection or refraction to deflect the beam side ways. Evaluation of these fibre tips shows that they are unique with regard to the angle at which the light is deflected off axis and to the size and shape of the spot at the urethra wall (van Swol *et al.* 1995) . These measurements have been used to determine an optimal dosimetry with the best fitted parameters for each device. For some tips it may be best to irradiate the prostate (non-contact) from a fixed position while others have to be moved to 'paint' the prostatic tissue to obtain a coagulation zone 10–20mm deep into the tissue. The necrotic tissue obstructing the urethra will slough off within a few weeks. To create an opening immediately, side-firing fibres can also be used in-contact vaporising the obstructing tissue. The explosive vapour bubble of the holmium laser has also been also investigated for an efficient removal of the abundant tissue (Johnson *et al.* 1992).

5.8   Cancer detection and treatment

Photo dynamic therapy

Besides thermal and mechanical effects, laser light can also induce chemical effects, which are used for photodynamic therapy (PDT). A photosensitive drug is administered to a patient and then it selectively accumulates in cancerous tissues. When the tissue is illuminated by a particular wavelength, the drug in the cancerous tissue will emit fluorescence light disclosing the presence of the malignant tissue (Fisher *et al.* 1995). In this way diffuse diseased tissue can be detected.

Depending on the drug, another particular wavelength can be used to induce a chemical process that releases a toxic component which kills the malignant cells that contain the drug. For this therapy, it is important that the target tissue is illuminated as uniformly as possible. Special diffusing fibre tips have been developed that are optimised for the geometry of the target organ. These are usually spherical or cylindrical. The power density at the tissue surface should exceed the threshold level for activation of the photodynamic drug in the tissue, but the heat input should not result in hyperthermia. Usually the tissue is exposed to several hundreds of milliwatts per square centimetre for several minutes. The power should also be limited to prevent damage to the diffusing tip since a part of the energy is absorbed in the diffusing material.

At this time, the only PDT drug approved in most countries is Photofrin. This drug is activated near 630nm which is commonly generated with argon-ion-laser-pumped dye lasers. However, dye lasers pumped with pulsed laser systems like Copper-vapour, frequency-doubled Nd:YAG (KTP) and Excimer lasers are also attractive because of their high efficiency (over 30%). A side effect of Photofrin is that the patient is not allowed in bright daylight for several days up to weeks since the drug also sensitises the skin. New generation PDT drugs with less side-effects are already under clinical investigation (Fisher *et al.* 1995). They can be excited with the near IR wavelengths of diode lasers. Such lasers are much preferred to the bulky lasers currently in use. Clinical trials with PDT drugs are being performed to treat cancers in the skin, lungs, oesophagus, bowels and bladder and the results are promising.

6 Future developments

From the discussion above we can see something of the design goals for new medical laser systems. The "ideal" system will be simple to operate, and will have well-researched modes of operation. It will allow tunability in wavelength, pulse-length and power in order to match the needs of the particular procedure. Power delivery will be via an optical fibre for ease of access to the point of application. The system will also be small, efficient and of low cost. This suggests that diode lasers and diode-laser powered lasers and optical parametric oscillators are the most attractive systems.

As well as using fibres to deliver the laser light, it is expected that increasing use will be made of the fibre to transmit fluorescent and Raman light back to the medical laser system for spectroscopic analysis. Such "remote-sensing" will give the surgeon a useful feedback on the stage of treatment being performed.

6.1 Diode lasers

In the near future, it can be expected that new laser systems will be developed using solid state technology. Besides the higher efficiency of light generation, these systems are compact and are expected to be reliable and maintenance free. The performance of the present generation of high power diode lasers already makes them an attractive alternative for Nd:YAG lasers and in future also for electro surgery instruments. The materials currently used, limit the production of high power to the near IR (700–1000nm). With new materials blue light can be produced. No 'diode' alternatives are available in the orange-yellow-green range as replacement for argon pumped dye lasers for example. However, in this wavelength range high efficiencies are already obtained by using diode lasers to pump solid state lasers (frequency-doubled Nd:YAG laser pumping dye lasers). Diode lasers have relatively poor beam properties. For coupling into fibre delivery systems with a diameter below 500μm either the incoupling optics or the numerical apertures of the fibres have to be improved.

6.2 Fibres

After many years of development, flexible waveguides for the IR have now become commercially available. Similar to silica fibres, these IR fibres will be further improved in flexibility and transmission characteristics over the next years. New applications in various medical fields are being explored (DeRowe *et al.* 1994. Snaijer *et al.* 1998) such as the ablation of hard tissues using the Erbium laser.

Besides high power delivery, IR fibres will be ideal for sensing the IR light coming from the irradiated tissue which will allow the local temperatures to be determined (Eyal and Katzir 1994). Spectroscopic sensing or imaging through delivery systems of silica or IR transmitting fibres will become an important method for identification of diseased tissues.

Another promising development is the fibre laser. By using rare-earth ions as impurities in the fibre 'host' material, the fibre itself becomes the laser medium. Powers over 100mW have been obtained in the visible range. Although fibre laser development has been driven by their use in high speed communication networks, they might have potential for medical applications.

In the quest to make treatments less invasive, imaging techniques are more frequently used. Fibre delivery systems can be easily incorporated in small flexible endoscopes consisting of thousands of fibres themselves. With these instruments laser energy can be delivered to the target tissue at almost any place in the human body. For treatments guided by magnetic resonance imaging (MRI, Bleier *et al.* 1991) fibre based systems are one of the only means available for delivering energies at a therapeutic level in a 'non-metal' environment.

6.3   Choices

It is obvious that laser delivery systems can be used for many applications. However only for a few of these applications are they the only method of treatment or is their superiority above other methods beyond doubt. In many areas the laser modality has to compete with other 'thermal' modalities like electro-surgery or cryo-surgery or 'mechanical' modalities like focused ultrasound or electro-hydrolysis. In some areas the introduction of lasers has been a catalyst for stimulating the improvement of the standard techniques. The initial success of laser angioplasty was overtaken by smooth balloons and guide-wires which can pass through almost any total occlusion. Similarly, the 'golden' standard , trans-urethral (electric) resection of the prostate (TURP), was only for a short time in the 'shadow' of laser light. At present, various devices such as 'vaportrodes' and 'rollerballs' have provided a come-back of electro-surgery.

The challenge for researchers and clinicians with a strong affinity to laser applications, is to improve and simplify the laser delivery systems and prove in randomised trials the benefit of laser applications. Even when laser modalities are superior, it has to be shown that they are cost effective. The threshold for acceptance of this high-technology has to be low.

7   Conclusions

A basic understanding of the dynamical processes of heat during the illumination of biological tissue with laser light will improve the direct surgical success and the long term clinical outcome. Intensive cooperation between surgeons, physicists and engineers within the laboratory and operating theatre will stimulate the improvement of application strategies and the improvement of delivery systems. The following two pages summarise the technical data and applications of many lasers now in use.

Acknowledgements

The author would like to acknowledge the contribution of his colleagues in the Medical Laser Center, Christiaan C P van Swol. PhD, and Matthijs CM Grimbergen, BSc, to the research and clinical developments discussed in this chapter.

laser	Excimer ArF	Excimer KrF	Excimer XeCl	Excimer XeF	Argon	Dye	Dye	KTP
wavelength(s) (range) [nm]	193	249	308	351	488, 514	(400 - 800)	(400 - 800)	532
cw / pulsed	pulsed	pulsed	pulsed	pulsed	cw	pulsed	cw	(semi) cw
pulse domain	5-25 ns	2-50 ns	10-150 ns	10-30 ns		1 - 10 us		10 ns
pulse energy [mJ]	300	300	300	300		1,000		1
rep rate [Hz]	100	100	100	100		10		kHz
max power [W]	10	10	10	10	20	10	5	40
penetration depth [um]	1	5	40	10	330	250	250	40
transportation	articulated arm				fibre			
absorbing chromophore	proteins water				blood	blood pigments	blood	blood
typical tissue effect								
selective coagulation					X	X	X	X
deep coagulation						X		X
dehydration/boiling					X	X	X	X
explosive vapour formation	X	X		X		X		
carbonisation					X	X	X	X
thermal vaporisation					X	X	X	X
stone fragmentation						X		
selective vaporisation						X		
superficial ablation	X		X	X				
Photo Dynamic Therapy							X	
clinical specialism								
Aesthetic Surg					X	X		
Cardiology			X		X	X		X
Dermatology					X	X	X	X
Ear, Nose and Throat								X
General Surgery								
Gynaecology								
Neuro Surgery			X					
Opthalmology	X	X			X	X	X	
Orthopedics								
Urology								

laser	Cu vapour	Diode	Nd:YAG	Nd:YAG	Ho:YAG	Er:YAG	CO2 cw	CO2 pulsed
wavelength(s) (range) [nm]	511, 578	630	1064	1064	2100	2940	10600	10600
cw / pulsed	(semi) cw	cw	pulsed	cw	pulsed	pulsed	cw	pulsed
pulse domain	10 ns	ms	10 ns, us		300 us	300 us		10 - 2000 us
pulse energy [mJ]	1		1,000		2000	1000		2000
rep rate [Hz]	kHz		kHz		20	20		kHz
max power [W]	30	60	100	100	80	20	100	100
penetration depth [um]		1300	1400	1400	300	1	20	20
transportation		fibre	fibre		art.arm (fibre)	art.arm (fibre)	articulated arm	articulated arm
absorbing chromophore	blood		pigments	pigments	water	water	water	water
typical tissue effect								
selective coagulation	X		X				X	
deep coagulation		X		X	X			
dehydration/boiling		X		X				
explosive vapour formation			X		X	X	X	X
carbonisation		X		X			X	
thermal vaporisation		X		X	X		X	X
stone fragmentation			X		X			
selective vaporisation			X				X	X
superficial ablation						X	X	X
Photo Dynamic Therapy	X	X						
clinical specialism								
Aesthetic Surg	X	X	X			X	X	X
Cardiology		X		X	X		X	X
Dermatology	X	X	X			X	X	X
Ear, Nose and Throat		X	X	X			X	X
General Surgery		X		X			X	X
Gynaecology		X		X			X	X
Neuro Surgery		X		X			X	X
Opthalmology		X	X		X			
Orthopedics					X			
Urology		X		X	X		X	X

References

Abel T, Harrington J A and Foy P R, 1994, Optical properties of hollow calcium aluminate glass waveguides. *Applied Optics* **33** 3919.

Apfelberg D B and Smoller B, 1993, Preliminary analysis of histological results of Hexascan device with continuous tunable dye laser at 514 (argon) and 577NM (yellow). *Lasers Surg Med* **13** 106.

Barnes A E, May R G, Gollapudi S and Claus R O, 1995, Sapphire fibres: optical attenuation and splicing techniques. *Applied Optics***34** 6855.

Barroso E G, Haklin M F and Staren E D, 1995, Characteristics of Nd:YAG sculptured contact probes after prolonged laser application. *Lasers. Surg Med* **16** 76.

Bleier A R, Jolesz F A, Cohen M S, Weisskoff R M, Dalcanton J J, Higuchi N, Feinberg D A, Rosen B R, McKinstry R C and Hushek S G, 1991, Real-time magnetic resonance imaging of laser heat deposition in tissue. *Magn Reson Med* **21** 132.

Borst C, Jansen E W L, Tulleken C A F, Grundeman P F, Mansvelt Beck H J, van Dongen J W, Hodde K C and Bredee J J, 1996, Coronary artery bypass grafting without cardiopulmonary bypass and without interruption of the flow using a novel anastomosis site restraining device ("Octopus"). *J Am Coll Card* **27** 1356.

Buchelt M, Kutschera H P, Katterschafka T, Kiss H, Schneider B and Ullrich R, 1992, Erb:YAG and Hol:YAG laser ablation of meniscus and intervertebral discs. *Lasers Surg Med.* **12** 375

Chang C T and Auth D C, 1978, Radiation characteristics of a tapered cylindrical optical fibre. *J Opt Soc Am* **68** 1191.

Chato J C, 1985, Selected thermophysical properties of biological materials, in Shitzer A, Eberhart RC (eds): *Heat transfer in medicine and biology.* (Plenum Press), p413.

DeRowe A, Ophir D and Katzir A, 1994, Experimental study of CO_2 laser myringotomy with a hand-held otoscope and fibreoptic delivery system. *Lasers Surg Med* **15** 249.

Eyal O and Katzir A, 1994, Thermal feedback control techniques for transistor-transistor logic triggered CO_2 laser used for irradiation of biological tissue utilizing infrared fibre-optic radiometry. *Applied Optics* **33** 1751.

Fisher A M R, Murphee A L and Gomer C J, 1995, Clinical and preclinical photodynamic therapy. *Lasers Surg Med* **17** 2.

Frazier O H, Cooley D A, Kadipasaoglu K A, Pehlivanoglu S, Lindermeir M, Barasch E, Conger J L, Wilansky S and Moore W H, 1995, Myocardial revascularization with laser: preliminary findings. *Circulation* **92** II-58.

Gallucci J G, Zeltsman D and Slotman G J, 1994, Nd:YAG laser scalpel compared with conventional techniques in head and neck cancer surgery. *Lasers Surg Med* **14** 139.

Gannot I, Dror J, Calderon S, Kaplan I and Croitoru N, 1994, Flexible waveguides for IR laser radiation and surgery applications. Lasers. *Surg Med* **14** 184.

Gerber G S, 1995, Lasers in treatment of benign prostatic hyperplasia. *Urology* **45** 193.

Harrington J A 1992, Laser power delivery in infrared fibre optics. In: Katzir, A., (Ed.) *Optical fibres in Medicine VII* **1649** p14. (Bellingham: SPIE.)

Hecht E, 1987, *Optics* (Prentice Hall)

Helfer D, Frenz M, Romano V and Weber H F, 1994, Fibre-end micro-lens system for endoscopic erbium-laser surgery application. *Applied Physics B* **58** 309.

Howes W L, 1984, Rainbow schlieren and its applications. *Appl Optics* **23** 2449.

Johnson D E, Cromeens D M and Price R E, 1992, Use of the holmium:YAG laser in urology. *Lasers Surg Med* **12** 353.

Katzir A, 1993, *Lasers and Optical fibres in Medicine*, (New York:: Academic Press.)

Lanzafame R J, 1990, Applications of lasers in laparoscopic cholecystectomy. *J Laparoendosc. Surg* **1** 33.

Magera J J and McCann B P, 1992, Silica-core fibres for medical diagnosis and therapy. In: Katzir, A., (Ed.) *Optical fibres in Medicine VII* **1649** p2.(Bellingham : SPIE.)

Mahlke G and Gossing P, 1987, *Fibre optic cables*,(New York: John Wiley & Sons.)

Matsuura Y, Abel T and Harrington J A, 1995, Optical properties of small-bore hollow glass waveguides. *Applied Optics* **34** 6842.

McKenzie A L, 1988, How to construct a bulb-tipped fibre: part II - Application. *Lasers Med Sc* **3** 273.

McKenzie A L, 1990, Physics of thermal processes in laser-tissue interaction. *Phys Med Biol* **35** 1175.

Merberg G N, 1993, Current status of infrared fibre optics for medical laser power delivery. *Lasers Surg Med* **13** 572.

Mordon S, Rotteleur G, Brunetaud J M and Apfelberg D B, 1993, Rationale for automatic scanners in laser treatment of port wine stains. *Lasers Surg Med* **13** 113.

Ossoff R H, Coleman J A, Courey M S, Duncavage J A, Werkhaven J A, Reinisch L, 1994, Clinical applications of lasers in head and neck surgery. *Las Surg Med* **15** 217.

Ross M, Watcher M A and Goodman M M, 1993, Comparison of the flashlamp pulsed dye laser with the argon tunable dye laser with robotized handpiece for facial telangiectasia. *Lasers Surg Med* **13** 374.

Russo V, Righini G C, Sottini S, and Trigari S, 1984, Lens-ended fibres for medical applications: a new fabrication technique. *Applied Optics* **23** 3277.

Sachinopoulou A, Beek J F, Tukkie R, Meijer D W, Grundeman P F, Verdaasdonk R M, Gemert M J C, 1996, Transmyocardial Revascularization: a study of literature. *Las Med Sci* **10** 83.

Snaijer A de, Verdaasdonk R M, Grimbergen M C M and van Swol C F P, 1998, Design and realization of a fibre delivery system for the continuous wave and pulsed CO2 laser. *SPIE proceedings* **3262** 108.

Torres J H, Springer T A, Welch A J, Pearce J A, 1990, Limitations of a thermal camera in measuring surface temperature of laser-irradiated tissues. *Lasers Surg Med* **10** 510.

Tulleken C A F, Verdaasdonk R M, Berendsen W and Mali W P T M1993, Use of the excimer laser in high flow bypass surgery of the brain. *J Neurosurg* **78** 477.

Tulleken C A F and Verdaasdonk R M, 1995, First clinical experience with Excimer laser assisted high flow bypass surgery of the brain. *Acta Neurochirurgica* **134** 66.

Tulleken C A F, Verdaasdonk R M, Mansvelt Beck H J, 1997, Nonocclusive excimer laser-assisted end-to-side anastomosis. *Ann Thorac Surg* **63** S138.

van Leeuwen T G, van-der-Veen M J, Verdaasdonk R M and Borst C, 1991, Noncontact tissue ablation by holmium:YSGG laser pulses in blood. *Lasers Surg Med* **11** 26.

van Gemert M J C, Welch A J, 1989, Time constants in thermal laser medicine. *Lasers Surg Med* **9** 405.

van Saarloos P P and Constable I J, 1993, Improved excimer laser photorefractive keratectomy system. *Lasers Surg Med* **13** 189.

van Staveren H J, Marijnissen H P, Aalders M C G, Star W M, 1995, Construction, quality assurance, calibration of spherical isotropic fibre optic light diffusers. *Las Med Sci* **10** 137.

van Swol C F P, Verdaasdonk R M, van Vliet R J, Molenaar D G and Boon T A, 1995, Side-firing devices for laser prostatectomy. *World Journal of Urology* **13** 88.

van Swol C F P, Slaa E, Verdaasdonk R M, Rosette J J M C, Boon T A 1996, Variation in the output power of laser prostatectomy fibres: a need for power measurements. *Urology* **47** 672.

Vangsness C T,Jr, Watson T, Saadatmanesh V and Moran K, 1995, Pulsed Ho:YAG laser meniscectomy: effect of pulsewidth on tissue penetration rate and lateral thermal damage. *Lasers Surg Med* **16** 61.

Verdaasdonk R M, Borst C and Gemert M J C, 1990, Explosive onset of continuous wave laser tissue ablation. *Phys Med Biol* **35** 1129.

Verdaasdonk R M, Jansen E D, Holstege F C and Borst C, 1991, Mechanism of CW Nd:YAG laser recanalization with modified fibre tips: influence of temperature and axial force on tissue penetration in vitro. *Lasers Surg Med* **11** 204.

Verdaasdonk R M, Holstege F C, Jansen E D and Borst C, 1991, Temperature along the surface of modified fibre tips for Nd:YAG laser angioplasty. *Lasers Surg Med* **11** 213.

Verdaasdonk R M and Borst C, 1991, Ray-tracing of optically modified fibre tips I: laser angioplasty probes. *Appl Optics* **30** 2159.

Verdaasdonk R M and Borst C, 1991, Ray-tracing of optically modified fibre tips II: laser scalpels. *Appl Optics* **30** 2172.

Verdaasdonk R M, Swol C F P, van Leeuwen A G J M, Tulleken C A and Boon T A 1994, Multifibre excimer laser catheter design strategies for various medical applications. In: Katzir, A. and Harrington, J.A., (Eds.) *Specialty fibres for biomedical and systems applications(* 2131th edn. pp 118–126. Bellingham: SPIE.)

Verdaasdonk R M, van Swol C F P and Coates J, (1995) Characterization of handpieces to control tissue ablation with pulsed CO2 laser. In: Jacques, S.L., (Ed.) *Laser-Tissue Interaction VI* (2391th edn. Bellingham: SPIE.)

Verdaasdonk R M and Borst C, 1995, Optics of fibres and fibre probes. In: Welch, A.J. and Gemert, M.J.C., (Eds.) *Optical-Thermal response of laser irradiated tissue* p.619 (New York: Plenum Press.)

Verdaasdonk R M, 1995, Imaging laser induced thermal fields and effects. In: Jacques, S.L., (Ed.) *Laser-Tissue Interaction VI* **2391** 165. (Bellingham: SPIE)

Vandertop W P, Verdaasdonk R M, van Swol C F P, 1997, Laser neuroendoscopy using Nd:YAG and Diode contact laser with pretreated fibre tips. *J Neurosurgery* **88** 82.

Vo Dinh T, Panjehpour M, Overholt B F, Farris C, Buckley F P and Sneed R, 1995, In vivo cancer diagnosis of the esophagus using differential normalized fluorescence (DNF) indices. *Lasers Surg Med.* **16** 41.

Wakabayashi A, 1991, Expanded applications of diagnostic and therapeutic thoracoscopy. *J Thorac Cardiovasc Surg* **102** 721.

Ward H, 1987, Molding of laser energy by shaped optic fibre tips. *Lasers Surg Med* **7** 405.

Weinstein C, 1994, Ultrapulse carbon dioxide laser removal of periocular wrinkles in association with laser blepharoplasty. *J Clin Laser Med Surg* **12** 205.

Wheeland R G, 1995, Clinical uses of lasers in dermatology. *Lasers Surg Med* **16** 2.

Zelickson B D, Mehregan D A, Zarrin A A, Coles C, Hartwig P, Olson S and Leaf Davis J, 1994, Clinical, histologic, and ultrastructural evaluation of tattoos treated with three laser systems. *Lasers. Surg Med* **15** 364.

Solid state lasers and nonlinear optics for LIDAR

P F Moulton

Q-Peak Inc., Bedford, USA

1 LIDAR overview

Radio and microwaves have long been used for the detection of remote objects through the use of radar techniques. These systems are very well developed, and are widely deployed. Similar techniques can be used with visible or near-visible wavelengths. Such Light Detection and Ranging (Lidar) techniques allow a number of measurements to be made that differ from those obtained by Radar and may be significantly better. The shorter wavelengths allow better angular resolution, the very short pulses that can be generated with light can allow better distance resolution, the wavelengths involved can allow significant backscatter from aerosols, and the photon energies involved can allow the unique property of detecting specific atomic or molecular species. These differences have resulted in considerable interest in atmospheric Lidar, which is developing rapidly using wavelengths ranging from some 200nm to 10μm and beyond.

This section reviews some of the science of these systems and considers the kind of coherent optical sources are that are required. More details can be found in the books by Measures (1992) and Jetalian (1992). NASA is heavily involved in Lidar technologies, and maintains a number of informative websites, including NASA Langley 1999.

1.1 Types of Lidar

Hard target ranging

The simplest form of lidar, and the form that is closest in principle to radar, is basic laser ranging. In this case a pulsed laser is used to send out a short pulse of light to a hard target. A measurement of the time taken till the return signal comes back gives the distance to the object. When used in a point-and-shoot mode, this is the basis of the military laser rangefinder discussed in more detail by Barr (this volume). If the directions of the laser beam and receiver are scanned, a three dimensional picture can be built up. Automatic vehicle recognition is an example of one commercially successful system built by Schwartz Electro-Optics (SEO). A pulsed laser diode and a scanning system are mounted above a road. The laser pulses are scanned across the vehicles as they travel underneath. The reflected signals can then be analysed to build up a three-dimensional representation of the vehicles, as shown in Figure 1. Some 600 of these "Autosense" systems have been installed throughout the world.

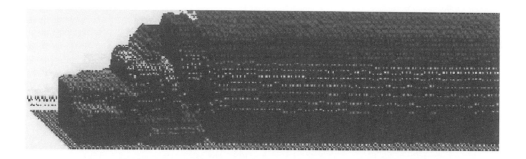

Figure 1. *Three dimensional image of a truck built up by the Autosense Lidar system. Distance from the transmitter was coded by colour, but here is reproduced in grey scale only. (Commercial Sensors Division, SEO)*

Additional information can be obtained if the frequency of the reflected light is analysed. A reflection from a moving object will be Doppler shifted in frequency. This can be exploited if a single frequency laser is used. Through the use of coherent detection (ie the return signal is mixed with a local oscillator) the change in frequency of the reflected signal can be measured, and hence the velocity calculated. This can be done with pulsed lasers to record range as well as speed, though for the most sensitive measurements of velocity a continuous wave source is used due to its potentially narrower spectral width. This technique can be used to measure the speed of vehicles, with the advantage over the more established radar techniques of a smaller beam divergence allowing better targeting of individual vehicles.

The same technique can be used for monitoring vibrations. The challenges in building such systems become evident when one calculates the frequency shift associated with small vibrations. If light of wavelength 1550nm is used to monitor a vibration of amplitude 1mm and frequency 230Hz, the maximum optical frequency shift is only 47kHz.

Aerosol mapping

Perhaps the best known lidar application is in the three-dimensional mapping of aerosols (water droplets, dust, smoke, etc). The lidar system is scanned in angular direction between pulses, and the backscattered signal of each pulse is recorded as a function of time. Each volume element of the aerosol scatters part of the laser pulse back to the lidar detector system. In this way the distribution of water droplets in a cloud, or smoke in a plume, can be determined remotely.

The study of aerosol transport in the atmosphere is of importance in studies of climatology due to its role in the "greenhouse effect". NASA, for example, has flown a

Q-switched flashlamp-pumped Nd:YAG laser in the space shuttle to probe aerosol distribution throughout the full height of the atmosphere. In this case, rather than scanning the angle of view of the lidar, the orbital motion of the shuttle allowed different shots to probe different vertical columns of the atmosphere. Named LITE (Lidar in Space Technology Experiment), this was the first space-borne lidar system. Figure 2 shows an image of a convective storm over Africa obtained by this system. At the lower latitudes the high altitude clouds preclude a study of aerosol scattering all the way to ground level, but at the higher latitudes in the example shown, haze in the troposphere between altitudes of two and five kilometres can be clearly seen. This was regarded as a big scientific success, and also demonstrated that laser and lidar technologies have progressed to the stage where they can be space qualified.

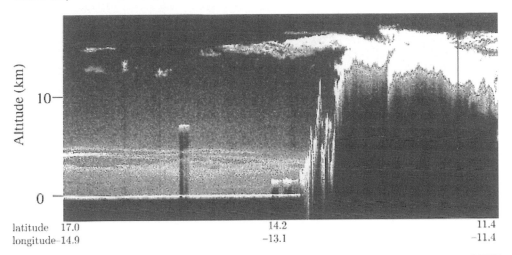

Figure 2. *Acrosol distributions over part of Africa determined by the space-borne LITE system. Picture from NASA (1998).*

In the same way that coherent detection and Doppler shifts allowed velocity mapping of hard targets, lidar based on aerosol scattering allows the remote measurement of wind velocities. Global mapping of winds is important in climatology and meteorology. Considerable efforts are also being expended in using coherent lidar to map the wind velocities at airports. The turbulence caused by the passing of a large aircraft has to damp out before the next aircraft can safely land. Currently, more than sufficient time is left between aircraft to allow this to happen. Terminal productivity could potentially be improved if a lidar system was able to show controllers when the turbulence caused by each aircraft had reduced to an acceptable level. Similar systems would be of use in aircraft in flight to detect clear-air-turbulence ahead of them before it was encountered.

Differential absorption Lidar (DIAL)

An important variant of lidar uses two pulses of different wavelengths to determine the concentration of a particular atomic or molecular species in the atmosphere. The return signal from one pulse is mapped out as above, using a laser with a wavelength that is not strongly absorbed by the species of interest. Soon after, a second pulse is sent out along the same direction, but this time with a wavelength of light that is strongly absorbed. The two wavelengths are normally chosen to be similar, so that the contributions to the two return signals due to Raleigh and Mie scattering are similar. The difference between the two return signals is then indicative of the energy absorbed by the probed species. Such a pair of return signals is shown schematically in Figure 3, for wavelengths λ_{off} and λ_{on} that are respectively off and on resonance. In the absence of absorption, we see a drop in return signal between distance R_1 and R_2 of ΔE_{off}. This decrease is due to conventional lidar losses such as the scattering process itself, and the inverse square law (see Section 1.2). However, when the laser is tuned into resonance with an absorption line, a smaller fraction of the power that reached R_1 will get to R_2, and so the difference in the return signal from the two distances, ΔE_{on}, is greater.

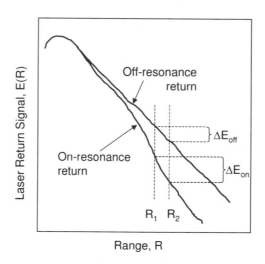

Figure 3. *Lidar return signal as a function of range (time) for a laser that has been tuned on and off resonance with an absorption line. Adapted from Measures (1992)*

It can be shown (Measures, 1992) that the concentration of the species averaged over the volume element between distances R_1 and R_2 from the lidar is

$$N = \frac{1}{2(R_2 - R_1)[\sigma_A(\lambda_{\text{on}}) - \sigma_A(\lambda_{\text{off}})]} \left[\ln \frac{E(\lambda_{\text{off}}, R_2)E(\lambda_{\text{on}}, R_1)}{E(\lambda_{\text{on}}, R_2)E(\lambda_{\text{off}}, R_1)} \right] \tag{1}$$

where $E(\lambda, R)$ are the detected signal energies, and $\sigma_A(\lambda)$ are the absorption cross sections at the two wavelengths. For typical absorption cross sections in the 10^{-18}cm^2 range, with a large difference between their values on and off resonance, concentrations as low as 1 ppm per metre of path length can be determined. If the measurement is taken over a kilometre, then measurements at the part per billion level can be made.

This type of system is now being explored for sensing a wide range of molecules. The laser source must be able to operate at the wavelength of an appropriate absorption feature, and be able to be tuned rapidly on and off resonance. Important examples of species to be probed include ozone at 290–300nm, water at 820nm, SO_2 at ∼300nm, and NO_2 at ∼ 450nm.

One example of a commercial DIAL system is the Elight 510M system. A flashlamp-pumped Ti:sapphire laser is frequency tripled to the UV to match the SO_2 absorption. The system is sufficiently compact and rugged to fit within a small van. The pulses are transmitted through a scanning telescope, and the return signals on and off resonance are processed to give an image of the SO_2 concentration. Figure 4 shows a two-dimensional map of the SO_2 density near a smoke stack, as determined by this system. The manufacturers specify for SO_2 detection a best accuracy of 4 parts per billion if a 500m spatial resolution is used. At the expense of a poorer sensitivity, a spatial resolution as high as 7.5m is achievable.

A notable example of a scientific DIAL system is the 820nm LASE instrument built by NASA to probe the distribution of water vapour in the atmosphere. This system is flown in an ER-2 aircraft at an altitude of 70,000 feet. The laser design will be described in Section 2.2.

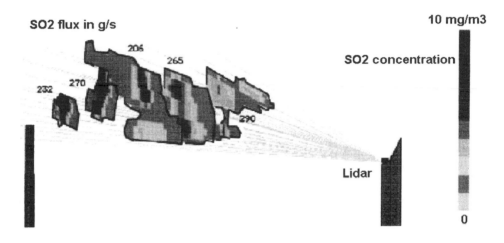

Figure 4. *Concentration of sulphur dioxide near a smokestack as measured by the Elight 510M system. Adapted from Elight Web site (1998).*

Raman and fluorescence Lidar

As discussed above, DIAL can be used with backscattering to determine the spatial distribution of molecules. This technique depends on the presence of scattering centres. However, there are two techniques that can be used to probe atomic and molecular species directly.

The Raman effect can be utilised by sending up a pulse at one wavelength, and observing the Raman shifted return signals from different atomic and molecular species. No atmospheric scattering by other entities is needed for measurement, and the nitrogen in the atmosphere provides a convenient calibration. As different species provide distinct Raman lines displaced in frequency from that of the laser, one lidar system can sense multiple species, and the lidar system need not be tunable as is needed for DIAL techniques. However, the Raman effect is generally rather weak. While this is an advantage for allowing the lidar beam to experience little attenuation over long distances, it has the disadvantage of resulting in a sensitivity that is orders of magnitude lower than that of DIAL.

An alternative technique is to use a laser tuned to an absorption line of the species of interest, and observe the fluorescence radiated by the atom or molecule as it decays back to its ground state. This can detect some entities that are not readily sensed otherwise, such as certain biological compounds. Spectral and temporal analysis can be used to distinguish amongst multiple species with overlapping absorption bands. However, range analysis can be complicated when the fluorescence decay time is long, and in many atmospheric conditions the fluorescence of molecules is strongly quenched. Only one in a million absorbed photons may result in the production of a photon by fluorescence. This can reduce the sensitivity of the system to below that attainable by DIAL techniques. Also, the spectral signature of the fluorescence may require a broad detector response, which leaves the system susceptible to background light, and to light produced by Raman scattering off other atmospheric constituents. This technique has, however, proved particularly useful for the study of the upper atmosphere where a layer of sodium atoms is located. This is probed using lasers at 590nm.

1.2   The return signal

In order to design a lidar system and its associated laser, one must know what sort of strength of return signal is expected. On propagation over a distance R towards a target, the energy in the emitted beam will be reduced to a fraction $T_a(\lambda, R)$ due to atmospheric attenuation. The wavelength dependence of this function will be discussed in Section 1.3.

When the return signal from a distance R away is analysed, the system considers a pulse that has propagated a distance $2R$. The thickness of atmosphere that is probed with a pulse of temporal length τ (or a detector with temporal response of this value) is thus $c\tau/2$. Let us consider the amount of light scattered back to the lidar system from one such length element. If we express the backscattering coefficient for the atmosphere in this element as $\beta(\lambda_L, \lambda, R)$ per centimetre per steradian, the total amount of light scattered by this element is then $4\pi\beta(\lambda_L, \lambda, R)c\tau/2$. The value of β for weak Mie scattering is quite modest, at about $10^{-8}\mathrm{cm}^{-1}\mathrm{steradian}^{-1}$. If we assume that this light is scattered uniformly in all directions, then a fraction $A_0/4\pi R^2$ will be heading in a direction that will be collected by the telescope mirror of effective area A_0. This return signal has to propagate back through the same atmosphere as the incoming pulse, so a second factor of $T_a(\lambda, R)$ is needed to account for attenuation on the return trip. Once the return signal reaches the lidar, a fraction $T_{\mathrm{sys}}(\lambda)$ will pass through the telescope and associated optics and filters.

So far we have assumed that the transmitting telescope and the receiving telescope are looking along exactly the same path. While the same telescope may sometimes be shared by the transmit and receive channels, separate telescopes are often used. At short ranges the two telescopes can be looking at significantly different parts of the atmosphere, which is accounted for by a function $O(R)$ which is the geometric overlap function. Such fall off in overlap at small ranges can be quite advantageous, as to some extent it can counterbalance the effects of the inverse square law, and allow the lidar system to work over a reduced dynamic range of return signals.

Factoring all these terms together, we obtain the following lidar equation, which gives the return signal energy as a function of range, initial pulse energy E_L, and the terms discussed above.

$$E(\lambda, R) = E_L T_{\text{SYS}}(\lambda) O(R) T_a(\lambda, R) T_a(\lambda, R) \frac{A_0}{R^2} \beta(\lambda_L, \lambda, R) \frac{c\tau}{2} \qquad (2)$$

This shows that the return signal rapidly gets weaker with increasing range, which accounts for the general shape of the curve shown in Figure 3. Consider how much light is scattered back to the lidar from a 100m length of atmosphere a distance of 1km from the lidar. If we make the optimistic assumptions of unity system transmission, atmospheric transmission, and overlap function, and the 10^{-8}cm^{-1}steradian^{-1} scattering coefficient mentioned above, we find that only about 10^{-12} of the emitted light is received by the system. One then has to consider what is the minimum detectable energy in order to calculate the energy of the initial laser pulse required. The receiver for most lidar systems is based on a photo-multiplier tube or a photodiode. The former has the advantage of significant gain, and so it is normally used if possible. Until recently this was limited to the 200–1000nm range, but recent advances in transferred electron photocathodes have extended this to 1700nm. The minimum detectable return signal is often limited by shot noise. Taking this into account we find the need for pulse energies in the multi-millijoule to joule range for pulses of a few nanoseconds. Clearly, lidar systems need relatively powerful sources. These will be discussed in detail in the following sections.

1.3   Atmospheric attenuation

Atmospheric absorption

The transmission of the atmosphere in the visible, UV, and IR regions is shown in low resolution in Figure 5. As expected, we see high transmission throughout the visible, but in many other regions the atmosphere is opaque due to absorption by ozone, water-vapour, and carbon dioxide. Any atmospheric lidar system must work in a region of relatively high atmospheric transmission. Thus there are "windows" in and near the visible, around 1.5μm, in parts of the 3–5μm range, and in the 8–13μm range. However, if these windows are looked at under higher resolution, one finds that there are many significant narrow absorption lines due to water and carbon dioxide. Thus relatively small changes in wavelength can result in quite significant changes in atmospheric absorption.

Another wavelength-dependent feature of the system is its capacity to damage eyesight. In this regard, wavelengths around 1.5μm are generally the least dangerous, while those in the visible and out to 1.1μm are the most dangerous. More details of the maximum permissible exposure levels as a function of wavelength can be found in ANSI (1993).

Atmospheric scattering

There are two major regimes of scattering, which depend on the size of the particles involved (Measures 1992). Rayleigh scattering is the term given to scattering off the non-uniformly distributed molecules in the atmosphere. The backscatter coefficient for this process is proportional to the molecular number density, and inversely proportional to the

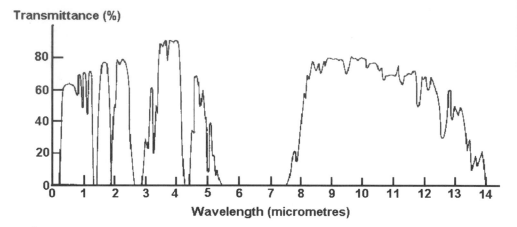

Figure 5. *A low resolution plot of the transmission of a 1.8km horizontal path at sea level containing 17mm of precipitable water. Adapted by permission of John Wiley and Sons Inc. from Measures R M, 1992, Laser Remote Sensing: Fundamentals and Applications.*

fourth power of the wavelength involved. It is this strong wavelength dependence which gives the clear sky its predominantly blue colour. At sea level for a standard atmosphere the coefficient is

$$\beta(\lambda) = 1.39 \left[\frac{550}{\lambda(\text{nm})}\right]^4 \times 10^{-8} \text{cm}^{-1}\text{sr}^{-1} \tag{3}$$

Scattering by macroscopic particles is called Mie scattering. For particle sizes much less than that of the wavelength the same λ^{-4} dependence is observed. The scattering behaviour becomes more complicated when the size of the particle becomes comparable to the wavelength of the light, and then for larger particles the scattering is independent of wavelength, with the amount of scattering being proportional to the number density of the particles and to their size. This is the reason why diffuse clouds look white; light of all incident wavelengths is scattered almost equally.

2   Laser sources for lidar

From the discussion presented in Section 1, it can be seen that the wavelength, pulse energy, and pulse duration of the laser source are of primary importance. Additionally, the system designer must take into account the size, reliability, ease-of-use, efficiency, and cost of the laser. These requirements limit most lidars to using solid-state or CO_2 gas lasers. The recent advances in diode-laser pumped solid-state lasers (Hanna, this volume) are also starting to be incorporated into lidar systems.

In this section we shall first address the most well-developed family of lasers for lidar, the neodymium lasers. (The use of nonlinear optics to shift the frequency of these and other lasers will be considered in Section 3.) We shall then discuss tunable lasers operating in the near IR, mid-infrared lasers, and finally UV lasers.

2.1 Neodymium-doped lasers

For many years, flashlamp-pumped Nd lasers have been the workhorses of the solid-state laser field. They can also be well suited to lidar, as in a Q-switched mode they can provide output pulses of appropriate energy (up to Joules) and duration (5–20ns). They are not significantly tunable, however, so are not directly applicable to DIAL or fluorescence lidar. Their 1064nm wavelength is suitable for scattering off aerosols, but is not in the eye-safe region. However, these disadvantages can be overcome by using the Nd laser as a driver for a variety of frequency conversion schemes using nonlinear optics and/or other laser media.

Lasers used in a lidar system must be particularly robust, and so the typical designs are very different from the average Nd laser used in a scientific laboratory. They must be turnkey systems, with no need for maintenance over long periods. However, a diffraction-limited output is not always required. An example of one such engineered system is shown in Figure 6. This shows an oscillator followed by an amplifier, with two Nd:YLF rods being pumped by one flashlamp. The use of YLF as a host material is advantageous as this material exhibits very low thermal lensing, so that the output energy varies very little with repetition rate between 10 and 30Hz. The mirrors are all glued in place on solid mounts, with angular alignment being made by rotating the Risley wedges in the cavity. The two prisms folding the path allow the entire system to be built into a metal case the size of a shoebox. This laser operates with an M^2 value of about 15, and produces 700mJ pulses at up to 30Hz when the lamp energy per pulse is 70J. In Section 4 we shall describe the use of this laser as the driver of a 1.5μm OPO for lidar applications.

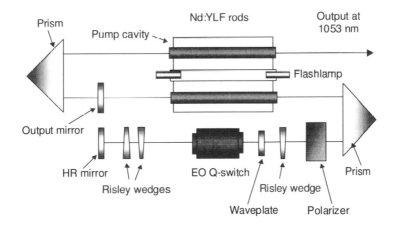

Figure 6. *A rugged flashlamp-pumped Nd:YLF laser suitable for lidar.*

Replacing flashlamps by pulsed diode lasers can lead to significant increases in efficiency, but currently also results in a major increase in cost. Side-pumping neodymium-doped rods with diode lasers tends not to produce such high gain as with the use of flashlamps. Fibertek have taken account of these issues to produce an oscillator-amplifier system that produces 1.2J pulses at 100Hz, with an M^2 of about 5. A similar oscillator design is used to that shown in Figure 6 but two rods are used to get sufficient gain. A two-rod pre-amplifier and a two-rod power amplifier are then used to amplify the laser pulses to the Joule level. This takes a total of 640 diode bars at 50W per bar. The cost for the diode lasers alone was approximately half a million dollars. However, this did produce an energetic and refined laser, and this cost may be justified for satellite installations, for example.

For some applications a much higher repetition rate is required, but with a lower pulse energy. Our company's solution to this design goal is to use cw laser diodes side-pumping a Nd:YLF crystal, in which the laser mode makes several passes through the "gain sheet" produced in the crystal. This allows a TEM_{00} beam to extract a large fraction of the stored energy, while the low excitation density minimises stress and beam distortion, and reduces the parasitic processes that act to deplete the population inversion before Q-switching. This geometry also allows efficient extraction of waste heat out of the top and bottom of the slab. The module shown in Figure 7 has a single pass gain of 4. When two of these are used in a Q-switched oscillator, followed by two further modules as amplifiers, 14ns pulses are generated at 5kHz with an average power of 40W. This indicates a peak power in the TEM_{00} mode of some 0.6MW. The intensities available with this laser have been shown to produce very efficient frequency upconversion through nonlinear optics.

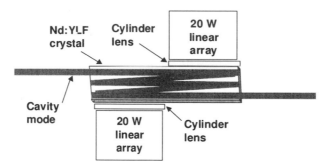

Figure 7. *A side-pumped Nd:YLF slab used as a gain module for a high repetition-rate laser system.*

2.2   Tunable near-IR lasers

DIAL systems also require short, high energy pulses, but additionally need to be tuned on and off an atomic or molecular absorption feature. The major features of five possible tunable laser systems are shown in Table 1, though to-date only Ti:sapphire and Cr:LiSAF have been widely used in Lidar. In this subsection the near infrared Ti:sapphire laser will be discussed. The Mid-IR and UV lasers Cr:ZnSe and Ce:LiSAF will be discussed in the following subsections.

The Ti:sapphire laser was first demonstrated by Moulton (1986). It has a broad tuning range in a region of considerable technological importance. This range includes wavelengths of interest in remote sensing. When the laser is combined with nonlinear optics this frequency tunability can be extended to other parts of the spectrum. Ti:sapphire can be grown in large crystal sizes and has excellent thermo-mechanical properties, allowing laser operation at high average powers.

The broad bandwidth of the laser transition leads to a relatively low gain, so the

Parameter	Ti:sapphire	Cr:LiSAF	Cr:YAG	Cr:ZnSe	Ce:LiSAF
Centre wavelength (nm)	800	850	1430	2400	292
Tuning range (nm)	680–1100	780–1050	1340–1570	2300–2500	280–297
Effective gain cross section $(10^{-19}\text{ cm}^{-2})$	2.5–3	0.32	3	8	60
Lifetime (μs)	3.2	67	4	7	0.028
Maximum efficiency	1.0	0.53	0.25	1.0	0.35–0.7
Thermal conductivity $(\text{Wm}^{-1}\text{K}^{-1})$	28	3.1(c)	13	18	3.1(c)
Expansion coefficient (10^{-6})	6.7	-10(c) 19(a)	6.7	7	-10(c) 19(a)
Thermal shock parameter (Wm^{-1})	3400	84(a) 160(c)	1450	1380	84(a) 160(c)
dn/dT (10^{-6})	+14	-4(e) -2.5(o)	+7.3	+60	-4(e) -2.5 (o)
Thermal lensing behaviour	Moderate	Weak	Moderate	Strong	Weak

Table 1. *Selected properties of five tunable laser materials.*

material is usually excited by other lasers. The peak absorption is at about 500nm. In cw operation the cw argon-ion laser was widely used, but frequency-doubled diode-pumped Nd lasers are gaining in popularity due to their all-solid-state nature (Harrison *et al.* 1991). For the pulsed operation required for lidar, a frequency-doubled Q-switched Nd laser is frequently used. Such a pumping scheme results in gain-switched operation of the Ti:sapphire laser, typically producing nanosecond pulses.

We take as an example the design (Rines and Moulton 1991) that was the basis for the laser flown in the LASE instrument mentioned in Section 1.1. Ten nanosecond pulses of almost half a Joule in energy were produced in a near-diffraction limited beam at 10Hz. The laser geometry is shown in Figure 8. Two Ti:sapphire crystals were used to off-set the spatial dispersion of the Brewster surfaces, and two-ended pumping helped reduce the pump fluence on the laser crystals. A relatively large beam diameter with good transverse coherence was obtained using an unstable resonator based on a graded reflectivity mirror. Four Brewster-angled fused-silica prisms were used as the tuning elements. The laser operated with a conversion efficiency close to 40% across the 790 to 910nm range.

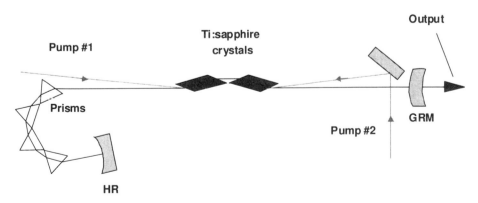

Figure 8. *A high brightness, laser-pumped Ti:sapphire laser developed at SEO (From Rines and Moulton, 1991)*

Despite the short upper-state lifetime of Ti:sapphire lasers, flashlamp-pumped operation can be achieved. Work at the Technical University of Berlin (Hoffstadt 1994) showed that a 0.1% doped rod of diameter 10mm and length 152mm could produce up to 2.5J of output for 200J into the flashlamps. When run at about 100Hz repetition rate, an average output power in excess of 200W was obtained, with an efficiency of 1.9%. A laser based on this design is the driver for the Elight DIAL laser described in Section 1.1.

2.3   Mid infra-red lasers

The 1064nm Nd laser is close to being an ideal system, except for the issue of eye safety. The intense pulses needed for lidar can be a major hazard for personnel in the vicinity of the transmitter. This is one reason for the push to develop lasers in the more eye-safe region beyond 1.4μm. The two most attractive established materials for lidar are thulium at 2.0μm and holmium at 2.1μm.

	Ho:YAG	Tm:YAG	Cr^{2+}:ZnSe	Nd:YAG
Wavelength (nm)	2100	2010	2350	1064
Stimulated emission cross section (cm^2)	1.5×10^{-20}	2×10^{-21}	8×10^{-19}	3×10^{-19}
Saturation Fluence (J/cm^{-2})	6	50	0.11	0.6

Table 2. *Selected laser parameters of Ho:YAG and Tm:YAG compared with Nd:YAG and the new material Cr^{2+}:ZnSe.*

Table 2 gives data which illustrate the challenges of design of these lasers relative to Nd:YAG. The stimulated emission cross sections are more than an order of magnitude down on that of Nd:YAG, which will lead to relatively low gain. Perhaps even more of a problem is the high saturation fluence of these lasers. For efficient extraction of stored energy a laser will be operating with an intracavity energy that is of the order of the saturation fluence. This means that the components in the 2μm laser or amplifier will experience fluences much greater than those in a Nd laser. This problem is compounded by the thicker layers that are necessarily used on the multi-layer mirror coatings. Furthermore, these laser transitions terminate on the ground state of the ion, resulting in a "$3\frac{1}{2}$ level" laser system. Significant pump energy must be absorbed just to bring the laser crystal to transparency. Thus to minimise the volumetric threshold pump energy, a relatively low ion doping is needed. However, the pump transitions are relatively weakly absorbing, so this demands relatively high doping levels. These incompatible demands mean that lamp-pumping is particularly difficult.

One way that laser designers have circumvented this problem with holmium lasers is by using diode pumping and a sensitizer scheme. Both holmium and thulium are doped into the same rod. A thulium ion absorbs one pump photon at 785nm. This excited ion interacts with another thulium ion in the ground state. This results in two Tm ions excited to a lesser degree. This excitation diffuses through the crystal by ion-ion interaction, until an excited Tm ion interacts with an Ho ion. The excitation at the Ho site then contributes to the population inversion that drives the laser. Thus one pump photon can actually provide two excited Ho ions, reducing the otherwise deleterious effects of a pump photon of very much greater energy than the laser photon. However, this form of excitation brings with it its own problems. There is thermal equilibrium between the excited Ho and Tm ions. The energy transfer process between these two species has a finite timescale, and so not all of the expected excitation density is available when the Ho laser is Q-switched. There are also other undesirable energy transfer mechanisms at work.

The thulium laser can work with diode pumping at 681nm or 785nm. A similar advantageous 2-for-1 energy transfer mechanism takes place, but the lower gains of the thulium laser bring with them their own challenges.

Table 2 also gives relevant parameters for the new Cr^{2+}:ZnSe laser, which has been developed at Lawrence Livermore National Laboratories. This is an extremely promising new material, which is discussed in more detail by Krupke (this volume).

2.4   UV lasers

The detection of ozone, SO_2 and biological agents requires lasers operating in the region of 300nm. Such wavelengths can be obtained using nonlinear frequency conversion, as will be reviewed in Section 3. Excimer lasers also access this wavelength range, but the toxic gases and high voltages involved with these systems do not make them particularly attractive. This has led to a search for tunable lasers operating directly in the UV. However, operation at these wavelengths is difficult. The Einstein relations show that obtaining high gain is more difficult as lasers move to shorter wavelengths. The high photon energies can also lead to problems with multi-step absorptions promoting electrons from the valence to conduction bands. This in turn leads to the possibility of colour-centre formation and dielectric breakdown.

Paramagnetic ion lasers based on the 4f–4f transitions (rare earths) or 3d–3d transitions (transition metals) have relatively small splittings between levels, leading to laser operation in the near-IR and visible. However, electronic transitions between the 5d and 4f levels in the rare earths are in the UV range. These transitions are parity-allowed, and so have short (nanosecond) lifetimes and high gains. They are spectrally broad, due to a large lattice shift between the 5d and 4f configurations, and the same shift is responsible for giving them a four-level nature.

Ce:LiCAF and Ce:LiSAF have been identified as two promising materials tunable in the 280–315nm range. They are pumped by a frequency-quadrupled neodymium laser at 266nm. These materials are resistant to degradation by the UV pump light, and their broad bandwidth allows tunability to particular atomic or molecular resonances. In one example, Sarukara *et al.* (1997) used a 15mJ UV pulse to pump a Ce:LiCAF master oscillator to obtain a 1mJ 1ns pulse at 289nm. This pulse was then amplified in a double-pass Ce:LiCAF amplifier powered by a 46mJ UV pump pulse. The input signal was amplified to 14mJ. More recent work at Lambda Physik (Govorkov *et al.* 1998) has created a prism-tuned oscillator-amplifier system providing a tunable output over the 280–316nm range with a linewidth of 0.3–0.5nm and an average power of 1.9W at a 1kHz repetition rate.

3   Nonlinear optics for lidar

It remains true that there are relatively few "good" pulsed laser types. The discussion in Section 2 has highlighted the Nd 1064nm laser and the Ti:sapphire laser as particularly useful systems for providing high peak-power pulses in a reliable and efficient manner. Although there are currently important developments in mid-IR and UV lasers, for many applications a better choice is to use nonlinear optics to shift the frequency of a Nd or Ti:sapphire laser into the desired range. The well-established techniques of second and third harmonic generation allow systems to access shorter wavelengths. These shorter wavelengths, or the fundamental wavelength itself, can be used as the pump sources for optical parametric oscillators (OPO). The OPO can then provide tunability at a wavelength longer than that of its pump (Dunn and Ebrahimzadeh, this volume).

3.1   Harmonic generation theory

The concepts of parametric conversion are covered by Dunn and Ebrahimzadeh (this volume), Yariv (1975), and Eimerl (1987). For a fundamental laser beam that is uniform in space and time we can describe the second harmonic conversion efficiency η by

$$\eta = \tanh^2 \left[\frac{1}{2} \tanh^{-1} (\mathrm{sn}\, [2\eta_0^{1/2},\ 1 + \delta^2/4\eta_0]) \right] \tag{4}$$

where $\mathrm{sn}(a, b)$ is the Jacobi elliptic function, $\delta = \Delta k L/2$ is the dephasing caused by the phase mismatch Δk in a single pass of the crystal of length L, and η_0 is the "drive":

$$\eta_0 = C^2 L^2 I \tag{5}$$

In this case d_{eff} is the effective nonlinear coefficient in pm/V, λ_f is the wavelength of the fundamental in μm, n_f and n_{sh} are the refractive indices of the crystal at the fundamental and second harmonic respectively, L is the crystal length in cm, I is the intensity of the fundamental in GWcm^{-2}, and C (in units of GW$^{-1/2}$) is given by

$$C = \frac{5.46 d_{\mathrm{eff}}}{\lambda_f (n_f n_{\mathrm{sh}} n_{\mathrm{sh}})^{1/2}} \tag{6}$$

For small conversion efficiencies and dephasing, the Jacobi function approximates to

$$\eta = \eta_0 \left(\frac{\sin \delta}{\delta} \right)^2 \tag{7}$$

but this regime is not desirable. Instead, the system should be designed to give significant conversion efficiency, and to be perfectly phase-matched ($\delta = 0$), giving a conversion efficiency of

$$\eta = \tanh^2(\eta_0^{1/2}) \tag{8}$$

In reality there will never be perfect phase matching, due to the beam range of wavelengths $\Delta\lambda$, the beam divergence $\Delta\theta$, and a non-uniform crystal temperature T that varies by an amount ΔT. When the parameters are chosen for the best phasematching, the phase mismatch in the crystal can then be described by

$$\Delta k \approx \beta_\theta \Delta\theta + \beta_T \Delta T + \beta_\lambda \Delta\lambda \tag{9}$$

where

$$\beta_\theta = \frac{\partial \Delta k}{\partial \theta}, \qquad \beta_T = \frac{\partial \Delta k}{\partial T}, \qquad \beta_\lambda = \frac{\partial \Delta k}{\partial \lambda}. \tag{10}$$

In the case of non-critical phase-matching, the first derivative may vanish, and the small second-order terms would then need to be taken into account. In order to determine the various values of β, one requires accurate information about the crystal in terms of its Sellmeier equations and thermo-optic coefficients (dn/dT). Data on dn/dT may not always be available.

If the crystal is being used in a critically phase-matched geometry with a diffraction-limited Gaussian beam, the range of angles $\Delta\theta$ in the beam is

$$\Delta\theta \approx \frac{0.4\lambda}{\pi w_0} \tag{11}$$

where w_0 is the beam waist. Thus the dephasing caused by the beam divergence alone is

$$\delta \approx \left[\frac{0.2\beta_\theta \lambda}{\pi}\right] \frac{L}{w_0} \tag{12}$$

If the laser is operating in multi-transverse mode, this value of dephasing should be multiplied by the M^2 value of the beam.

Eimerl (1987) has calculated the conversion efficiency under a range of different dephasing and drive conditions, as shown in Figure 9a. The contours are of conversion efficiency. For near phase-matched conditions an 80% conversion efficiency is achievable with the drive parameter at the relatively modest value of 1.4. It can be seen that an increase in the drive will result in back-conversion and lowering of the efficiency of the second harmonic generation process. At any single low value of the drive parameter, higher values of dephasing result in a lower conversion efficiency. This graph also shows that the conversion efficiency using the $\delta \cong 0$ peak becomes increasingly sensitive to dephasing at higher drives.

The contours in Figure 9a were calculated for plane-wave "top-hat" circular beams, which also had a "top-hat" intensity profile with respect to time. Real laser pulses have a spatial variation close to Gaussian, and often have a temporal variation that is near-gaussian. This means that if a pulse energy and crystal length are chosen to give near-unity conversion efficiency for the peak of the pulse, the conversion in the wings will be much less than one. If, however, the intensity is increased to allow the conversion efficiency in the wings to approach unity, the peak of the pulse may have experienced substantial back-conversion. Cousins (1993) has taken account of these effects and produced a more realistic set of efficiency contours by integrating over time and space results similar to those produced by Eimerl. Now, for second harmonic generation close to being phase matched, 80% conversion efficiency is not achieved until a drive parameter of 12 is used, as shown in Figure 9b. One can see some similarity between the two contour graphs,

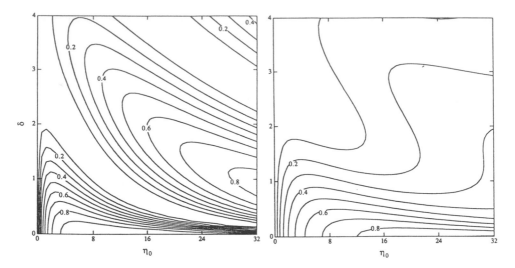

Figure 9. *Contours of efficiency of second harmonic generation as functions of the drive parameter η_0 and the dephasing δ. Part (a) is for plane waves with a uniform intensity in time and space, and is taken from Eimerl (1987). Part (b) is for a laser pulse with Gaussian distribution in time and space, and is taken from Cousins (1993).*

but the curves that have been obtained from the more realistic space and time variations show that it is more difficult to obtain high conversion efficiencies in this regime than in the idealised case.

This discussion shows that desirable properties for nonlinear crystals include (i) a large effective nonlinear coefficient d_{eff}, as the drive varies with d_{eff}^2, (ii) small values of β, indicating a need for non-critical phase matching (this may be even more important than high d_{eff}), (iii) large crystal lengths and wide apertures, (iv) low absorption at all the wavelengths involved (including low multi-photon absorption), and (v) high damage thresholds. The first two points are sometimes taken together into a figure of merit known (confusingly) as the "threshold" parameter. This parameter takes into account the d_{eff} and the angular acceptance bandwidth. It is commonly expressed in units of GW, with lower values being preferable.

Although this section has centred on second harmonic generation, similar concepts are true in optical parametric oscillators.

3.2   Nonlinear materials

Amongst the first commercially successful nonlinear crystals were KDP (potassium dihydrogen phosphate), and ADP (ammonium dihydrogen phosphate). These could be grown in large sizes, but had relatively low effective nonlinear coefficients, were hygroscopic, and had small acceptance angles. Lithium niobate and lithium iodate have also been used for a long time and, while they have relatively high nonlinear coefficients, they suffer from low damage thresholds.

In the past two decades a number of new nonlinear materials have been developed, which overcome some of these difficulties. KTP (potassium titanyl phosphate) is particularly widely used for the second harmonic generation of the 1064nm Nd laser. It has a high nonlinearity ($d_{eff} = 3.18$ pmV^{-1} compared with KDP at 0.37 pmV^{-1}), a large acceptance angle for this process, and has recently become available in relatively large sizes (10 x 10 x 20mm). KTP and its more recently developed isomorphs KTA (potassium titanyl arsenate) and RTA (rubidium titanyl arsenate) are also being exploited in optical parametric oscillators. The relatively high damage threshold of the material is often important. KTP has a damage threshold for 1064nm 10ns pulses of 1.5GWcm^{-2}, whereas LiNbO$_3$ has a damage threshold of only $\sim$0.1GWcm^{-2}). Indeed, it is this increase in damage thresholds of new nonlinear materials that has played a major part in the successful development of optical parametric oscillators. Some relevant data for the crystals are given in Table 3. Related information is also presented in Table 6 of the chapter in this volume by Krupke.

	KTP	BBO	LBO	CLBO
Nonlinear coefficients d (pm/V)	$d_{24}= 3.9$ $d_{15} = 2.04$	$d_{21} = $ -2.3	$d_{31} = 0.85$ $d_{32} = 0.67$	$d_{36} = 0.8$
Refractive indices	$n_x = 1.74$ $n_y = 1.75$ $n_z = 1.83$	$n_o = 1.66$ $n_e = 1.54$	$n_x = 1.57$ $n_y = 1.59$ $n_z = 1.61$	$n_o = 1.5$ $n_e = 1.45$
Transparency range (nm)	350–3500	189–3500	160–2600	180–2750
d2/n3	2.7	1.3	0.18	0.19
Thermal conductivity (Wcm^{-1}K^{-1}	13	1.6 (c) 1.2 (a)	3.5 (av)	
Thermal expansion coefficient (10^{-6})		4(a) 36 (c)	10.8 (x) -8.8 (y) 3.4 (z)	
dn/dT (10^{-6})	11 (x) 13 (y) 16 (z)	-9.3 (o) -16.6 (e)	-1.5 (x) -13.4 (y) -8.2 (z)	
Damage threshold (GWcm^{-2}) (1064 nm, 10 ns pulses)	1.5	5	10	

Table 3. *Selected properties of four important new nonlinear crystals.*

A new family of nonlinear crystals based on borates has been developed over the last fifteen years or so. β-barium borate (BBO) was the first of these new crystals, and excelled due to its excellent UV transparency, high nonlinear coefficient, and a large birefringence that allowed a wide range of phasematching. Lithium borate (LBO) has a particularly high damage threshold and low absorption, and can be non-critically phase-matched for generating the second harmonic of 1064nm Nd lasers. It has an angular acceptance bandwidth β_θ that is often significantly higher than that of BBO. CLBO is the most recently developed of the three crystals and it has a lower two-photon absorption than BBO.

The relative merits of the borates for two different tasks will be evaluated. Table 4 compares the three materials for second harmonic generation of 921nm from a pulsed Ti:sapphire laser. The 460nm light produced in this process is of interest in a third harmonic scheme to reach 307nm for ozone detection. The low angular acceptance bandwidth

Crystal	LBO	LBO	BBO	CLBO
Type interaction	I	II	I	I
λ_1 (μm)	0.921	0.921	0.921	0.921
λ_2 (μm)	0.921	0.921	0.921	0.921
Phasematch angle (degrees)	68.68	48.6	25.54	31.3
Walkoff (mrad)	12.3	9.26	61.2	32.5
d_{eff} (pm/V)	0.77	0.55	2.1	0.49
Angular BW (mrad cm)	2.06	5.52	0.4	0.84
Temperature BW (pm cm)	6.87	5.92	25.5	14.6
Spectral BW (pm cm)	784	726	473	953
Crystal temperature (C)	20	20	25	20
n_1	1.608225	1.567121	1.65742	1.487707
n_2	1.608225	1.599775	1.65742	1.487707
n_3	1.608225	1.583448	1.65742	1.487707
β_0 (cm^{-1}/ mrad)	2.702	1.008	13.916	6.626
C (GW$^{-1/2}$)	2.237	1.635	5.830	1.600
P_{th} (GW)	0.012	0.003	0.048	0.146

Table 4. *Comparison of borate crystals generating the second harmonic of 921nm.*

of BBO (and to a lesser extent CLBO) is a significant drawback. This value means that with a 10mm long crystal a beam divergence of only 0.4mrad would significantly lower the conversion efficiency. The lower d_{eff} of LBO is less of a problem, resulting in a rather better figure of merit (theshold power) for LBO than BBO. BBO also has a rather high walkoff angle, which would also act to lower the conversion efficiency.

To reach the UV at 262nm for, say, the pumping of Ce-doped lasers, the second harmonic of a Nd:YLF laser can itself be frequency doubled. In this case both BBO and CLBO could be used. Here CLBO has an effective nonlinear coefficient that is half that of BBO, but an angular acceptance bandwidth that is three times that of BBO. This results in CLBO having the lower threshold, but the defining issue is probably one of two-photon absorption. This is significantly less in CLBO than BBO, and so the CLBO is definitely preferable to BBO at high average powers. The problems of multiphoton absorption in the UV are similar to those outlined earlier for laser materials, and again the values in different materials are not well known.

To be of any use, the crystal must be able to be phase-matched for the required nonlinear process. Figure 10 shows the phase matching angle for type I second harmonic generation from the visible and near IR for three nonlinear crystals. Although BBO and LBO appear to show desirable non-critical phase matching at the lowest end of their wavelength ranges, this is not useful as at these angles the effective nonlinear coefficient goes to zero.

In the range of the Ti:sapphire laser both BBO and LBO are worthy of consideration. Table 4 has shown that LBO would seem to be the preferred choice, but a longer crystal will be needed to reach the maximum conversion efficiency than would be the case in BBO. Our company has evaluated these two crystals for the second harmonic generation of 911nm radiation with nanosecond pulses from a gain-switched Ti:sapphire laser. At a drive parameter of 2 both LBO and BBO reached 50% conversion efficiency. We were

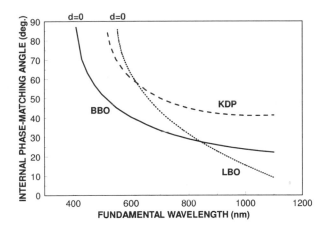

Figure 10. *Phase matching angles for second harmonic generation for type I processes.*

able to drive BBO to a parameter of 8, when we obtained 65% conversion efficiency, corresponding to a blue-pulse energy of a quarter of a Joule. This is consistent with the theory presented earlier.

The final family of nonlinear crystals to mention is that of periodically poled materials. Myers (this volume) reviews these materials. They have yet to demonstrate their ability to withstand the high energies and average powers encountered in lidar systems, but are expected to become important in the future. This section has concentrated on harmonic generation. Similar ideas hold for optical parametric oscillators, the science of which is covered in detail in the chapter by Dunn and Ebrahimzadeh.

4 Design choices

The previous sections have reviewed lidar techniques, laser sources, and nonlinear optics. This section will show how such knowledge allows the designer to make informed choices among different possible systems, and will detail some of the systems built at Schwartz Electro-Optics (now Q-Peak Inc).

4.1 Eye-safe aerosol Lidar

Although Q-switched Nd:YAG lasers are well suited for the measurement of aerosol clouds, eye-safety is a major concern, especially for ground-based or aircraft-based lidar systems. A similar source, but with a wavelength greater than $1.4\mu m$ is needed to decrease the hazard while maintaining the functionality.

Possibilities include Raman shifting a Nd laser using a methane cell at 1500psi, but such schemes are unattractive due to breakdown of the gas depositing carbon on the windows, and the explosive hazards associated with methane. Erbium, thulium, or holmium lasers could generate appropriate wavelengths directly. Scientists at NASA (Yu *et al.* 1998), for example, have demonstrated a diode-pumped Ho:Tm:YLF laser producing up to 125mJ per pulse at 6Hz. The Ho laser is side-pumped by ten GaAlAs diode arrays, with each array providing a peak power of 360W near 792nm. Injection seeding of the ring oscillator by a Ho:Tm:YLF microchip-laser provided unidirectional and single-frequency operation. This laser is intended to fly as the "SPARCLE" experiment on a future space shuttle mission. It will undertake remote sensing of wind velocities using coherent lidar.

An alternative to the laser system described above is an optical parametric oscillator using a Nd laser as the pump source. Disadvantages of the SPARCLE system include an optical-to-optical conversion efficiency of only 3%, and a relatively complicated cavity. The OPO option, which will be described below, benefits from a relatively simple and efficient "bolt-on" component to a well-established neodymium laser.

The nonlinear crystals KTA and KTP have the attractive property of being non-critically phase matched for the parametric generation of ~1500nm light from a 1064nm pump. This gives a large angular acceptance bandwidth of ~65mrad-cm. As well as allowing a signficant beam divergence, this large figure means that the alignment of the beam through the crystal is not particularly sensitive, and that a beam with a large value of M^2 should still be able to be used effectively. In 1993 we explored the use of a KTP OPO to shift the wavelength of our 1064nm, 1.1J, 10Hz compact Nd:YAG laser to the eye-safe region. A simple OPO cavity was constructed with a 20mm long KTP crystal surrounded by a pair of mirrors which double-passed the pump through the crystal. For a pump laser pulse energy of 1.1J we obtained a 41% conversion efficiency to generate in one direction pulses of 450mJ at 1550nm with an M^2 of about 40. The double-pass of the pump was advantageous in achieving high efficiency, but the OPO had to be well separated from the pump to stop the reflected pump light interfering with the operation of the Nd:YAG laser system. In order to produce a compact laser we engineered a ring OPO, as shown in Figure 11a. The lack of double-passing the pump through the nonlinear

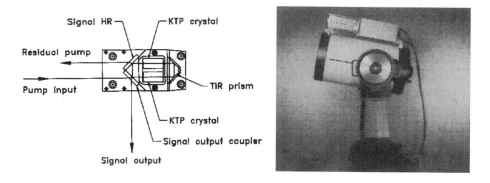

Figure 11. *(a) Schematic of the KTP OPO and (b) the compact laser head with OPO mounted on a 16 inch telescope.*

crystal was partially compensated by using four crystals to provide a gain-length of 40mm. The conversion efficiency was reduced to 34%, but with the lack of pump feedback we were able to construct a very compact system that showed no signs of damage. The system is particularly robust, with all the components dropped into place and glued to the solid surround. The OPO acts almost as a passive converter of the 1064nm radiation to the eye-safe region. Figure 11b shows the compact laser head and OPO mounted on top of a 16 inch Mead telescope for lidar measurements.

Another aerosol lidar required a much higher average power than the one described above. In this case the 100Hz repetition rate diode-pumped Nd:YAG laser produced by Fibertek was used as the driver. With 120W average power of pump light, absorption of the pump, signal, and idler radiation may be an issue. Any such absorption would create a thermal lens. When the 1064nm pumped OPO has a signal wavelength at 1530nm, the idler is near 3300nm. A 20mm thick slice of KTP transmits only 23% of incident 3297 radiation, while the otherwise similar KTA crystal has a transmission of 75%. KTA and KTP have essentially the same nonlinear coefficients. Apart from the replacement of KTP by KTA, the high power OPO was similar to that shown in Figure 11, with the exception of a longer cavity to provide a lower divergence beam from the OPO. With an average pump power of 110W, this system generated 33W average power at 1530nm (Webb *et al.* 1998). This is believed to be the highest average power OPO yet produced. A system similar to this has now been engineered into a helicopter-based lidar system for the detection of biological agents.

4.2   UV laser for ozone DIAL

Tunable and energetic UV sources are needed for DIAL sensing of ozone (290–350nm) and the excitation of biologically active species at $\sim$ 290nm. Wavelengths in this region are also useful for DIAL sensing of mercury and a variety of molecules such as SO_2.

Ozone has a broad unstructured absorption peak at 290nm, with an absorption cross section that falls by five orders of magnitude to a minimum at around 370nm. Thus the DIAL system need not have a narrow linewidth, but it does need significant tunability to tune on and off the absorption peak. The magnitude of tuning is in contrast to that required for sensing SO_2, where a wavelength difference of only $\sim$2nm is required.

Older solutions to this design challenge have included the use of excimer lasers with stimulated Raman scattering (SRS) to access a number of lines, quadrupled Nd:YAG lasers with SRS also to give line tunability, and frequency doubled dye lasers to give continuous tunability. However, for an aircraft platform, an all-solid-state solution is preferred. Such a source could be based on one of three technologies: (i) Ce-doped lasers pumped by a frequency quadrupled Nd:YAG laser; (ii) the frequency-doubling of a BBO OPO that is pumped by a frequency-tripled Nd:YAG laser; and (iii) frequency-tripled Ti:sapphire lasers pumped with a frequency-doubled Nd:YAG laser.

The OPO and Ti:sapphire approaches have the advantage over the Ce-doped lasers of a more mature technology, and a wider tuning range. The relative merits of the OPO and the Ti:sapphire laser will now be presented.

- The OPO has an advantage in its broad tunability, with frequency doubling of the OPO giving a tuning range of 225–340nm. Only second harmonic generation of the tunable source is needed, compared with the third harmonic required for the Ti:sapphire laser.

- However, because of the phase matching requirements outlined in Section 3, for efficient operation the OPO needs a diffraction limited pump beam from the frequency-tripled Nd:YAG laser. While such sources exist for the laboratory, there are no powerful sources that are sufficiently robust and reliable to be considered for space-borne applications. The Ti:sapphire laser, on the other hand, can readily be designed to work efficiently with a multimode pump source.

- The broad tuning of the OPO is a result of a rapid change in phase-matched wavelength with angle of propagation through the crystal. This means that any small movement of the angle of propagation of the pump beam through the crystal will result in a change of the OPO wavelength. Thus the aligment stability of the OPO and the UV pump are both critical. This is not the case for the Ti:sapphire laser, where a large shift in pump-beam position will merely reduce the output power slightly.

- The OPO can be line-narrowed using a grating, but micro-radian control would be needed. There are no suitable sources for injection seeding in the visible wavelengths at which the OPO would operate. The Ti:sapphire laser, on the other hand, can be tuned in the manner discussed in Section 2.2, and can be injection seeded by a single-mode diode-laser.

- While it may be possible to scale the OPO to the relatively high energies needed, the Ti:sapphire laser has already been proven to work well at the 400mJ level.

- The Ti:sapphire laser has the disadvantage of prior difficulties in efficient frequency tripling.

The comparisons above led our company to develop a source based on a Ti:sapphire laser, primarily on the basis that this technology was the more practicable to engineer into a space-qualified system.

The fundamental source is a flashlamp-pumped Nd:YLF oscillator-amplifier that undergoes two further stages of amplification in flashlamp-pumped rods. This light is doubled into the green to pump a gain-switched Ti:sapphire laser of the type discussed in Section 2.2. One of the challenges remains the efficient frequency tripling of the Ti:sapphire laser. This is currently achieved by single pass frequency doubling the laser to blue wavelength regions in the manner described in Section 3.2. The blue light is then sum-frequency mixed with the residual Ti:sapphire radiation. This can be achieved in LBO, BBO, or CLBO. Our calculations show that CLBO is the lowest threshold material but problems with the hygroscopic nature and mechanical fragility of CLBO remain to be solved. LBO has a higher threshold power but does not suffer from the practical problems of CLBO. BBO has the highest threshold of the three but has the advantage that one crystal can be used over a wide range of wavelengths. We chose to use a 20mm long LBO crystal and with this obtained record levels (>35%) of third harmonic conversion efficiency (Dergachev et al. 1999), producing up to 40mJ of light tunable from 254 to 315nm. This is a higher energy than any other nanosecond narrow-linewidth UV source, and has exceeded the energies reported for OPO based systems.

4.3   Broadly tunable Mid-IR for DIAL

A source covering the atmospheric windows in the 2 to $10\mu m$ region allows the sensing of a wide range of molecules. There are many fundamental and overtone transitions in this range. The ideal source would be able to access all of the mid-IR atmospheric windows, with a narrow enough linewidth to resolve the individual lines of molecules of low mass (< 1GHz). The first design choice is amongst (i) a carbon dioxide laser, which with an isotopic mix could access many lines in this region, (ii) a Raman-shifted tunable laser, which could give wide coverage, but would need high peak powers to obtain a reasonably efficient Raman conversion efficiency, and (iii) an infrared OPO, which may allow broad tunability, but which may suffer optical damage at the intensities involved. The desire for the compact and robust operation of solid-state lasers leads to the OPO being chosen.

The next choice is in the materials to use for the mid-IR OPO. The discussion in Section 3 confined itself to the classic dielectric materials such as the KTP family and the borates. However, these materials tend to have high phonon-frequencies, and so their transmission is poor at wavelengths above 2–5μm, depending on the material. Semiconductors such as CdSe and AgGaSe generally have lower phonon-frequencies, and so their IR transmission cut-offs tend to be up in the 10–20μm region. Cadmium selenide, for example, is transparent up to 25μm. These semiconductors have effective nonlinear coefficients that are comparable with those of KTP for operation at shorter wavelengths. However, the bandgaps of the semiconductor materials are in the near-IR, which means that they would suffer strong one- or two-photon absorption if pumped by a neodymium laser. Apart from the reduced efficiency that this would entail, the free carriers so induced would be liable to cause material damage. Thus these materials need a pump source in the 2μm range. The birefringence of the crystals allows phase matching of the OPO wavelengths with this type of pump.

Given that CdSe can be used to generate 3–10μm with a 2μm pump, the choice is now one of the pump laser. Although some work has been carried out elsewhere using thulium or holmium lasers to power mid-IR OPOs, our company has chosen to base a system on the more established Nd:YAG laser, shifted in frequency using a KTA OPO. This has been called a "tandem OPO" (Isyanova *et al.* 1996).

Section 4.1 showed how a KTA OPO could be used as a "bolt-on" attachment to a Q-switched Nd:YAG laser head. For this application we developed a flashlamp-pumped Nd:YLF oscillator and amplifier that produced up to 200mJ on a single longitudinal mode through injection seeding. These pulses were 16ns long. The KTA OPO was angle tuned to obtain signal output tunable in the 1.5 – 2μm range. As the signal was being produced over these wavelengths, the idler tuned in the 3.5–2μm range, as shown in Figure 12. The oscillator produced up to 90mJ of energy split between the signal and idler. The KTA OPO could in principle be injection seeded if a particular single-frequency was required.

The CdSe OPO was arranged for non-critical phase matching to allow best operation. The tunability of the non-critically phase-matched CdSe OPO comes from the tuning of its pump source, the KTA OPO. Figure 12 shows the CdSe OPO producing an idler tunable from 7.8 to 10.2μm, and a signal tunable from 3 to 5.2μm. One or other OPO is then able to access almost all of the wavelengths in the atmospheric windows at 3–5μm and 8–12μm. The desire for broadly tunable and robust pulsed sources across the mid-IR is close to being realised.

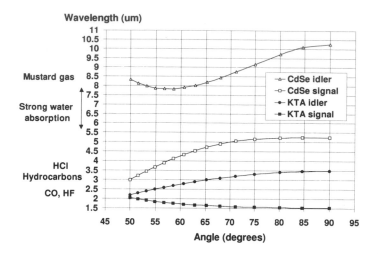

Figure 12. *Tuning curve for the KTA OPO as a function of angle, and the wavelengths that were produced by the CdSe OPO as its pump source (the KTA OPO) was tuned.*

5   Conclusions

Lidar provides a unique tool for making measurements of such diverse items as automobile traffic, the heights of clouds and global distributions of water vapour and ozone. Recent advances in the technology of solid state lasers and nonlinear optics have provided new and improved sources for lidar systems, promising improved performance and wider applications.

Acknowledgements

Much of the company work cited was done at the Research Division of SEO, now organised as Q-Peak, Inc. The author would like to recognise the particular efforts of Glen Rines, Richard Schwarz, James Harrison, Yelena Isyanova and Alex Dergachev related to the development of lidar sources. In addition, we would like to recognise the support of NASA and U.S.Air Force.

References

ANSI, 1993, American National Standard Institute document Z-136.1

Cousins A, 1993, "Power conversion efficiency in second harmonic generation with nonuniform beams", *IEEE J. Quantum Electron* **29**, 217.

Dergachev A Y, Pati B and Moulton P F, 1999, "Efficient third-harmonic generation with a Ti:sapphire laser", *OSA Topical Meeting on Advanced Solid State Lasers, Boston, MA* Paper PD12

Eimerl D, 1987, "High average power harmonic generation", *IEEE J. Quantum Electron.* **23**, 575

Elight 1998, Website http://www.elight.de/lidard.htm

Govorkov S V, Weissner A O, Schroder Th, Stamm U, Zchocke W and Basting D, 1998, "Efficient high average power and narrow spectral line width operation of Ce:LiCAF laser at 1kHz repetition rate", *OSA Trends in Optics and Photonics, Advanced Solid State Lasers, Vol 19* W R Bosenberg, M M Fejer, eds. (Optical Society of America, Washington, DC)

Harrison J, Finch A, Rines D M, Rines G A and Moulton P F, 1991, "Low-threshold, CW, All-solid-state Ti:Al$_2$O$_3$ Laser," *Opt. Lett.* **16**, 581

Harrison J, Moulton P F and Scott G A, 1995, "13W, M^2 ¡ 1.2 Nd:YLF laser pumped by a pair of 20W diode-laser bars", *CLEO '95, Baltimore,MD* Paper CPD20.

Hoffstadt A, 1994, "High-Average-Power Flash-Lamp-Pumped Ti:Sapphire Laser", *Opt. Lett.* **19**, 1523

Isyanova Y, Rines G A, Welford D and Moulton P F, 1996, "Tandem OPO generating 1.5–10 mm wavelengths", *Trends in Optics and Photonics Volume 1, Advanced Solid State Lasers,* S A Payne and C R Pollock, eds. (Optical Society of America, Washington, DC 1996), Vol. 1, 174.

Jetalian A V, 1992, *Laser Radar Systems,* (Artech House, Boston)

Measures R M, 1992, *Laser Remote Sensing: Fundamentals and Applications* (Krieger Publishing Company, Malabar, Florida)

Moulton P F, 1986, "Spectroscopic and Laser Characteristics of Ti:Al$_2$O$_3$", *J. Opt. Soc. Am. B* **3**, 125

NASA 1998, Website http://www-arb.larc.nasa.gov/lite/

NASA Langley, 1999, Website http://aesd.larc.nasa.gov/GL/glf.html

Rines G A and Moulton P F, 1991, "Performance of Gain-Switched Ti:Al$_2$O$_3$ Unstable-Resonator Lasers", *OSA Proceedings on Advanced Solid State Lasers,* Hans P. Jenssen and George Dubé, eds. (Optical Society of America, Washington, DC 1991), Vol. 6, 88.

Sarukura N, Liu Z, Ohtake H, Segawa Y, Dubinskii M A, Semashko V V, Naumov A K, Korableva S L and Abdulsabirov R Y, 1997, "Ultraviolet short pulses from an all-solid-state Ce:LiCAF master-oscillator-power-amplifier system", *Opt. Lett.* **22** 994

Webb M S, Moulton P F, Kasinski J J, Burnham R L, Loiacono G and Stolzenberger R, 1998, "High-average-power KTA OPO", *Optics Lett.* **23** 1161

Yariv A, 1975, *Quantum Electronics,* 2nd Edition, (Wiley).

Yu J, Singh U N, Barnes N P and Petros M, 1998, "125 mJ diode-pumped injection-seeded Ho:Tm:YLF laser", *Opt. Lett.* **23** 780

Challenges for new laser sources in the defence industry

John R M Barr

Pilkington Optronics, Glasgow, Scotland

1   Introduction.

Probably the main challenge for new laser sources is to be accepted and exploited by industry. Of the many types and variants of lasers that are developed and described in the scientific literature, very few find more than niche applications. The purpose of this article is to summarise the acceptance of lasers for military applications and to identify reasons why lasers succeeded in displacing alternative technologies.

Military applications of lasers have been one of the main drivers for the development of laser technology. This was particularly true during the early years following the first demonstration of a laser, a ruby laser by T H Maiman (Hughes Research Laboratories, Malibu, Ca) in 1960. At this time, due to the immaturity of the field, the applications envisaged were not circumscribed by technical and financial reality and included science fiction ideas such as the "death ray", a concept which is only just being realised today in a limited and much less dramatic form as a defence against ballistic missiles. Those applications which have won through into practice were generally developed during the early years of the 1960's following the demonstration of the ruby laser.

The laser has been fielded in quantity for a very limited number of military applications. Primarily these are rangefinding and laser target marking and designation. An increasing use of lasers for illumination of targets and for training applications is also apparent. The laser is being developed for battlefield communications and as a weapon in its own right. Rangefinding and designation will be looked at in some detail in this chapter as examples of successful applications for lasers. The reasons for the adoption of lasers for these applications will be explored. Unclassified details of other applications may be found in Hewish, 1997.

2   Rangefinding.

The military need for rangefinding has been well established ever since projectiles were introduced into warfare. The procurement of specific instruments to measure range without exposing the user to unnecessary risk probably arose with the introduction of long range artillery on land and at sea. By the start of this century, optical rangefinders, based on the measurement of angles of sight, were in common use by the military. The first product introduced in 1888 by Barr & Stroud Ltd was an improved optical rangefinder (Moss and Russell, 1988a). These optical rangefinders all relied on measuring the angular size of an observed distant object. Radar techniques for establishing direction and range of targets became well established by the middle of this century. However, the relatively long radar wavelength results in significant beam-spreading with difficulty in selecting one object amongst many. Similar techniques at optical wavelengths allow much lower beam divergences and the possiblity of isolating much smaller objects from the background.

Within a year or so following the development of the laser the concept of laser rangefinding was demonstrated (in February 1961, see Stitch 1972). The laser rangefinder (LRF) had a number of unique attributes that solved many of the problems found with optical rangefinders and it is worth exploring these in some detail, since these improvements were the key to the successful insertion of laser rangefinders into real world applications.

2.1 A simple laser rangefinder.

The most common type of military rangefinder uses a time of flight technique to measure the range. An optical pulse is transmitted towards the target and light scattered from the target is collected. The time delay, τ, between transmission of the laser pulse and the detection of the backscatter is used to calculate the range, R, using the range equation:

$$R = \frac{c\tau}{2} \tag{1}$$

where c is the speed of light (300m per microsecond). Figure 1 shows the component parts of a laser rangefinder. The function of each part of the system will now be described.

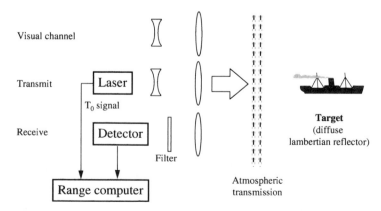

Figure 1. *A functional architecture for a laser rangefinder.*

The visual channel.

The visual channel allows the user to point the laser at the target. In a real system, the visual optics may not be part of the rangefinder equipment, but in this case, means will be provided to allow the user to boresight the LRF to the visual channel of the host system. The visual channel does not need to rely solely on visible light, but could be a thermal imager to allow day and night capability.

The laser transmit channel.

The laser is transmitted through a telescope toward the target. The laser wavelength, pulselength and energy are important characteristics which will be discussed after an outline of the theory of laser rangefinders is presented. Ruby, Nd:YAG, Er:glass, semiconductor lasers and frequency shifted Nd:YAG lasers have all been used for this application. Th telescope tailors the raw divergence of the laser to the value required by the application. Military lasers are designed to be robust and reliable and as insensitive as possible to the effects of temperature change and vibration. Achieving good beam quality has not generally been a primary design goal. Typically, the beam quality may be in the range 10–15mm.mrads (approximately 10 times the diffraction limit at 1064nm) so the raw laser divergence could be as large as 3mrad for a 5mm beam size. The telescope has to magnify the beam size by six times, say, to reduce the divergence to $\theta < 0.5$mrad.

The divergence of the laser beam from the LRF is important because it impacts on the usability of the system. Assuming that the maximum range of operation for the LRF is 10km and the target dimensions are about 2m then the divergence has to be less than 200μrad to ensure that the laser beam does not overspill the target. Overspill reduces the energy that is available for the return from the target. Clutter is the name given to multiple returns from objects that impinge on the laser beam, for example, trees, buildings or vehicles. From the user's perspective, clutter leads to uncertainty as to which range is the correct range for his target. A small laser divergence allows the user to minimise potential clutter returns. Most LRF designs include allowance for what is called first/last logic. In this case the reported range will be either the first or last of the returns. The user is supposed to apply sufficient intelligence to deduce which is more likely for his specific situation. For example, if the target is viewed through trees then selecting the last return is reasonable whereas if the target is clearly visible against a background then the first return is most likely to be correct.

The receive channel.

The receive channel collects the laser light scattered from the target and focuses this onto the detector. The larger the receive aperture, D, the more energy will be detected. The receive channel may well contain a number of filters to reduce background light, for example a narrow band-pass filter to reduce the effect of unwanted light from the sun or other light sources within the field of view (FOV) of the detector. The FOV is defined either by the size of the sensitive area of the detector or by an internal aperture or spatial filter if one is included.

Ideally the FOV should be minimised to reduce the effects of background light. In practice, the FOV will be larger than the the laser divergence, θ, otherwise returning laser energy will be wasted. Further, increasing the FOV slightly simplifies the opto-mechanical problem of maintaining the relative boresight between the laser transmit channel and the receive channel. Boresight refers to the deceptively difficult problem of keeping two optical channels pointing in the same direction and is one of the key tolerances that a full analysis of an LRF will reveal.

The various channels need not be kept separate and various combinations are seen in practice. It is common to combine the receiver and visual channel or the laser and visual channel, but it is less common to combine all three channels. The reason for trying to combine all channels is to minimise the total optical aperture for the system and hence hopefully reduce the system size and complexity. Normally there is a penalty in combining all channels resulting from lower optical transmission or increased backscatter of the outgoing laser light from the output telescope into the receiver.

2.2   Why were laser rangefinders adopted?

The main reasons for one technology superseding another in a given application are that the new technology offers:

1. A reduction in cost to the customer. This can either be a lower capital cost or a lower cost of ownership taking through-life costs into account, or some combination that means that the new technology is cheaper than the competition.

2. Additional functionality that solves more of the customers problems.

3. Better performance at the same task.

4. Improved physical characteristics: Smaller, more efficient, more physically robust, easier to use and so on.

Table 1 provides a brief comparison between optical and laser rangefinders. Laser rangefinders score in virtually all categories when compared with optical rangefinders. The accuracy of measuring the range is better, especially at longer ranges. The LRF is simpler to use and requires less skill from the operator and operates as well at night as during the day.

Feature	Optical rangefinders	Laser rangefinders
Weight	5kg	<2.0kg
Size	1m (long dimension)	<20x20x15cm^3
Cost (estimated at 1998 £'s)	£125 (in 1900) approx £18,000 (in 1998)	£5,000–15,000 depending on volume and specification.
Controls	Entirely manual	Automatic; digital interfaces.
Night operation	Assumes illuminated targets	Yes
Range error	Increases with R^2	Fixed ($< \pm 5$m)

Table 1. *A comparison between optical rangefinders and laser rangefinders. The data for the optical rangefinders is taken from Moss and Russell, 1988b.*

Probably the main reason that LRFs replaced optical rangefinders was that the electronic nature of the LRF made it straightforward to integrate its operation with a fully electronic fire control computer. Such a device could take into account all the various factors that influence accurate fire control. As a result less shells could be expended before a target is destroyed with a higher probability of first shot on target. In this way the survivability of the tank was improved and more targets destroyed for a given load of shells. The customer can offset the increased cost of ownership of the LRF compared to the optical rangefinder against the better performance of the platform, generally an armoured fighting vehicle. Additionally, from a 1960's view point, the fact that LRFs could operate at night meant that the anticipated development of thermal imagers could be fully integrated into the fire control system, providing a full day and night fighting capability.

During the mid 1960's, LRFs were developed by a number of companies. Barr and Stroud Ltd, as Pilkington Optronics was then known, worked with the Hughes Aircraft Company to develop a ruby LRF . A version of this device, the LF1 was offered for sale in 1965 and was the first commercially available laser rangefinder. A variant integrated with a tank sight and known as the LF2 was put into service on the Chieftain tank in 1969 and was the first such device to be fielded (Moss and Russell, 1988c).

2.3   The theory of laser rangefinders

A simplified analysis of a standard military laser rangefinder will be presented. Rangefinders of this type generally use a single-pulse time-of-flight technique to measure the range. Other techniques have also been demonstrated using modulated continuous wave lasers or coherent detection, but a complete description is outside the scope of this article and details may be found in the literature (see Forrester and Hulme 1981, Kamerman 1993).

The purpose of a functional analysis is to identify those areas in a rangefinder design that impact on the performance of the instrument. For a simple rangefinder one might expect that the laser energy, receiver sensitivity and receiver telescope aperture are all important parameters that influence how well the device will work. The analysis allows a trade-off study to be carried out among all the various parameters leading to an optimum design for the given application and, ideally, will allow a complete specification of the operation of the laser rangefinder together with the tolerances that have to be met in the detailed opto-mechanical design. Examples of this process will be given later.

The performance of a laser rangefinder depends on many parameters, a simple subset of which is given in Table 2.

Consider the situation where a laser illuminates a target at a range R. The energy on target can be calculated as:

$$E_{\text{target}} = E T_L e^{-\sigma R} \eta \qquad (2)$$

where the overlap efficiency, η, includes the effect of target size, laser divergence and FOV of the detector (clearly no energy that lies outside the FOV can contribute directly to the return). In general η is a relatively complex parameter to calculate for a realistic scenario where a general far field beam distribution, a real target and boresight errors between the visual, laser and receiver channels of the LRF are considered. For simplicity, a model

Parameter	Definition	Value	Units
A	Target area.	5.3	m^2
c	Speed of light.	300	m/μs
D	Aperture of the receive channel.	0.05	m
E	Laser energy	8	mJ
E_{det}	Laser energy returned to the detector at R=1000m.	110	fJ
E_{min}	Minimum energy detected by the detector for a 50% probability of detection and 0.001 false alarm rate.	0.5	fJ
r	Reflectivity of the target.	0.1	-
R	Range to the target.	1	km
T_{L}	Transmission of the laser channel.	0.8	-
T_{det}	Transmission of the receive channel to the detector.	0.3	-
V	Atmospheric visibility.	23.5	km
θ	Divergence of the laser after the transmit telescope (full angle, 90% contained energy).	0.5	mrad
θ_{FOV}	Field of view of the receiver (full angle).	4	mrad
ϕ	Normal to the target surface	0	rads
σ	Atmospheric absorption cross section.	0.029	km^{-1}
ε	Extinction ratio against an extended target at R=1km with r=100%.	34	db
η	Overlap efficiency of the laser beam on the target and within the FOV of the receiver at R=1000m	1	-
Ω_{det}	Solid angle of the receive channel at a range R=1000m	2x10^{-9}	Sr
λ	Laser wavelength	1.5	μm

Table 2. *Summary of the main parameters describing the performance of a rangefinder. The values are appropriate to an "eyesafe" 1.5μm LRF. Values which depend on a specific range have been calculated for R=1000m.*

based on a flat topped laser beam and no boresight errors will be considered. The target will either be smaller than the laser beam size at the range of interest (this is the case that normally limits the range of the LRF since much of the energy overspills the target) or the target will be larger than the laser beam. Atmospheric transmission is included through the factor $e^{-\sigma R}$. Other symbols are defined in Table 2.

If the laser beam is smaller than the target (extended target) then $\eta = 1$ otherwise

$$\eta = \frac{4A}{\pi \left(R\theta \right)^2} \tag{3}$$

for the target smaller than the laser beam (finite target). In the flat topped laser beam approximation, the laser energy should be interpreted as the energy within the divergence angle θ, for most cases this will either be 90% or 86.5% of the total laser energy.

Next, the amount of energy reflected from the target into the FOV of the receiver has to be estimated. Obviously the surface finish and detailed shape of the target will influence the return in the most general case; equally obviously this will be very difficult to calculate and impossible to predict in advance. The usual solution is to assume that the target is a square flat board painted with an appropriate material to simulate the worst case likely to be met in practice. Normally the reflectivity of the surface is assumed to be diffuse with a Lambertian profile since this provides a calculable solution to the problem.

The radiance (units: $\text{Wm}^{-2}\text{Sr}^{-1}$) from an illuminated Lambertian diffuse reflector is assumed to be a constant over the surface (see Bass, Van Stryland, Williams and Wolfe, 1995). Consequently the energy scattered into a given solid angle depends only on the projected area of the illuminated Lambertian diffuser. The projected area decreases with angle, ϕ, to the surface normal as $\cos\phi$. The reflected energy per unit solid angle is a maximum at the surface normal and decreases to zero at $\phi = \pi/2$ radians. Let the surface radiance be L, and assume that the reflectivity of the surface is 100% so the total energy scattered by the surface is equal to the incident energy and is calculated as:

$$E = \int_S \int_\Omega L \cos\phi \, dA_s d\Omega \tag{4}$$

where the first integral is over S, the illuminated surface of the target (A_s), and the second integral is over the hemisphere into which the laser light is scattered by the surface. The solid angle is:

$$d\Omega = 2\pi \sin\phi \, d\phi \tag{5}$$

and the integration limits are $0 \le \phi \le \pi/2$. Evaluating the integrals in Equation 4 yields:

$$L = \frac{E}{A_s \pi} \tag{6}$$

Equation 6 allows an estimate for the reflectivity of a Lambertian diffuse reflector (expressed as energy scattered per unit solid angle) to be given as:

$$LA_s \cos\phi = \frac{rE}{\pi} \cos\phi \tag{7}$$

where r is the reflectivity of the surface. The integration over the surface in Equation 4 has been carried out. For simplicity, the receiver will be assumed to lie along a normal to the surface ($\phi = 0$). The solid angle of the receive channel at a range R is:

$$\Omega_{\text{det}} = \frac{\pi D^2}{4R^2} \tag{8}$$

An expression for the energy collected by the receiver can now be found using Equations 2, 7 and 8:

$$E_{\text{det}} = \frac{r}{\pi}(E_{\text{target}}) \, \Omega_{\text{det}} \, (T_{\text{det}} \, e^{-\sigma R}) \tag{9}$$

$$= E \, T_L T_{\text{det}} r e^{-2\sigma R} \eta \frac{D^2}{4R^2} \tag{10}$$

The maximum range of the rangefinder against the target can be found by replacing E_{det} by E_{min} and using the expressions for η, for an extended target:

$$Re^{\sigma R} = \left(\frac{ErT_L T_{\text{det}} D^2}{4E_{\text{min}}} \right)^{1/2} \tag{11}$$

and for a finite target:

$$Re^{\sigma R/2} = \left(\frac{ET_L T_{\text{det}} r A D^2}{E_{\min} \pi \theta^2} \right)^{1/4}.$$ (12)

If atmospheric absorption and scattering may be neglected, Equation 12 is usually the best approximation to the maximum range of the LRF. As the right hand side of this equation is to the inverse fourth power, the range increases only slowly with design parameters. It is probably more sensible to increase the aperture D and decrease the LRF divergence than build a more powerful laser.

Receiver characteristics

The receiver is a very sensitive detector followed by an amplifier and a trigger circuit. The trigger is normally set to respond to a certain voltage level. The noise from the detector and amplifier is usually approximated as a Gaussian distribution characterised by an rms voltage, the noise level. The trigger level is set as a compromise between being too large (i.e. increasing $E_{\min}$) or by being too low and risking being triggered by the detector noise and increasing the false alarm rate. When the voltage generated by $E_{\min}$ is equal to the trigger level then the Gaussian noise assures that a return of this energy is detected with a probability of 50%. Most receiver suppliers will quote a minimum detectable energy for 50% probability of detection and a 0.1% false alarm rate or near equivalent. These lie between 0.03–3fJ depending on wavelength and detector type. A fuller description of receiver performance may be found in Byren 1993, Analogue Modules Inc 1997, Barbeau 1996.

Atmospheric effects.

The atmosphere has very strong influence on laser rangefinding. Basically light only propagates through the atmosphere in a limited number of wavelength regions (visible, 1.5–2μm, 3–5μm, and 8–12μm). Even in the atmospheric transmission windows, light can be strongly absorbed by molecules such as CO_2 and H_2O. The details of the atmospheric transmission is accurately modelled by a number of commercially available software packages and the discussion of the details is beyond the scope of this article. A working approximation for the absorption cross-section, σ, is (see, for example, BSEN 60825–1:1994a):

$$\sigma = \frac{3.91}{V} \left(\frac{0.55}{\lambda} \right)^{0.585 V^{0.33}} \quad \left[\text{km}^{-1} \right].$$ (13)

In this formula, the visibility V is in km and wavelength is in μm. The approximation of Equation 13 is only valid in the atmospheric transmission bands and for any important work a more accurate value taking into account the detailed absorption properties of the atmosphere should be used. Note that the absorption cross-section decreases with increasing wavelength. This suggests that LRF operation at longer wavelengths might be beneficial. In practice this potential benefit is often negated by technical problems with the laser and receiver.

Atmosphere	Visibility (km)	σ (km^{-1})(Equation 13)	Model (km^{-1})
Light fog	1	2.090	0.9510
Hazy conditions	2	0.890	0.4790
Hazy conditions	5	0.270	0.1960
Moderate visibility	10	0.102	0.101
Standard clear	23	0.029	0.0478
Very clear	50	0.008	0.0194

Table 3. *A comparison of the simple prediction for cross section for various atmospheric transmissions to a code output. The laser wavelength is 1.5μm.*

A comparison for the predictions for this model and some code predictions is given in Table 3. Equation 13 is normally suitable for most initial approximations.

In addition to the effects of scattering and absorbtion of light by the atmosphere, rain and smoke can also reduce the effectiveness of LRFs. A value of $\sigma = 1.2$km^{-1} for a rain rate of 12mm/hour has been quoted (Forrester and Hulme, 1981). This is a severe rain fall corresponding to an inch of rain every two hours. The scattering from rain does not depend strongly on wavelength.

The atmosphere can also cause perturbations to the laser beam profile arising from the refractive index inhomogeneities caused by temperature variations of air. Quite large intensity fluctuations can arise known as scintillation. Further modifications to the intensity profile can be generated by the target surface profile which induce speckle into the return beam. The analysis of these effects can be very complex and will not be considered further here.

Target reflectivity.

The target reflectivity, r, is somewhat wavelength dependent. Data for for a variety of paints used on military vehicles is provided in Byren, 1993, and for 1.064μm the reflectivity r varies from 0.074–0.257, and for 1.54μm the reflectivity r varies from 0.06 to 0.21. For normal calculation purposes, a value of $r = 0.1$ is sufficiently accurate.

Extinction ratio.

The extinction ratio is the performance of the LRF against an extended target with $r=100\%$ and a 1km range with visibility neglected. It is defined as a ratio of received energy to E_{min} often expressed in dB:

$$\varepsilon = 10\log_{10}\left(\frac{E_{det}}{E_{min}}\right) = 10\log_{10}\left(\frac{ErT_LT_{det}D^2}{4R^2E_{min}}\right) \tag{14}$$

where $r=1$ and $R=1000$m. Knowledge of ε allows the manufacturer to describe the performance of the LRF without being too specific about the internal details of its construction.

Part of the usefulness of the extinction ratio is that it is easily measured by the manufacturer and user. Most suppliers will have standard targets at ranges between 500m and 1000m of known reflectivity against which the LRF can be tested using a calibrated

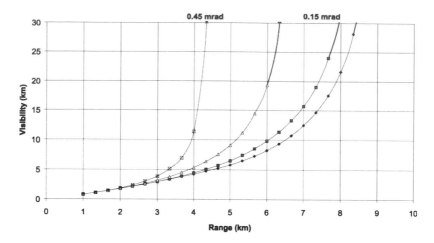

Figure 2. *Predictions of the simple LRF model for the parameters listed in Table 1. The effect on ranging performance of a variety of boresight errors between the aiming mark and the laser is shown.*

set of neutral density filters from which an extinction ratio in a standard format may be derived. Equations 11 and 12 can be rewritten using the extinction ratio.

Predictions from the simple rangefinding model.

Figure 2 shows the results of the simple rangefinding model using data listed in Table 1. The maximum effective range of the LRF for a given atmospheric visibility can be read from this figure. The predictions are for perfect alignment and for various levels of boresight errors. Boresight errors occur when the visual channel, laser and receiver channel are not pointing in exactly the same direction. This process is schematically shown in Figure 3. The net result of a boresight error is that the LRF performance may not be as good as expected. From a system perspective, the modelling of the LRF performance including boresight errors is very important since knowing the magnitude of typical errors allows the system specifications to be estimated. Furthermore, the system can be constructed to minimise the impact of boresight errors. Finally, the requirements on optomechanical mounting of the component parts of the system can be calculated. This important step allows the engineers to embark on the detailed design of the LRF.

2.4   Laser rangefinder technology.

A wide variety of different laser technology has been applied to rangefinders. Some pulsed rangefinders using direct detection of reflected light and solid state gain media are listed in Table 4. The cited examples are only a selection from the large number of products offered in the market place. The LRF technology has evolved in line with improvements in laser gain media and receiver technology. As an example, Nd:YAG superseded ruby as a gain medium during the 1970s since it is more efficient.

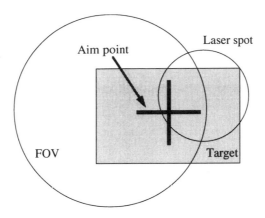

Figure 3. *A schematic diagram showing the relationship between the laser beam distribution to the aiming point and FOV of the LRF. In this case substantial boresight errors are apparent.*

Parameter	Ruby LRF LF2 Pilkington Optronics	Nd:YAG LRF LF18 Pilkington Optronics	Er:glass LRF	Nd:YAG & Raman LRF Harlem 2 Zeiss-Eltro Optronic	Nd:YAG & OPO LRF Mark VII Litton Lasers	Laser diode LRF Vector 1500 Leica
Wavelength	694.3nm	1.064μm	1.535μm	1.543μm	1.58μm	860nm
Energy	40mJ	40mJ	8mJ	10mJ (Class 3A)	<8mJ (Class1)	Class 1
Pulse length	40ns	8ns	17ns	5ns	N/A	N/A
Divergence	0.5mrad	0.7mrad	0.5mrad	1mrad	N/A	1.6mrad
Repetition	every 30s	4Hz	1Hz	0.5Hz	6/minute	12/minute
Receiver aperture	50mm	32mm	50mm	56mm	18mm	42mm
Receiver type	pmt	Si pin	InGaAs APD	InGaAs pin	N/A	N/A
R_{max}	10km	10km	19.995km	25km	19.995km	2.5km
R_{min}	400m	200m		50m	50m	25m
ΔR	10m	5m	5m	5m	5m	2m
Performance	Note 1	N/A	Note 2	N/A	N/A	Note 3
Weight	20kg	6kg	<1.5kg	2.5kg	1.6kg	

Table 4. *A comparison between the performance of various commercially available laser rangefinders. The data contained in this table has been extracted from data sheets issued by the companies. N/A denoted that data is not available. The performance notes are: (1) Ruby: could range to objects beyond visibility due to receiver sensitivity. (2) Er glass: more than 7km for 20km visibility and target reflectivity >0.1. (3) Laser diode: 1.5km (for visiblity of 10km, target reflectivity 0.4)*

More recently, the driving force in changing LRF technology has been the desire to use so called eyesafe lasers . These are lasers that operate at wavelengths between 1.5μm and 1.8μm. In this wavelength band the maximum permitted exposure level of the eye to laser radiation is at its highest (BSEN 60825–1:1994b). From the point of view of the user the laser can be Class I and can be used freely without the requirement to ensure that the target and other potential observers are suitably protected. Training and other uses of the eyesafe LRF are simplified thus reducing support costs and increasing the utility of the device. In this case the user may be willing to pay a higher price for the laser to minimise training costs and the costs of litigation resulting from accidents. Please note that this discussion is simplified and should not be taken to mean that all lasers between 1.5μm and 1.8μm are unconditionally safe or that other wavelength regions are not eyesafe under suitable conditions.

Ruby laser rangefinders.

The initial development of LRFs centred on flashlamp pumped ruby lasers (Cr^{3+} in Al_2O_3) operating at 694.3nm. By Q-switching, using rotating mirrors or prisms, pulse durations in the required range (10–100ns) are obtained. Ruby lasers can conveniently be detected using photomultiplier tubes with very good sensitivity. The problem with ruby lasers are:

- Efficiency. The energy level structure of ruby is composed of three levels which leads to inefficient performance (Koechner 95).

- Covertness. Ruby light is visible and can easily be seen by the unaided eye.

Ruby LRFs are now being phased out of service.

Nd:YAG laser rangefinders.

Nd:YAG (Nd^{3+} in $Y_3Al_5O_{12}$) is a four level laser system operating at 1064nm. Compact LRFs can be built using passive Q-switching based on saturable absorbers such as plastic films containing absorbing dyes or Cr^{4+} doped into hosts such as YAG. The Nd:YAG wavelength is compatible with very sensitive Si pin detectors. Apart from the eyesafe issue there is no technical reason for preferring other laser gain media for LRFs.

Er:glass laser rangefinders.

The only reason to use Er:glass is to obtain eyesafe performance. Er:glass is a three level laser host in a thermally poor host. It therefore suffers from low efficiency and the effects of thermal lensing. As a result Q-switching is normally achieved using rotating mirrors or prisms to minimise round trip losses and the repetition rate is limited to less than 1Hz. The Er:glass LRF has been adopted by many armed services and fielded in large volumes.

Raman shifted Nd:YAG lasers.

Exploitation of the Raman effect in methane (CH_4, Stokes shift of 2914cm^{-1}) allows efficient shifting of 1.064μm radiation into the eyesafe band. The Raman cell has to contain high pressure gas in order that the Raman process is efficient. Compared to Nd:YAG LRFs there is a reduction in system efficiency by about 50% both through losses in the Raman effect and through lower receiver sensitivity. The Raman route does provide higher repetition rates than Er:glass.

Nd:YAG pumped optical parametric oscillators.

Optical parametric oscillators offer many of the same advantages for generating eyesafe light as Raman cells. The normal material is non critically phasematched KTP (potassium titanyl phosphate) that generates light at $1.57\mu m$ when pumped by Nd:YAG. Typical optical to optical conversion efficiencies are about 30–50%. The basic advantage over Raman cells is that OPOs are smaller and easier to fabricate than high pressure cells.

Semiconductor laser diodes.

Semiconductor laser diodes are well established for ranging over short distances (<1km) against co-operative targets (i.e. targets with high specular reflectivities enhanced by corner cubes). Due to the small pulse energy (typically about $1\mu J$ or less) long distance rangefinding is very difficult. Some progress has been made using signal averaging techniques where the return is summed over a large number of pulses, N say. Under these conditions, the minimum detectable energy E_{min} is reduced by $N^{1/2}$ but this advantage is gained at the expense of complicating the operation and electronics of the device. It is fair to say that this technology has many potential advantages of size, efficiency and reliability. However, by comparison with conventional flashlamp pumped LRFs the technology has to be developed further. The technology is likely to be fielded first for use with short range weapons.

3 Laser target marking and designation.

Laser designation of targets and so called smart munitions (some of which are laser guided bombs) first came to the public's attention dramatically during the Gulf war. In fact, the development of this technique dates from the Vietnam war and resulted from the inability of traditional bombing techniques to destroy strategic targets such as bridges. The idea of using a laser as a torch to illuminate a target and allow the bomb to home in on the laser splash was first suggested in 1965 and first demonstrated around 1967. The production units were put into service during the Vietnam war in the early 1970s (Clancy 1995).

3.1 The designator system.

Unlike laser rangefinders, laser designators form only a single part of a more complex whole. There is the laser source called the designator, the delivery vehicle, normally an aircraft or helicopter, but laser seekers have also been added to artillery shells, and the seeker which detects the scattered laser radiation. During delivery, these three components have to work together correctly. Figure 4 shows an engagement where the designator is ground based. Designators can be used to allow a pilot to identify the correct target in a cluttered environment during a fast low level approach when there is only a short time to see, recognise and engage a target (Forrester and Hulme 1981). The scattered laser light is detected on board the aircraft using a laser spot tracker (LST). An indication of the target position is then provided to the pilot through a head up display (HUD) or helmet mounted display allowing unambiguous identification of the target and a normal attack using conventional bombs. Alternatively a laser guided bomb can be deployed to attack the illuminated target.

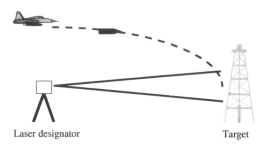

Laser designator   Target

Figure 4. *The principles of operation of a ground based laser designator.*

Complexity is built in because there are many LST and laser seekers in service around the world. On a given battlefield several designation events could take place simultaneously thus it is necessary to code the laser signal so that the LST or laser seeker locks onto the correct target. However, if operation between any designator and any seeker is to take place then the laser characteristics have to be completely specified. In practice this means that the laser wavelength and pulse duration are only allowed to vary between fairly narrow limits. The laser wavelength is fixed by the requirement to place a narrow band filter in front of the seeker detector to limit the detection of background radiation while the pulse duration is fixed to match the electrical bandwidth of the detector.

Laser designation has gained widespread acceptance because of the flexibility of the process. The designator can be ground based allowing immediate exploitation of an evolving tactical situation on the battlefield, or mounted on board the aircraft to allow attack of targets beyond the fringes of the ground battle. The seeker units for the laser guided bombs are very cheap compared with alternative electro-optical seekers. Finally, when used in conjunction with the LST, improved performance with conventional bombs is possible.

The net result of the various constraints upon the designator is that only Nd:YAG Q-switched lasers can be used. This limitation arises because of the other components that make up the complete system. Economics caused by the widespread investment in this technology and the large numbers of units fielded mean that it is very difficult to expect that alternative laser sources can displace the current laser designators. As a result all laser designators share a common laser technology and have done so for more than 30 years. It is unlikely to change in the near future. The arguments concerning eye-safety do not operate in the context of designation. This situation of apparent technical stagnation should be contrasted with the LRF history where 4 different technologies have been fielded in volume (ruby, Nd:YAG, Er:glass, Raman shifted Nd:YAG) with two additional technologies on the brink of acceptance (OPO shifted Nd:YAG, laser diodes).

3.2 Modelling of designation.

The modelling of laser designation can be based upon the simple analysis of laser rangefinders presented in Section 2.2. The only complication arises from the fact that the receiver and transmitter may, in general have different ranges since the transmitter may be ground based while the receiver is on the aircraft or bomb.

3.3   Laser designator technology.

Laser designators have always been based on flashlamp pumped Nd:YAG lasers operating at 1064nm. This is not to say that there have been no improvements in performance of these instruments. Successive generations of laser designator system have seen improvements in technology leading to a better product for the user. For the purposes of this article attention will be focussed on manportable laser designators and the airborne variants will be neglected.

Manportable laser designators, as the name implies, are carried by the user and not transported on a vehicle. As a result attention is focussed on size and weight of the complete system which normally comprises laser designator, battery, tripod and connecting cables. The number and size of the batteries is determined by the energy requirements from the laser and efficiency of the complete system.

The first generation of manportable laser designators tended to be large, heavy and composed of multiple sub units which had to be assembled before use. Typically weights ranged from 9 to 24kg for the minimum usable configuration (laser designator and batteries) of systems available prior to 1990. The equivalent figures for current generation of manportable laser designators is 7.6 to 12.7kg so from the point of view of the user a significant improvement in portability has been achieved. The reduction in weight has generally been achieved by improvements to all parts of the designator system: batteries, control electronics, coolant, and optomechanical mounting of optics; rather than any specific technical breakthrough.

One of the more modern laser designators available now is the LF28 designed and built by Pilkington Optronics. This device offers a number of developments over existing equipments. The following sub-sections describe each development in detail.

Slab technology.

The laser host is Nd:YAG in the form of a zig-zag slab rather than the more conventional rods (Eggleston *et al.* 1984). The reason for this approach is inherent in the mode of operation of laser designators. Normally the laser is only fired for a short time during the engagement. Under these conditions the laser has to perform well, at the correct energy and divergence, from immediately after switch on. Due to the time taken to establish a thermal lens in a rod (5-10s) this ideal has not been achieved to date in current laser designators and the limitation has reluctantly been accepted by system designers. Some designers use an intracavity compensation telescope to minimise the effects of thermal lensing. Such an approach adds significant complexity to the system and the potential for a reduction in reliability. Slab technology illustrated in Figure 5 avoids the problem and allows instant in-specification operation.

The zig-zag path through the slab shaped gain material allows the thermal lens to be averaged out. All input rays sample the complete temperature profile so that optical path difference is minimised across the beam. Any residual thermal effects tend to have much smaller magnitude and less impact on the laser performance.

Conduction cooling.

Most existing laser designators use liquid cooling loops or fans to remove waste heat from the pumphead and laser material. Owing to the flat nature of the slabs conduction cooling is used within LF28. It is a relatively simple matter to attach the slab to a

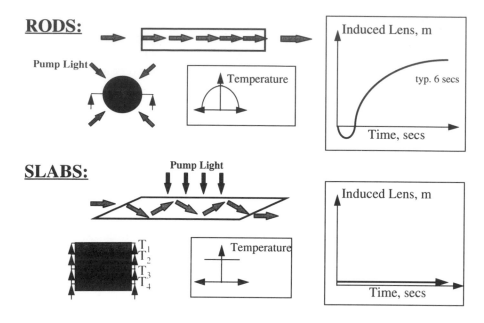

Figure 5. *A comparison of rod shaped gain media with slab shaped gain media.*

heatsink and control the temperature by conduction to the ambient atmosphere. A more difficult problem is to conductively cool the flashlamp and again this problem has been solved in this system.

The main advantage of conduction cooling is that less maintenance is required by the system. Fluid based cooling systems normally require frequent routine attention to maintain system performance. The reason for this is that the pumphead is flooded so that the pump light passes through the coolant. Any reduction in transmission of light through the coolant reduces the laser energy. Conduction cooling should allow the system to be free of routine maintenance for more than 5 years compared with the current yearly coolant replacement cycle.

Modern system design.

The laser designator features a clip on battery and modern easy-to-use man-machine interface. The clip on battery reduces the number of parts required for system operation and makes the system simple and rapid to deploy. The use of a microprocessor to control the system function allows a flexible reconfigurable display that is easy to use. The complete laser designator is shown in Figure 6.

Future compatible.

The modular architecture of LF28 and the use of slabs has been adopted with a future upgrade path to laser diode pumping in mind. It is technically straightforward to replace the pumphead of LF28 with an array of laser diodes. Laser diode pumping is well known to have better efficiency and provide a longer lifetime than flashlamps. Such a step would improve the efficiency of the unit threefold and reduce the heatload as well. It has not been adopted since the cost of laser diodes is significant compared with flashlamps. At

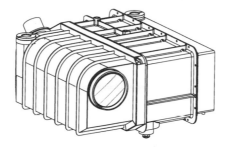

Figure 6. *A modern manportable laser designator*

current prices the adoption of laser diodes could increase the unit price by about 50% with negligible reduction in operating costs.

4 Conclusion.

A brief review of two of the main military laser applications has been presented. The technical information has been presented with the objective of explaining why a particular technology has been adopted. Various considerations apply but the main driver has always been cost, not just of the laser but the complete device in which the laser is embedded and the way in which it is used. Eyesafe laser rangefinders are a case where the LRF costs more or is less effective than non eyesafe LRFs, but the user recognises that the adoption of this technology reduces his overall costs. The many variants of LRFs is explained by the fact that there is little to influence the choice. In laser designation on the other hand, the laser is a small part of a larger, more complex system and the laser performance is constrained by the whole system. The laser has to be compatible with these parts and this has limited the technology to Nd:YAG lasers. However, changes in the laser system, particulary reduction in weight and improvements in reliability, are taking place.

5 Acknowledgements.

The author would like to acknowledge discussions with many colleagues at Pilkington Optronics on the subject of laser rangefinders and designators, especially Keith Stott, Mike Darlow, Willie Alexander, John Robertson and Frank Haran.

References.

Analogue Modules Inc, 1997, *Sensors and Amplifiers Data sheet.*
Barbeau N R, 1996, *Opt. Eng* **35** 2396.
Bass M, Van Stryland E, Williams D R, Wolfe W L, 1995, *Handbook of Optics* **2**, Chapter 24.
BSEN 60825–1:1994a, *Safety of laser products*, page 78.
BSEN 60825–1:1994b, *Safety of laser products*, Tables 1–4 and 6.

Byren R W, 1993, *The infra-red and electro-optic systems handbook,* **6** chapter 2   ed. Clifton Fox.

Clancy T, 1995 *Fighter Wing,* 147.

Eggleston J M, Kane T J, Kuhn K, Unternahrer J, Byer R L, 1984, The slab geometry laser-part 1: theory, *IEEE J. Quantum Electron.* **20** 289.

Forrester P A, Hulme K F, 1981, *Optical and Quantum Electronics,* **13** 259.

Hewish M, 1997, *Jane's International Defense Review,* **30** 38.

Kamerman G W, 1993 *The infra-red and electro-optic systems handbook,* **6** chapter 1 ed. Clifton Fox.

Koechner W, 1995, *Solid state laser engineering,* edition 4. Chapter 1.

Moss M, Russell I, 1988a, *Range and Vision,* Chapter 1.

Moss M, Russell I, 1988a, *Range and Vision,* Chapter 2.

Moss M, Russell I, 1988b, *Range and Vision,* Chapter 7.

Stitch M L, 1972, *Laser Handbook,* chapter F7, 1746, eds. F T Arecchi and E O Schulz-DuBois.

Lasers for interferometric gravitational wave detectors

S Rowan[1] and J Hough[2]

[1]University of Stanford, USA and
[2]University of Glasgow, Scotland

1 Introduction

The largest optical interferometers ever to be built on earth are currently under construction in the following projects:

- LIGO Project, USA (Barish, 1997),

- VIRGO Project, Italy/France (Vinet et al. 1997),

- GEO 600 Project, Germany/UK (Hough et al. 1997a)

- TAMA 300 Project, Japan (Tsubono et al. 1997).

When completed, this detector array should have the capability of detecting gravitational wave signals from violent astrophysical events in the Universe.

Some early relativists were sceptical about the existence of gravitational waves ; however, the 1993 Nobel Prize in Physics was awarded to Hulse and Taylor for their experimental observations and subsequent interpretations of the evolution of the orbit of the binary pulsar, PSR 1913 +16 (Hulse 1994, Taylor 1994), the decay of the binary orbit being consistent with angular momentum and energy being carried away from this system by gravitational waves (Will 1983, 1995).

Gravitational waves are produced when matter is accelerated in an asymmetrical way, but due to the nature of the gravitational interaction, significant levels of radiation are produced only when very large masses are accelerated in very strong gravitational fields. Such a situation cannot be found on earth but is found in a variety of astrophysical systems. Gravitational wave signals are expected over a very wide range of frequencies; from $\sim 10^{-17}$Hz in the case of ripples in the cosmological background to $\sim 10^3$Hz from the formation of neutron stars in supernova explosions. The most predictable sources are binary star systems. However there are many sources of much greater astrophysical interest associated with black hole interactions and coalescences, neutron star coalescences, stellar collapses to neutron stars and black holes (supernova explosions), pulsars, and the physics of the early Universe. For a full discussion of sources refer to the material contained in 'First International LISA Symposium' (LISA 1997) and 'Gravitational Waves Sources and Detectors' (Ciufolini and Fidecaro F, 1997).

Why is there currently such interest worldwide in the detection of gravitational waves? Partly because observation of the velocity and polarisation states of the signals will allow a direct experimental check of the wave predictions of General Relativity; but more importantly because the detection of the signals should provide observers with new and unique information about astrophysical processes.

2   Detection of gravitational waves

Gravitational waves are most simply thought of as ripples in the curvature of space-time, their effect being to change the separation of adjacent masses on earth or in space; this tidal effect is the basis of all present detectors. The problem for the experimental physicist is that the predicted magnitudes of the strains in space caused by gravitational waves are of the order of 10^{-21} or lower. Indeed current theoretical models suggest that in order to detect a few events per year (*e.g.* from coalescing neutron star binary systems) a sensitivity level close to 10^{-22} is required.

The necessity of being able to detect such small signal levels means that noise, resulting from a variety of sources, such as the thermal motion of molecules in the detector (thermal noise), from seismic or other mechanical disturbances, and from noise associated with the detector readout, whether electronic or optical, must be reduced to a very low level. For signals above ~10Hz ground based experiments are possible, but for lower frequencies where local fluctuating gravitational gradients and seismic noise on earth become a problem, detectors which will operate in space are being studied. (Danzmann *et al.* 1996).

2.1   Initial detectors

Early experiments in the field were based on looking for tidal strains induced in aluminium bars which were at room temperature and were well isolated from ground vibrations and acoustic noise in the laboratory (Weber 1969, Weber 1970). These detectors were very sensitive, being able to detect strains of the order of 10^{-15} over millisecond timescales, but their sensitivity was still far away from what was predicted to be required theoretically. Development of bar type detectors has continued with the emphasis being on cooling to reduce the noise levels, and currently systems at the Universities of Rome (Astone *et al.* 1996), Louisiana (Amaldi *et al.* 1989) and Perth (Western Australia) (Heng *et al.* 1996) are achieving sensitivity levels better than 10^{-18} for millisecond pulses. Bar detectors have a disadvantage, however, of being sensitive only to signals that have significant spectral energy in a narrow band around their resonant frequency.

2.2 Long baseline detectors on earth

An alternative design of gravitational wave detector offers the possibility of very high sensitivities over a wide range of frequency. This design uses test masses a long distance apart and freely suspended as pendulums to isolate against seismic noise and reduce the effects of thermal noise. Laser interferometry provides a means of sensing the motion of the masses as they interact with a gravitational wave (Figure 1).

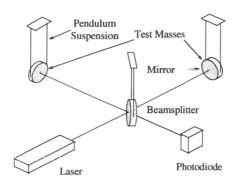

Figure 1. *Schematic of a gravitational wave detector using laser interferometry*

This technique is based on the Michelson interferometer and is particularly suited to the detection of gravitational waves as they have a quadrupole nature. Waves propagating perpendicular to the plane of the interferometer will result in one arm of the interferometer being increased in length while the other arm is decreased and vice versa. Gravitational wave strengths are characterised by the gravitational wave amplitude h, given by

$$h = 2\Delta L/L \qquad (1)$$

where ΔL is the change in the length L of one arm of the detector. Thus h is a measure of the strain in space induced by a gravitational wave.

The induced change in the length of the interferometer arms results in a small change in the interference pattern of the light observed at the interferometer output. A typical design specification to allow a reasonable probability for detecting sources requires a noise floor in strain smaller than $2 \times 10^{-23}/\sqrt{\text{Hz}}$ to be achieved. The distance between test masses possible on earth is limited to a few kilometres by geographical and cost factors. If we assume an arm length of 3km the above specification sets the requirement that the residual motion of each test mass is smaller than 3×10^{-20} m/$\sqrt{\text{Hz}}$ over the operating range of the detector, which may be from $\sim$10Hz to a few kHz, and requires that the optical detection system at the output of the interferometer must be good enough to detect such small motions.

2.3   Noise sources which limit the sensitivity of detectors

Fundamentally it should be possible to build systems using laser interferometry to monitor strains in space which are limited only by the Heisenberg Uncertainty Principle; however there are other practical issues which must be taken into account. Fluctuating gravitational gradients pose one limitation to the interferometer sensitivity achievable at low frequencies and it is the level of noise from this source which dictates that experiments to look for sub-Hz gravitational wave signals have to be carried out in space (Spero 1983, Saulson 1984). In general, for ground based detectors the most important limitations to sensitivity are a result of seismic and other ground-borne mechanical noise, thermal noise associated with the test masses and their suspensions and shot noise in the photocurrent from the photodiode which detects the interference pattern. The significance of each of these sources will be briefly reviewed.

Seismic noise

Seismic noise at a reasonably quiet site on the earth follows a spectrum in all three dimensions close to $10^{-7}/f^2$ m$\sqrt{\text{Hz}}$ and thus if the motion of each test mass has to be less than 3×10^{-20} m/$\sqrt{\text{Hz}}$ at a frequency such as 30Hz then the level of seismic isolation required at 30Hz in the horizontal direction is greater than 10^9. Since there is liable to be some coupling of vertical noise through to horizontal a significant level of isolation has to be provided in the vertical direction also. Such isolation can be provided in a relatively simple way; *e.g.* by suspending each test mass as the last stage of a multiple pendulum system which itself is hung from a plate mounted on passive 'rubber' isolation mounts or on an active (electro-mechanical) anti-vibration system.

Thermal noise

Thermal noise associated with the mirror masses and the last stage of their suspensions is likely to be the most significant noise source at the lower end of the operating range of the detector (Saulson 1990). The operating range of the detector lies between the resonances of the test masses and their pendulum suspensions, and thus it is the thermal noise in the tails of the resonant modes which is important. In order to keep this as low as possible the mechanical quality factors of the masses and pendulum resonances should be as high as possible. To achieve the level of sensitivity discussed above the quality factor of the test masses ($\sim$15–20kg) must be $\sim 3 \times 10^7$ and the quality factor of the pendulum resonances should be greater than 10^8. Discussions relevant to this are given in Rowan (1996). Obtaining these values puts significant constraints on the choice of material for the test masses and suspension fibres. One viable solution is to use fused silica masses hung by fused silica fibres (Braginsky *et al.* 1996, Rowan *et al.* 1997) although the use of other materials such as sapphire may be possible (Ju *et al.* 1996, Rowan *et al.* 1996).

Photoelectron shot noise

For gravitational wave signals to be detected, the output of the interferometer must be held at a particular point on an interference fringe. An obvious point to choose is halfway up an interference fringe since the change in photon number produced by a given differential change in arm length is greatest at this point. The interferometer may be stabilised to this point by sensing any changes in intensity at the interferometer output with a photodiode and feeding the resulting signal back, with suitable phase, to a transducer capable of changing the position of one of the interferometer mirrors. Information about changes in the length of the interferometer arms can then be obtained by monitoring the signal fed back to the transducer.

As mentioned earlier it is very important that the system used for sensing the optical fringe movement can sense strains in space of $2 \times 10^{-23}/\sqrt{\text{Hz}}$ or differences in the lengths of the two arms of less than 10^{-19}m, a minute displacement compared to the wavelength of light $\sim 10^{-6}$m. A limitation to the sensitivity of the optical readout scheme is set by shot noise in the detected photocurrent. From a consideration of the number of photoelectrons measured in a given time it can be shown that the detectable strain sensitivity depends on the level of laser power P of wavelength λ used to illuminate the interferometer of arm length L, and on the time τ, such that the detectable strain in time τ is given by

$$\tau = \frac{1}{L[\lambda hc/2\pi^2 P\tau]^{0.5}} \tag{2}$$

or

$$\frac{\text{Detectable strain}}{\sqrt{\text{Hz}}} = \frac{1}{L[\lambda hc/\pi^2 P]^{0.5}}, \tag{3}$$

where c is the velocity of light and h is Planck's constant. Thus to achieve the necessary level of strain sensitivity requires a laser, operating at a wavelength of 10^{-6}m, to provide 6×10^6W of power at the input to the interferometer. This is a formidable requirement.

However the situation can be helped greatly if a multipass arrangement is used in the arms of the interferometer, as this multiplies up the apparent movement by the number of bounces the light makes in the arms. The multiple beams can either be separate as in an optical delay line, or may lie on top of each other as in a Fabry-Perot resonant cavity, as shown in Figure 2.

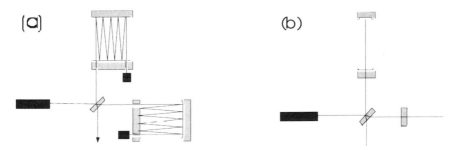

Figure 2. *Michelson Interferometers with (a) delay lines and (b) Fabry-Perot cavities in the arms of the interferometer.*

The maximum number of bounces which are desirable is set by the fact that the light should not stay in the arms longer than the characteristic time-scale of the signal, otherwise some cancellation of the detected signal may occur. Thus if signals of characteristic time-scale 1ms are to be searched for the number of bounces should not exceed 50 for an arm length of 3km. With 50 bounces the required laser power is reduced to 2.4×10^3W, still a formidable requirement.

It can be shown that the optimum signal to noise ratio in a Michelson interferometer is obtained when the arm lengths are such that the output light is very close to a minimum. (This is not intuitively obvious and is discussed more fully by Edelstein *et al.* (1978)). Thus rather than lock the interferometer to the side of a fringe as discussed above, it is usual to make use of a modulation technique to operate the interferometer close to a null in the interference pattern. An electro-optic phase modulator placed in front of the interferometer can be used to phase modulate the input laser light. If the arms of the interferometer are arranged to have a slight mismatch in length this results in a detected signal which when demodulated, is zero with the cavity exactly on a null fringe and changes sign on different sides of the null providing a bipolar error signal; this can be fed back to the transducer controlling the interferometer mirror to hold the interferometer locked near to a null fringe.

In this situation if the mirrors are of very low optical loss, nearly all of the light supplied to the interferometer is reflected back towards the laser. In other words the laser is not properly impedance matched to the interferometer. The impedance matching can be improved by placing another mirror of carefully chosen transmission, a power recycling mirror, between the laser and the interferometer so that a resonant cavity is formed between this mirror and the rest of the interferometer and no light comes back towards the laser (Drever *et al.* 1983). There is then a power build-up inside the interferometer, a build-up which can be high enough to create the required kilowatt of laser light at the beam-splitter from $\sim$ 10W or so out of the laser.

Using appropriate optical configurations the required laser power may thus be reduced to a level where laser sources are available; however stringent requirements on technical noise and reliability must be satisfied.

2.4   Laser specifications

Power fluctuations

As described above gravitational wave interferometers are typically designed to operate with the length of the interferometer arms set such that the output of the interferometer is close to a minimum in the output light. If the interferometer were operated exactly at the null point in the fringe pattern, then in principle it would be insensitive to power fluctuations in the input laser light. However, in practice, there will be small offsets from the null position causing some sensitivity to this noise source. In this case it can be shown (Hough *et al.* 1991) that the required power stability in the frequency range of interest for gravitational wave detection may be estimated to be

$$\delta P/P \sim h(\delta L/L)^{-1} \tag{4}$$

where $\delta P/P$ are the relative power fluctuations of the laser, and δL is the offset from the null fringe position for an interferometer of arm length L. From calculations of the effects of low frequency seismic noise on long baseline detectors (Hough *et al.* 1991) it can be estimated that δL is likely to be of the order of 10^{-13}m RMS, thus to detect a strain, h, of 10^{-22} over millisecond time-scales requires

$$\frac{\delta P}{P} < \frac{10^{-7}}{\sqrt{\text{Hz}}} \quad \text{at} \quad 1\text{kHz.} \tag{5}$$

To achieve this level of power fluctuations typically requires the use of active stabilisation techniques.

Frequency fluctuations

It can be shown that a change δx in the differential path length, x, of the interferometer arms causes a phase change $\delta\phi$ in the light at the interferometer output given by

$$\delta\phi = \frac{2\pi}{c}(\nu\delta x + x\delta\nu)$$

where $\delta\nu$ is a change in the laser frequency ν and c is the speed of light. From this it can be seen that if the lengths of the interferometer arms are exactly equal then $x = 0$ and the interferometer output is insensitive to fluctuations in the laser frequency, provided that, in the case of Fabry-Perot cavities in the arms, the fluctuations are not so great that the cavities cannot remain on resonance. In practice however, differences in the optical properties of the interferometer mirrors mean that the effective arm lengths will be slightly different, perhaps by a metre or so. Then the relationship between the limit to detectable gravitational wave amplitude and the fluctuations $\delta\nu$ of the laser frequency ν is given by (Hough *et al.* 1991)

$$\delta\nu/\nu \sim h(x/L)^{-1} \tag{6}$$

Hence for a detector with arms of length 3km to achieve a sensitivity of 10^{-22} for millisecond pulses requires

$$\frac{\delta\nu}{\nu} < \frac{10^{-20}}{\sqrt{\text{Hz}}}. \tag{7}$$

This level of frequency noise may be achieved by the use of appropriate laser frequency stabilisation systems involving high finesse reference cavities (Hough *et al.* 1991).

2.4.1 Beam geometry fluctuations

Fluctuations in the lateral or angular position of the input laser beam, along with changes in its size and variations in its phase-front curvature may all couple into the output signal of an interferometer and reduce its sensitivity. For example, fluctuations in the lateral position of the beam may couple into interferometer measurements through a misalignment of the beam-splitter with respect to the interferometer mirrors. A lateral movement of the input beam δz, coupled with an angular misalignment of the beamsplitter of $\alpha/2$ results in a phase mismatch $\delta\phi$ of the interfering beams, such that (Rüdiger *et al.* 1981)

$$\delta\phi = (4\pi/\lambda)\alpha\,\delta z \tag{8}$$

Assuming a typical beamsplitter misalignment of $\sim 10^{-7}$ radians, means that to achieve sensitivities of the level described above using a detector with 3km arms, and 50 bounces in each arm, a level of beam geometry fluctuations of approximately 2×10^{-12} m/$\sqrt{\text{Hz}}$ at 1kHz is required. Typically this will mean that the beam positional fluctuations of the laser need to be suppressed by several orders of magnitude. The two main methods of reducing beam geometry fluctuations are (1) passing the input beam through a single mode optical fibre and (2) using a resonant cavity as a mode cleaner.

Passing the beam through a single mode optical fibre helps to eliminate beam geometry fluctuations as deviations of the beam from a Gaussian TEM_{00} mode are equivalent to higher order spatial modes, which are thus attenuated by the optical fibre. However there are limitations to the use of optical fibres mainly due to the limited power handling capacity of the fibres; care must also be taken to avoid introducing extra beam geometry fluctuations from movements of the fibre itself.

A cavity may be used to reduce beam geometry fluctuations if it is adjusted to be resonant only for the TEM_{00} mode of the input light. Any higher order modes should thus be suppressed (Rüdiger *et al.* 1981). The use of a resonant cavity should allow the handling of higher laser powers as well as having the additional benefits of acting as a filter for fast fluctuations in laser frequency and power (Skeldon 1996, Willke 1998). This latter property is extremely useful for the conditioning of the light from some laser sources as will be discussed below.

Laser design

From equation (2) it can be seen that the photon-noise limited sensitivity of an interferometer is proportional to $(P)^{-0.5}$ where P is the laser power incident on the interferometer and $\lambda^{0.5}$ where λ is the wavelength of the laser light. Thus single frequency lasers of high output power and short wavelength are desirable to obtain the best interferometer sensitivity. With these constraints in mind, laser development has concentrated on argon-ion lasers and Nd:YAG lasers. Argon-ion lasers emitting light at 514nm have been used to illuminate several interferometric gravitational wave detector prototypes, see for example Shoemaker *et al.* (1988), Robertson *et al.* (1995). They have an output power in the required single spatial mode of operation (TEM_{00q}) typically of around several watts, sufficient for this type of laser to have been proposed as the initial laser for a full-scale interferometric detector (Vogt *et al.* 1989). For advanced detectors higher laser powers would be desirable and it has been demonstrated that the output of several argon-ion lasers could be coherently added for this purpose (Kerr and Hough, 1989). However the disadvantages of argon-ion lasers include the increased optical absorption and more pronounced effects due to scattering for light of this shorter wavelength. In addition argon-ion lasers are relatively inefficient.

Nd:YAG lasers, emitting at 1064nm, present an alternative candidate. The longer wavelength is less desirable than the 514nm of the argon-laser, as more laser power is needed at the Nd:YAG wavelength to obtain the same sensitivity; in addition, the resulting increase in beam diameter leads to a need for larger optical components. Nd:YAG sources do however have some compelling advantages and in particular the potential for scaling Nd:YAG laser designs up to levels of 100W or more (Shine 1995a) combined with their superior efficiency, has led all the long baseline interferometer projects to choose some form of Nd:YAG light source.

Compact sources of lower powers of Nd:YAG light have been available for several years in the form of monolithic diode-pumped ring lasers (Kane and Byer 1985). Investigations have shown that the technical noise associated with these lasers may be well controlled and reduced to levels comparable to those needed for gravitational wave interferometer sources. (Kane 1990, Fritschel *et al.* 1989, Campbell *et al.* 1992, Rowan *et al.* 1993, Harb *et al.* 1994). Different approaches to obtaining high powers of low noise Nd:YAG light have been studied, these approaches having in common the use of a stable lower power laser as a master oscillator.

One approach is to use a lower power Nd:YAG master oscillator to injection lock a higher power Nd:YAG slave laser, with the length of the slave laser cavity being locked to the frequency of the light from the master oscillator (Cregut *et al.* 1989, Nabors *et al.* 1989, Golla *et al.* 1993,). Up to 20W of single frequency laser light have been obtained using this method (Shine *et al.* 1995b), which has the desirable feature that the higher power slave laser light has noise properties which are for the most part dominated by those of the master laser light. (Farinas *et al.* 1995). This is desirable since it is typically easier to apply active noise reduction techniques to stabilise the lower power master lasers. Injection locked systems of this type are being developed for use by the VIRGO, TAMA 300 and GEO 600 projects, each of which requires ~10W of laser light for initial operation.

However, adapting this technique for producing still higher powers from the slave laser requires care, since the light power needed from the master oscillator also increases. To meet this requirement, systems in which a series of lasers are successively injection locked have been proposed.

An alternative scheme has been developed for use by the LIGO project (Weichmann 1998). Light from a master laser is passed through diode-pumped Nd:YAG amplification stages in a master oscillator/power amplifier (MOPA) configuration. This approach has the advantage of having the potential to allow very high powers of cw light to be obtained using multiple amplification stages, but with the need for multiple cavity locking schemes having been removed. However, the effects of this design configuration on the noise properties of the amplified light must be addressed.

In particular, to obtain high performance from the modulation techniques discussed in Section 2.3 it is necessary that at the modulation frequency, the power fluctuations of the laser light used must be shot noise limited in the amount of light detected at the interferometer output (typically ~1W)

Previous studies of the noise properties of optical amplifiers have shown that in a given output power of light from an optical amplifier, power fluctuations exist which are in excess of those obtained from a laser of the same output power, limited by shot noise (Harris *et al.* 1992). This gain dependent excess noise arises from the beating of the spontaneous emission from the amplifier with the light being amplified. Measurements of this excess noise at rf modulation frequencies have been made using a free space Nd:YAG linear optical amplifier system (Tulloch *et al.* 1998). For this type of light source to be suitable for use in an interferometric gravitational wave detector, it is necessary to reduce these high frequency power fluctuations; a suitable technique is to pass the light through a resonant cavity similar to that used to spatially filter the input laser light as described in Section 2.4. Above the corner frequency f_c of the cavity, power and frequency

fluctuations of the laser light are reduced by a factor f/f_c where f is the frequency at
which the fluctuation occurs, and

$$f_c = (\text{cavity free spectral range})/(2 \times \text{finesse})$$

Thus the excess power noise introduced by the amplification process may be reduced to an
appropriate level. The noise properties of saturated free space Nd:YAG optical amplifiers
remain to be experimentally evaluated.

A light source with the potential to combine the increased efficiency of solid-state lasers
with the advantages of using shorter wavelength light is a frequency doubled Nd:YAG
laser. Whilst injection locked Nd:YAG systems have been used to produce up to 11.2W
of frequency doubled light (Yang 1993), reliable sources of high powers of single frequency,
controllable, doubled light are as yet not widely available. Development of these sources
requires further improvements in available non-linear doubling materials.

Signal recycling

To enhance further the sensitivity of an interferometric detector and to allow some narrowing of the detection bandwidth, (valuable in searches for continuous wave sources of
gravitational radiation), another technique known as signal recycling can be implemented
(Meers, 1988, Strain and Meers 1991, Heinzel *et al.* 1998). This relies on the fact that
sidebands created on the light by gravitational wave signals interacting with the arms do
not interfere destructively and so do appear at the output of the interferometer. If a mirror of suitably chosen reflectivity is put at the output of the system as shown in Figure 3,
then the sidebands can be recycled back into the interferometer where they resonate and
hence the signal size over a given bandwidth (set by the mirror reflectivity) is enhanced.

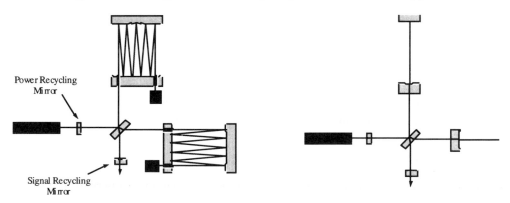

Figure 3. *The implementation of power and signal recycling on the interferometers
shown in Figure 2.*

2.5   Long baseline detectors under construction

Prototype detectors using laser interferometry have been constructed by:

- Max-Planck-Institut für Quantenoptik in Garching (Shoemaker *et al.* 1988),
- University of Glasgow (Robertson *et al.* 1995),
- California Institute of Technology (Abramovici *et al.* 1996),
- Massachusets Institute of Technology (Fritschel *et al.* 1998),
- Institute of Space and Astronautical Science in Tokyo (Araya *et al.* 1997) and
- the astronomical observatory in Tokyo (Mizuno *et al.* 1996).

These detectors have arm lengths varying from 10m to 100m and have or had either multibeam delay lines or resonant Fabry-Perot cavities in their arms. Several years ago the sensitivities of some of these detectors reached a level, better than 10^{-18} for millisecond pulses, where it was sensible to decide to build detectors of much longer baseline which should be capable detecting gravitational waves. Thus an international network of gravitational wave detectors is now under construction.

The American LIGO project comprises the building of two detector systems with arms of 4km length, one in Hanford, Washington State, and one in Livingston, Louisiana. The vacuum system, laser and input optics and first suspension system for the detector in Hanford are now installed and the vacuum system is in place in Louisiana. The French/Italian VIRGO detector of 3km arm length at Cascina near Pisa is at the stage where the central buildings are close to completion and vacuum tanks to house the interferometry are being installed. The TAMA 300 detector, which has arms of length 300m, is at a relatively advanced stage of construction at Tokyo. This detector is being built mainly underground; the vacuum system is complete and initial operation with light in the arms has started. All the systems mentioned above are designed to use resonant cavities in the arms of the detectors and use standard wire sling techniques for suspending the test masses. The German/British device is somewhat different. It makes use of a four-pass delay-line system with advanced optical signal enhancement techniques, utilises very low loss fused silica suspensions for the test masses, and should have a sensitivity at frequencies above a few hundred Hz comparable to the first phases of VIRGO and LIGO when they become operational. Construction is advancing well with the necessary buildings and vacuum pipes for the arms being in place. Installation of the suspensions for the optical elements is now beginning. Figure 4 shows the site and the directions of the two arms.

In two years time initial operation of the detector should commence and during the following years we can expect some very interesting coincidence searches for gravitational waves, at a sensitivity of approximately 10^{-21} for pulses of several milliseconds duration.

2.6   Longer baseline detectors in space

Perhaps the most interesting sources of gravitational waves, those resulting from black hole formation and coalescence, lie in the region of 10^{-4}Hz to 10^{-1}Hz and a detector whose strain sensitivity is approximately 10^{-23} over relevant timescales is required to search for these. The most promising way of looking for such signals is to fly a laser interferometer in space *i.e.* to launch a number of drag-free space craft into orbit and to compare the

Figure 4. *A bird's eye view of the GEO 600 detector, sited in Ruthe near Hannover.*

distances between test masses in these craft along arms making significant angles with each other using laser interferometry.

Two such experiments have been proposed. The first, LISA (see for example "First International LISA Symposium", (LISA, 1997)) is being proposed by an American/European team; it consists of an array of 3 drag-free spacecraft at the vertices of an equilateral triangle of length of side 5×10^6km. This cluster is placed in an Earth-like orbit at a distance of 1 AU from the Sun, and 20 degrees behind the Earth. Proof masses inside the spacecraft (two in each spacecraft) form the end points of three separate but not independent interferometers. Each single two-arm Michelson type interferometer is formed from a vertex (actually consisting of the proof masses in a 'central' spacecraft), and the masses in two remote spacecraft as indicated in Figure 5.

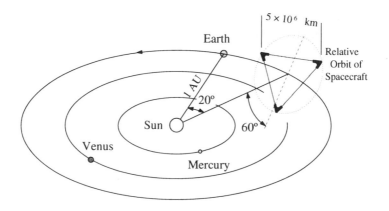

Figure 5. *The proposed LISA detector.*

The three-interferometer configuration provides redundancy against component failure, gives better detection probability, and allows the determination of polarisation of the incoming radiation. The spacecraft in which they are accommodated shield each pair of proof masses from external disturbances (*e.g.* solar radiation pressure). Drag-free control servos enable the spacecraft to follow the proof masses to a high level of precision, the drag compensation being effected using proportional electric thrusters. Illumination of the interferometers is by highly stabilised laser light from Nd:YAG lasers at a wavelength of 1.064 microns, laser powers of ~ 1W being available from monolithic, non planar ring oscillators which are diode pumped. For each interferometer, consisting of a central spacecraft and two distant spacecraft, two lasers in the central spacecraft, each pointing along one of the arms, are phase locked together so they effectively behave as a single laser.

For LISA to achieve its design performance, adjacent arm lengths have to be sensed to an accuracy of better than 30 pm/$\sqrt{\text{Hz}}$. Because of the long distances involved and the spatial extent of the laser beams, the low photon fluxes make it impossible to use standard mirrors for reflection; thus active mirrors with phase locked laser transponders on the spacecraft will be implemented. Telescope mirrors will be used to reduce diffraction losses on transmission of the beam and to increase the collecting area for reception of the beam. Given that the available laser power in each arm is of the order of 1W, and that arguments similar to those already discussed for ground based detectors can be made, photoelectron shot noise considerations suggest that the diameters of the transmitting and receiving mirrors on the space craft need to be $\sim$30cm. Further, just as in the case of the ground based detectors, the presence of laser frequency noise is a limiting factor. It leads to an error in the measurement of each arm length. If the arms are equal, these errors cancel out but, if they are unequal, the comparison of lengths used to search for gravitational waves may be dominated by frequency noise. For the 5×10^9m long arms of LISA, a difference in arm length of 10^8m is likely. Then for a relative arm length measurement of 2×10^{-12} m/$\sqrt{\text{Hz}}$, using the error budget level allowed in the LISA design for this noise source, application of equation (6) suggests that a laser stability of $\sim 6 \times 10^{-6}$Hz/$\sqrt{\text{Hz}}$ is required. This is a level much better than can be achieved from the laser on its own. Thus frequency stabilisation has to be provided.

The primary method of stabilisation is to lock the frequency of one laser in the system on to a Fabry-Perot cavity mounted on one of the craft, (see for example McNamara *et al.* (1997)) and then to effectively transfer this stability to other lasers in the system by phase locking techniques. With the temperature fluctuations inside each craft limited in the region of 10^{-3}Hz to approximately 10^{-6} K/$\sqrt{\text{Hz}}$ by three stages of thermal insulation, a cavity formed of material of low expansion coefficient such as ULE allows a stability level of approximately 30Hz/$\sqrt{\text{Hz}}$. This level of laser frequency noise is clearly much worse than the required 6×10^{-6}Hz/$\sqrt{\text{Hz}}$ and a further correction scheme is required.

Further frequency correction is provided by comparing the phase of the light returning in each arm with the phase of the transmitted light. The phase difference, measured over the time of flight in the arm, allows an estimate of laser frequency noise to be made (Shoemaker *et al.* 1988; Faller 1991; Giampieri *et al.* 1996). For the arm of length L

$$\delta\phi = (4\pi/c)L\,\delta\nu \qquad (9)$$

and thus if the spectral density $\delta\phi$ is measured, the spectral density can be estimated. This estimate can then be used to correct the signal obtained by subtracting the phase difference measurements in two adjacent arms, allowing the search for gravitational waves to be carried out. This correction is made easier if each arm length is known to a few km and the difference in arm length is known to a few tens of metres. These quantities should be available from radar and optical ranging measurements. If, however, they are not well enough known, then they can be found by searching through a range of possible values to minimise the effect of frequency noise on the "gravitational wave" signal.

There are many other issues associated with the laser interferometry for LISA which are not dealt with here and the interested reader should refer to Hough *et al.* (1997b) for a discussion of some of these.

LISA has been adopted by ESA as a Cornerstone project in their post Horizon 2000 programme. However because of financial uncertainties the timescale of this programme is somewhat long and the possibility of an earlier launch as a NASA led or ESA led collaborative medium scale mission is being enthusiastically addressed at present. The second experiment mentioned earlier is OMEGA (Hellings 1996) and calls for three craft to be placed in a geocentric orbit, the arm length in this case being 10^9m. OMEGA is currently being proposed to NASA as a MIDEX mission.

3   Conclusion

A large amount of effort worldwide is now being invested in the development of both ground and space based searches for gravitational radiation and we are entering a new era where the signals from neutron star and black hole interactions will widen our understanding of the Universe. Beyond this, however, there is the very exciting prospect that the dawning of gravitational wave astronomy will be similar to the birth of radio astronomy and X-ray astronomy and will allow the discovery of very active sources currently unknown to us.

Acknowledgements

We wish to thank G. Cagnoli for assistance with the preparation of this paper, and PPARC and the University of Glasgow for support.

References

Abramovici A, Althouse W, Camp J, Durance D, Giaime J A, Gillespie A, Kawamura S, Kuhnert A, Lyons T, Raab F J, Savage Jr R L, Shoemaker D, Sievers L, Spero R, Vogt R,Weiss R, Whitcomb S, and Zucker M, 1996, *Phys. Lett. A* **218** 157.

Amaldi E, Aguiar O, Bassan M, Bonifazi P, Carelli P, Castellano M G, Cavallari G, Coccia E, Cosmelli C, Fairbank W M, Frasca S, Foglietti V, Habel R, Hamilton W O, Henderson J, Johnson W, Lane K R, Mann A G, McAshan M S, Michelson P F, Modena I, Palletino G V, Pizzella G, Price J C, Rapagnani R, Ricci F, Solomonson N, Stevenson T R, Taber R C and Xu B-X, 1989, *Astron. and Astro.* **216** 325.

Araya A, Mio N, Tsubono K, Suehiro K, Telada S, Ohashi M, Fujimoto M-K, 1997, *Appl. Opt.* **36** 1446.

Astone P, Bassan M, Bonifazi P, Carelli P, Coccia E, Cosmelli C, Fafone V, Frasca S, Marini A, Mazzitelli G, Minenkov Y, Modena I, Modestino G, Moleti A, Palletino GV, Papa MA, Pizzella G, Rapagnani P, Ricci F, Ronga F, Terenzi R, Visco M and Votano L, 1996, International Conference on Low Temperature Physics, Prague, *Czech. J. Phys.* **46** suppl. pt S5 2907

Barish B C, 1997, *"LIGO: Status and Prospects"* Proc. of Conference on Gravitational Wave Detection Tokyo, 163, eds. K. Tsubono, M K Fujimoto, K. Kuroda, (Universal Academy Press Inc, Tokyo).

Bertotti B, Ambrosini R, Armstrong J W, Asmar S W, Comoretto G, Giampieri G, Iess L, Koyama Y, Messeri A, Vecchio A and Wahlquist H D, 1995, *Astron. and Astr.* **296** 13.

Braginsky V B, Mitrofanov V P and Tokmakov K V, 1996, *Phys. Lett. A.* **218** 164.

Campbell A M, Rowan S and Hough J, 1992, *Phys. Lett. A* **170** 363.

Ciufolini and Fidecaro F, 1997, "Gravitational Waves, Sources and Detectors" *Edoardo Amaldi Foundation Series* **2** (eds. , World Scientific)

Cregut O, Man C N, Shoemaker D, Brillet A, Menhert A, Peuser P, Schmitt N P, Zeller P and Wallermoth K, 1989, *Phys. Lett. A.* **140** 294.

Danzmann K and the LISA Study Team,1996, *Class. Quant. Grav.* **13** 11A A247-50.

Drever R W P, Hough J, Munley A J, Lee S-A, Spero R, Whitcomb S E, Ward H, Ford G M, Hereld M, Robertson N A, Kerr I, Pugh J R, Newton G P, Meers B, Brooks E D III and Gursel Y, 1983, *"Quantum optics, experimental gravitation and measurement theory"* 503, eds. Meystre P, Scully M.O., (Plenum Press, New York).

Edelstein W A, Hough J, Pugh J R and Martin W, 1978, *J. Phys. E.* **11** 710.

Faller J, 1991 in Workshop Proceedings : Technologies for a Laser Gravitational Wave Observatory in Space (LAGOS), JPL D-8541, 2.

Farinas A D, Gustafson E K and Byer R L, 1995, *J. Opt. Soc. Am.B* **12** 2.

Fritschel P, Jeffries A and Kane TJ, 1989, *Opt. Lett.* **14** 993.

Fritschel P, Gonzalez G, Lantz B, Saha P and Zucker M, 1998, *Phys.Rev.Lett.* **80** 3181.

Giampieri G, Hellings R, Tinto M, and Faller J, 1996,*Opt. Comm.* **123** 669.

Golla D, Freitag I, Zellmer H, Schone W, Kropke I and Welling H, 1993, *Opt. Comm.* **98** 86.

Harb C C, Gray M B, Bachor H-A, Schilling R, Rottengatter P, Freitag I and Welling H, 1994, *IEEE J. Quant. Elec.* **30** 2907.

Harris M, Loudon R, Shepherd T J and Vaughan J M, 1992, *J. Mod. Opt.* **39** 1195.

Heinzel G, Strain K A, Mizuno J, Skeldon K D, Willke B, Winkler W, Schilling R and Danzmann K, 1998, *Phys. Rev. Lett.* In press.

Hellings R W, 1996,*Cont. Phys.* **37** 457.

Heng I S, Blair D G, Ivanov E N and Tobar M, 1996, *Phys. Lett. A.* **218** 190.

Hough J, Ward H, Kerr G A, Mackenzie N L, Meers B J, Newton G, Robertson D I, Robertson N A, and Schilling R, 1991, in *The detection of gravitational waves* ed. D. Blair,(Cambridge University Press.)

Hough J, for the GEO 600 Team, 1997a, *Proc. of Conference on Gravitational Wave Detection, Tokyo*, 175, eds. K. Tsubono, M K Fujimoto and K Kuroda, (Universal Academy Press Inc, Tokyo).

Hough J, for the LISA Science Team, 1997b *Fundamental Physics in Space*, Alpbach Summer School 1997, ESA SP **420** 253.

Hulse RA, 1994, *Rev. Mod. Phys.* **66** 699.

Kerr G A and Hough J, 1989, *Appl. Phys.B* **49** 491.

Ju L, Notcutt M, Blair D, Bondu F and Zhao C N, 1996, *Phys. Lett. A.* **218** 197.

Kane T J, 1990, *IEEE Phot. Tech. Lett.* **2** 244.

Kane T J and Byer R L, 1985, *Opt. Lett.* **10** 65.

Kaspi V M, Taylor J H and Ryba M, 1994, *Astr. J.* **248** 712.

LISA, 1997, "First International LISA Symposium", 1997, *Class. Quant. Grav.* **14** 6.

McNamara P W, Ward H, Hough J, Robertson D, 1997, *Class. Quantum Grav.* **14** 1543.

Meers B J, 1988, *Phys. Rev.D* **38** 2317.

Mizuno E, Kawashima N, Miyoke S, Heflin E G, Wada K, Naito W, Nagano S and Arakawa K, 1996, *Proc. VIRGO 96,* 108, (Cascina, World Scientific).

Nabors C D, Farinas A D, Day T, Yang S T, Gustafson E K and Byer R L, 1989, *Opt. Lett.* **14** 21.

Robertson D, Morrison E, Hough J, Killbourn S, Meers B J, Newton G P, Strain K A and Ward H, 1995, *Sci. Rev. Instr.* **66** 4447.

Rowan S, Campbell A M, Skeldon K and Hough J, 1994, *Phys. Lett. A.* **41** 1263.

Rowan S, Twyford S M and Hough J, 1996, *Proc. 2nd E. Amaldi Conf. on Gravitational Waves, CERN, Geneva,* in press.

Rowan S, Twyford S M, Hutchins R, Kovalik J, Logan J E, McLaren A C, Robertson N A and Hough J, 1997, *Phys. Lett. A.* **233** 303.

Rüdiger A, Schilling R, Schnupp L, Winkler W, Billing H and Maischberger K, 1981, *Optica Acta* **26** 641.

Saulson P R, 1984, *Phys. Rev. D.* **30** 732.

Saulson P R, 1990, *Phys. Rev. D* **42** 2437.

Shine Jr. R J, Alfrey A J and Byer R L, 1995a, *Opt. Lett.* **20** 459.

Shine Jr. R J, Alfrey A J and Byer R L, 1995b, *OSA Proc. on Advanced Solid-State Lasers,* **24** 216.

Shoemaker D, Schilling R, Schnupp L, Winkler W, Maischberger K and Rüdiger A, 1988, *Phys. Rev. D* **38** 423.

Skeldon K D, Strain K A, Grant A I and Hough J, 1996, *Rev. Sci. Inst.***67** 2443.

Spero R E, 1983, in *"Science Underground"* ed. M.M. Nieto *et al.* AIP New York.

Strain K A and Meers B J, 1991, *Phys. Rev. Lett.* **66** 1391.

Taylor J H, 1994, *Rev. Mod. Phys.* **66** 711.

Tsubono K and the TAMA collaboration,1997, *"TAMA project" Proc. of Conference on Gravitational Wave Detection, Tokyo,* 183, eds. Tsubono K, Fujimoto M.K., Kuroda K., (Universal Academy Press Inc, Tokyo).

Tulloch W M, Rutherford T S, Huntingdon E H, Ewart R, Harb C C, Wilke B, Gustafson E K, Fejer M M, Byer R L, Rowan S and Hough J, 1998, *Opt. Lett.* **23.** 1852

Vinet J-Y *et al.* , Proc. of Gravitation and Cosmology, ICGC-95 Conference, Pune, India 13-19 Dec 1995 Astrophysics and Space Science Library, Kluwer Academic Publishers 1997 211 89.

Vogt R E, Drever R W P, Thorne K S, Raab F J and Weiss R, 1989, "A laser interferometer gravitational wave observatory", Proposal to the National Science Foundation.

Weber J, 1969, *Phys. Rev. Lett.* **22** 1320.

Weber J, 1970, *Phys. Rev. Lett.* **25** 180.

Weichmann W, Kane T J, Haserot O, Adams F, Truong G and Kmetec J D, 1998,*CLEO 98 Optical Digest* **6** 432.

Willke B, Uehara N, Gustafson E K, Byer R L, King P J, Seel S U and Savage Jr R L, 1998, *Opt. Lett.* **23** 21.

Will C M, 1983, *"Theory and experiment in gravitational physics"* (revised edition, C.U.P., Cambridge)

Will C M, 1995, *"The confrontation between General Relativity and Experiment: an update" in General Relativity, the proceedings of the 46th Scottish Universities Summer School in Physics* (Institute of Physics Press)

Yang S T, 1993, PhD thesis, Stanford University.

Lasers in material processing

Peter Loosen

Fraunhofer-Institut für Lasertechnik, Aachen, Germany

1 Introduction

Over the past two decades lasers have developed into reliable, productive and widely used tools in industrial manufacturing. Even more use would be made of lasers if the cost, size and weight of the laser systems could be reduced. This may happen with the use of diode lasers. High power diode lasers are compact, light-weight and have the potential for low cost. This may open up new markets and applications for laser material processing.

The next section of this chapter will give an overview of the applications of lasers in material processing. Section 3 will discuss the issues and present technical solutions of the use of diode lasers in this field.

2 Industrial materials processing with lasers

2.1 Examples of laser materials processing

Evidence for the growing importance of lasers in industrial manufacturing is given in Figure 1 which shows the annual sales of lasers used in materials processing. In this context the term "laser" is used for laser beam sources. "Laser systems" refers to an integrated system involving the laser, beam steering and workpiece handling. The picture indicates a steady growth over the past several years with an average annual growth rate of about 20% and a total world market volume in 1998 of 1.8 billion DM for lasers and 4.9 billion DM for laser systems.

The broad range of applications which is driving this market development is illustrated in Figure 2 using automotive production as an example. In today's car numerous parts are produced using lasers, mainly in the field of welding and cutting. Advantages of using lasers instead of conventional tools are: high processing speed and thus high productivity, high quality and the possibility of 100% in-line quality control and high flexibility. These general aspects will be put in concrete form in the rest of this section along with examples for standard applications such as cutting, welding and surface treatment.

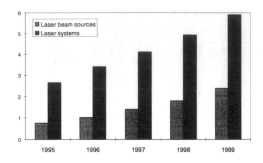

Figure 1. *Development of the world market for lasers and laser systems used in industrial materials processing (in DM billions, 1998 and 1999 estimated) (Mayer 1998).*

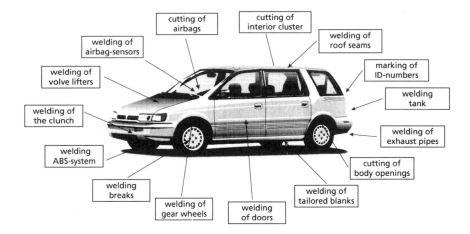

Figure 2. *Illustration of the broad range of laser applications: laser manufactured parts and laser applications in a typical car (by courtesy of Rofin-Sinar, Hamburg).*

Cutting flat sheet steel is the biggest market segment for today's high-power laser systems. Figure 3a shows the impressive range of steel thicknesses which can be cut fully automatically with systems such as these shown in Figure 3b. Sheet thicknesses ranging from tens of micrometres up to about 50mm can be processed, leaving cut edges of high rectangularity, low surface roughness and with negligible distortion of the material at the cut edges. Cutting speeds are up to 150m/min (0.2mm steel) or 40m/min (1mm steel). Systems capable of high accelerations can be used to cut more than 6 holes per second in a 1mm thick steel sheet with hole diameter about 10mm. This productivity is comparable with conventional punching systems. However the laser system has much higher flexibility, and is thus well adapted for today's needs for flexible and productive manufacturing of small lot sizes. The majority of lasers used for cutting are continuously working (CW) CO_2-lasers with an output power from 1 to 5kW (see section 2.3).

Welding with high-power lasers is an application that is gaining increased importance

Figure 3. *(a)(left) Examples of laser cut steel parts, illustrating the broad application range; (b)(right) typical laser cutting system, by courtesy of Trumpf, Ditzingen.*

Figure 4. *Welding of metals with lasers: low weight steel beam (left); typical weld seam cross section (middle); welding of an automotive Al-space frame with a fibre-guided Nd:YAG-laser (right).*

in the automotive industry as well as in steel and ship-building. Figure 4 shows two examples: a low-weight steel beam, which can be flexibly tailored to the specific end-use, and a space-frame for a car, made from aluminium. The main advantages of using lasers in both cases are the small width of the weld seam and the high processing speed. These are due to the tiny focused beam. This goes along with low energy input into the material compared with conventional welding techniques, thus reducing thermal distortion and tolerances of the components. For welding applications, CO_2-lasers in the power range from 2 to 20kW are used in combination with handling systems. These either move the beam by means of mobile folding mirrors or move the workpiece itself. A flexible alternative is to use optical fibres for beam delivery in conjunction with a robot-arm and

Figure 5. *Steering rack (left): the teeth have been hardened with laser; the micrograph (right) shows a cross-section of one tooth with the hardened zone.*

focussing head. This is shown in Figure 4. However, because of their wavelength, CO_2 lasers cannot normally be used with conventional optical fibres and in this example a 3kW Nd:YAG laser is in use.

Surface treatments such as laser-based transformation hardenings have been investigated and developed for many years (see example in Figure 5). These laser-based techniques have not yet been widely used in industry despite their numerous advantages. The laser allows the rapid heating of a well-localised work-piece zone to bring it above the austenitic temperature of carbon steel. Subsequent rapid cooling, by the conduction of heat into the cold bulk of the metal, creates the hardened surface. The relatively low heat input to the work-piece leads to low distortion, high flexibility and in many cases, higher hardness figures than with conventional techniques. However, lasers are very much more expensive than the conventional heat sources for this treatment and this restricts laser hardening to niche applications at present. When the laser cost reduces with the introduction of high-power diode lasers to this field, a much greater use of this technique is expected.

2.2   Fundamental processes in laser materials processing

The starting process in laser materials processing is always the absorption of a raw or focused laser beam at the surface of the workpiece. The beam is usually incident normal to the surface as indicated in Figure 6. This absorption is determined by the absorptivity and reflectivity of the material. Metals, for example, absorb an increasing fraction of the beam power as the wavelength decreases from the 10.6μm of CO_2 lasers to the 1.06μm of Nd:YAG lasers and to the 0.8–1.0μm of diode lasers. For material hardening the surface shape does not change during processing and data taken from Figure 6 can give a direct measure of how efficiently laser power can be used. In this case the shorter wavelength lasers are preferable due to their higher absorptivity.

However, for laser cutting and welding the environment of the absorption zone changes significantly during processing. Much higher absorption is achieved due to absorption at higher angles of incidence and multiple reflections in the absorption zone (Dausinger 1995).

The second stage of laser material processing is the transfer of heat into the workpiece by conduction. This is illustrated in Figure 7. In the initial phase the heat-flow is nearly one dimensional into the depth of the workpiece. Lateral heat conduction , which can be viewed as a loss mechanism, is negligible. However, at longer times the isotherms turn into a three dimensional form. The absorption zone can then be thought of as a point source with large lateral heat conduction losses. The larger the laser spot size and the smaller the thermal diffusivity, the longer the process can be regarded as being in the

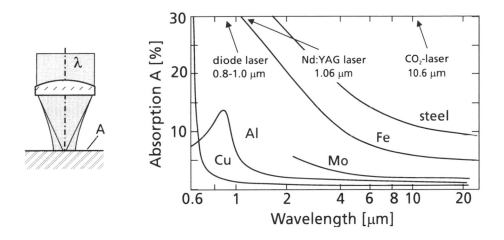

Figure 6. *Typical data for the absorption of laser radiation at different metals with ideal surfaces as a function of the laser wavelength.*

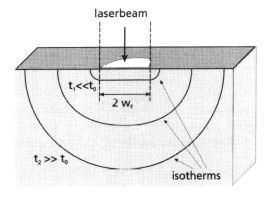

Figure 7. *Conduction of the heat, being absorbed at the surface, into the bulk material. It is 1-dimensional for $t_1 \ll t_0$ and 3-dimensional for $t_2 \gg t_0$*

one dimensional case. The characteristic time scale t_0 associated with the transition from largely 1-D heat flow to 3-D is by

$$t_0 = \frac{w_F^2}{\kappa} \tag{1}$$

where w_F is the radius of the focussed laser beam spot$[\mu m]$ and κ is the thermal diffusivity $[\text{m}^2\text{s}^{-1}]$, itself defined by $\kappa = \lambda/\rho c$ in terms of λ, the thermal conductivity $[\text{Wm}^{-1}\text{K}^{-1}]$, ρ, the density $[\text{kgm}^{-3}]$ and c, the heat capacity $[\text{Jkg}^{-1}\text{K}^{-1}]$.

Different types of materials processing, work on different timescales relative to t_0. For most cases of laser transformation hardening the laser power and the interaction zone width are chosen to allow the absorption zone to be heated up to the required 700–1000^{circ}C in a time scale much less than t_0. During this time the heat flow is largely

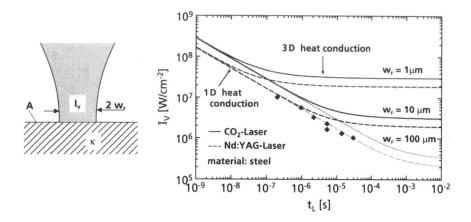

Figure 8. *Threshold intensities for the onset of vaporisation at a steel surface. The calculations have been performed for a static beam of Gaussian intensity distribution.*

one-dimensional and little energy is lost. To achieve this rapid heating the laser beam intensities are in the range of 10^2–10^4 Wcm^{-2}. Once the required temperature is reached, the laser power is switched off and the material cools down rapidly by three dimensional heat conduction.

In the case of welding, drilling and cutting however, the surface shape does not stay unaffected: the material has to be heated above the melting point or, in case of drilling and welding, up to the vaporisation temperature. Depending on the interaction time, processing speed, focal diameter and absorption properties, a certain threshold intensity I_V is required to reach the vaporisation temperature T_V. Treusch (1985) found that for the case of a static intensity distribution of Gaussian shape the beam-centre intensity required for vaporisation [in W/cm^2] is

$$I_V \propto \frac{T_V \kappa}{A w_F \arctan\left(\frac{8\kappa t_L}{w_F^2}\right)^{1/2}} \tag{2}$$

where A is the absorption coefficient of the material and t_L is the interaction time (and κ and w_F as before). In Figure 8 this relation is given as a function of interaction time t_L. For a beam with a diameter of $2w_F$ moving with a constant speed v, the time axis can be qualitatively transformed into a speed axis by the relation: $v = 2w_F/t_L$.

If sufficient metal vapour is produced by a laser operating above the threshold intensity, a "keyhole" is formed in the material. This hole is kept open by the vapour pressure opposing the capillary forces that tend to close the hole. This keyhole is vital to laser welding and drilling and is the reason for the high efficiency and the narrow interaction zone of these processes. The incident laser energy is able to penetrate the vapour-filled keyhole and deposit heat into the keyhole throughout the depth of the material. This is in contrast to conventional heat sources which have to transport energy from the surface through the material thickness by slow and inefficient heat conduction into the bulk material. The laser can be considered as a "hot needle" which is pulled through the workpiece, efficiently depositing energy through the depth of the material.

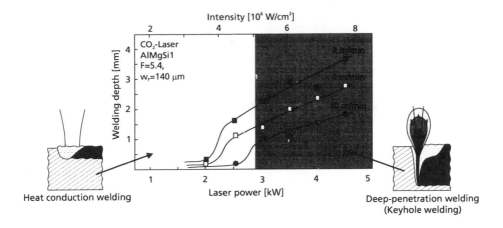

Figure 9. *Increase of the welding depth with laser intensity due to the formation of a keyhole.*

The advantage of keyhole welding becomes evident in Figure 9. Below the critical threshold intensity, no keyhole is formed. The welding process is dominated by heat conduction and is quite inefficient: welding speeds are low. Above the threshold intensity however, the beam penetrates deeply into the workpiece and welding speeds are much higher. In order to achieve this regime, called "deep penetration" or "keyhole welding", a laser with sufficiently high beam quality is needed in order for it to be focussed a small enough spot so that the required threshold intensity is reached (see Section 3.3).

In the case of laser cutting no vapour-filled keyhole is produced. Instead, the molten material is blown out behind the interaction zone by a high speed jet of reactive or inert process gas, fed by an appropriate nozzle. A typical geometry is shown in Figure 10. High

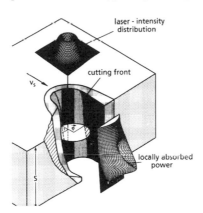

Figure 10. *Principle of laser cutting (Petring 1994).*

beam quality is needed because the cutting kerf has to be small. All the material in the cutting kerf has to be molten at the applied power; high speed thus requires small kerf,

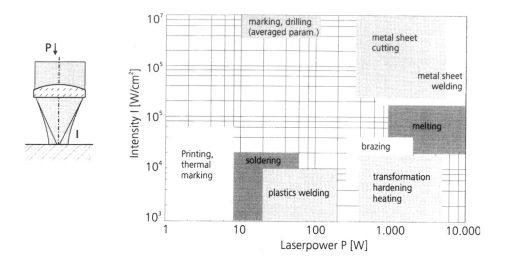

Figure 11. *Typical parameter ranges for laser power and process intensities for common laser applications.*

small focal spot and high beam quality.

In addition to the examples discussed above, many other laser applications are shown in Figure 11 which indicates the laser power and intensity typically required for the specific processes. Apart from the "high-power" and "high-intensity" applications of cutting and welding, lasers have a variety of diverse uses, especially in the low-intensity range. Examples are hardening, welding of plastics and soldering. Cutting, welding and marking presently represent the largest market segment. The other processes are also highly developed from the process point of view but are not widely used due to the high cost of conventional lasers.

The process oriented data in Figure 11 can be transformed into the laser device oriented requirements on laser power and beam quality that are necessary to produce the desired intensity. Beam quality here is used as a term defining how finely a laser can be focused. It is measured by the beam-parameter product Q (in mm·mrad). The smaller the Q value, the smaller the focal radius (for a detailed discussion see also Section 3.3). As indicated in Figure 12 the beam parameter product Q can be directly derived from the process intensity by assuming a certain numerical aperture of the focusing optics Θ_f. The straight lines in Figure 12 represent lines of constant intensity, applications requiring low intensity are located at the top of the diagram, those with high intensity at the bottom.

The result of the transformation of the data in Figure 11 into a power/beam-quality diagram is shown in Figure 13 for a typical focusing system with an NA of 0.12 (F#4). The data are put together with two curves which indicate where today's industrial CO_2- and Nd:YAG-lasers are located. The picture makes evident that from the point of view of laser power and beam quality any process can be served by these lasers, especially the important applications of the cutting and welding of metals.

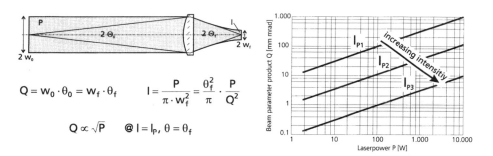

$$Q = w_0 \cdot \theta_0 = w_f \cdot \theta_f \qquad I = \frac{P}{\pi \cdot w_f^2} = \frac{\theta_f^2}{\pi} \cdot \frac{P}{Q^2}$$

$$Q \propto \sqrt{P} \qquad @ \, I = I_p, \, \theta = \theta_f$$

Figure 12. *Relation between the process intensity and the beam quality Q (beam parameter product), required to achieve this intensity.*

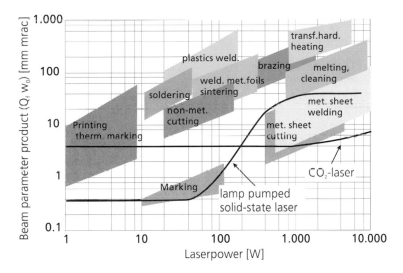

Figure 13. *Laser beam qualities and laser powers for a range of laser applications. Present commercial CO_2- and Nd:YAG-lasers are located above the respective curves.*

2.3 Comparison of laser systems

The present workhorses in industrial laser manufacturing are CO_2 lasers and lamp-pumped Nd:YAG-lasers. Two typical high-power cw systems are shown in Figure 14. Over the past 10 years such lasers have grown into reliable and rugged tools and have proven their industrial usefulness in a broad variety of different applications and industrial environments. Some drawbacks of these systems are

- long wavelength of CO_2-lasers which results in low absorption at metal surfaces,
- low efficiencies, large energy consumption, heavy and bulky power supplies,
- no fibre transportation possible for CO_2-lasers,
- short periods between maintenances.

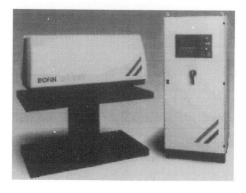

Figure 14. *Examples of industrial lasers for materials processing. Left: 2kW CO₂-laser (by courtesy of Rofin-Sinar, Hamburg). Right: 3kW lamp-pumped Nd:YAG-laser (by courtesy of Haas, Schramberg).*

Intensive developments are going on worldwide with the aim of providing new lasers to overcome these drawbacks. Diode pumping of solid state lasers instead of lamp pumping will raise their efficiency, reliability and beam quality considerably — see Table 1. High-power diode lasers for direct applications will push efficiency and compactness even further and will contribute to reduced production costs of lasers as well. However, the main development issue with high-power diode lasers for direct materials processing is their poor beam quality and beam shaping. These topics will be addressed in detail in the next section.

	CO_2-laser	Nd:YAG-laser (lamp pumped)	high-power diode laser
Wavelength	10.6μm	1.06μm	0.8–1.0μm
Output power	up to 20kW	up to 5kW	up to 4kW
Efficiency	5–10%	2–5 %	20–40%
Typical beam-parameter product [mm×mrad] @ 2kW	0.3	25	600
Fibre transportation	no	yes	yes
Maintenance intervals [h]	2,000–3,000	500–1,000	5,000–10,000

Table 1. *Comparison of typical performance data of industrial lasers.*

3   High-power diode lasers for material processing

3.1   Direct applications of high-power diode lasers

High power diode laser systems in most cases are based on diode laser bars of the type shown in Figure 15. In order to achieve high output powers from such a bar, a number of individual diode laser structures are arranged in parallel on the bar, which is typically 10mm wide.

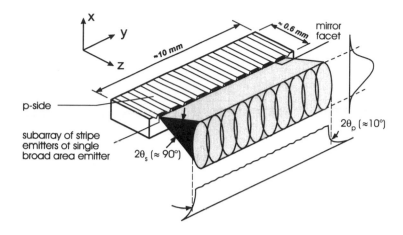

Figure 15. *Diode laser bar, as typically used in industrial diode laser systems*

The individual diode laser structures on the bar may consist of an array of stripe emitters or be "broad area" emitters. In the direction perpendicular to the plane of the pn-junction (x-dimension in Figure 15) the beam quality in general is close to diffraction limited as the beam is emitted from a region about 1μm wide. The corresponding divergence angle is quite large and may achieve figures of up to NA 0.8 (approx. 100 degrees full angle). Since the beam is diverging rapidly in this dimension, the x-axis in Figure 15 is usually called the "fast axis".

In contrast to the fast axis, the emitting width in the y-direction of an individual diode laser structure is usually much larger, up to some 100μm. The beam quality is much lower compared to the fast axis and typically amounts to some ten times the diffraction limit. Typical divergence angles are thus in the range of several degrees: this slowly diverging behaviour is the reason why this axis usually is called the "slow direction". In most current diode laser bars, adjacent diode laser structures are completely incoherent with each other. Coupling between the individual structures is often suppressed by v-grooves as shown in Figure 15 in order to suppress "super modes" and lasing in transverse direction.

In all discussions of the current section the coordinate system is labelled as shown in Figure 15 : the propagation coordinate of the beam is the z-direction; x and y denote the lateral coordinates.

In order to achieve the required high output powers from diode bars, proper cooling of the components has to be ensured. This is done very efficiently by means of microchannelled heat sinks, which are used as diode laser mounts (Beach *et al.*1992). Figure 16 illustrates a specific type based on micro-structured copper sheets (Loosen *et al.* 1998).

Fine channel structures are laser cut into the copper sheets and cooling water is forced through them. This creates a large inner surface and a good thermal contact between the heat conducting solid and the coolant. The coarse structures seen in the picture serve as water manifolds, guiding the water from the inlet openings to the microchannels and back to the outlet openings. Typical water flow rate is around 500ml/min per cooler at feeding pressures in the range of 1–4bar.

Figure 16. *Left: Microchannel coolers, built up from thin copper sheets (typical thickness 0.3mm). Fine channel structures down to a width up 100μm are laser cut into the sheets, which are subsequently diffusion bonded. Right: Final cooler with the water-inlets and outlets, going through the cooler in order to enable stacking of several components.*

A complete diode laser sub-assembly is shown in the right-hand part of Figure 16. The diode laser bar has been soldered on the front edge of the cooler and has been supplied with a top-electrode as n-contact.

In many systems, the diode lasers are put on top of each other in order to increase the output power to the desired level. Such a stack is shown in the lefthand part of Figure 17. The diode lasers are put in parallel with regard to the cooling circuit and in series with respect to the electrical supply. The water connections of the cooler have to go through as shown in Figure 16. As the stack is built up, appropriate water seals have to be made between sub-assemblies and good electrical connections must be made to allow the drive currents of up to 50A to pass with low loss.

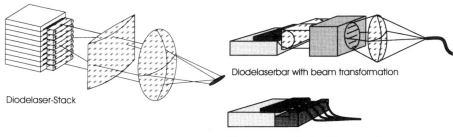

Figure 17. *Schematics of different technical set-ups for incoherent beam combination: Stacking (left), fibre coupling of a transformed beam (upper right) and direct coupling into a fibre bundle (lower right).*

3.2   Principles of incoherent beam combination

Incoherent beam combination of individual diode lasers is widely used in high-power diode lasers for materials processing, medical applications and for the pumping of solid-state

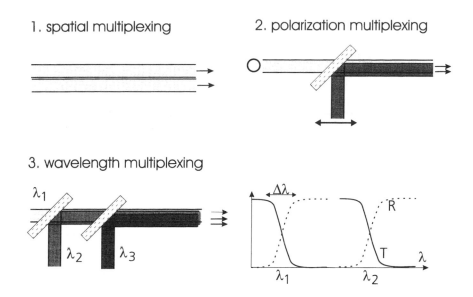

Figure 18. *Basic principles of incoherent beam combination.*

lasers. It is a straightforward way to achieve relatively high output powers at high system efficiency, although the potential to achieve high beam quality is limited (section 3.3). Schematic pictures of present technical setups are given in Figure 17. At low output powers of up to approximately 100W, where only a couple of diode lasers is needed, the techniques of direct coupling into a fibre bundle or the utilisation of a beam transformation system (Section 3.4) are common. At higher output powers of up to several kilowatts another set up is used: the individual diode lasers are stacked on top of each other or arranged in parallel.

The basic physical principles used in these systems are shown in Figure 18. The most common arrangement utilises "spatial multiplexing" where the individual laser beams are simply set side-by-side in a one- or two-dimensional array, thus increasing output power as well as the size of the beam. Two complementary techniques, which allow an increase in output power at constant beam size are also shown: in polarisation multiplexed systems two mutually perpendicularly polarised beams are coupled via a polarisation-coupler; in wavelength multiplexed systems several beams with different wavelengths are coupled with edge-filters so that the combined beams are collinear. The discussion in the following sections refers mainly to the spatial multiplexing; only where explicitly mentioned, will polarisation and frequency multiplexing be addressed.

Figure 19 illustrates for the one-dimensional case how N adjacent beams overlap in the far-field, for example, in the focal plane of a focusing lens. In the fully incoherent case the intensity of the total beam is N times the intensity of one individual beam. If the coherence among the beams is increased, the individual beams start to interfere with each other, thus producing an intensity distribution with a much smaller width and having a peak intensity that is proportional to the square of the number of beams N. This is the case of a "classical" laser with a well defined phase relation across the full cross-section.

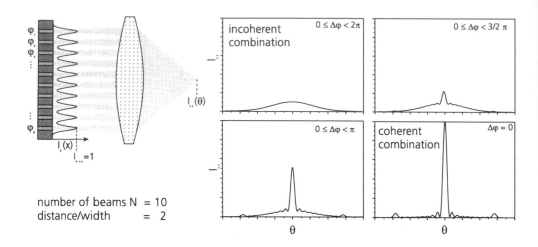

Figure 19. *Comparison between incoherent and coherent beam combination: In a numerical simulation several coherent beams with similar intensity distribution are superimposed in the focal plane of a lens. The coherence between the beams is increased from the top-left to the lower-right by reducing the statistical phase differences among the beams.*

In contrast, laser systems with incoherent beam superposition are in-between a classical laser and a pure incoherent light source. However, well designed systems with incoherent beam combination are comparatively simple to realise and allow the production of beams with properties and output powers that cover nearly all of the commercially relevant applications.

3.3 Beam quality of incoherently combined beams

Beam quality is a measure of how tightly a beam can be focused. The higher the beam quality, the smaller the spot size and the higher the maximum achievable laser intensity. According to the ISO-standard (ISO 1997) this property can be characterised by the beam-parameter product (Q), defined as the product of the waist radius (w_0) and the far-field divergence (θ_0) of the beam. In the best case, that is, without any aberrations, this beam-parameter product remains constant if the beam is transformed by passive optical components such as lenses or mirrors (Siegman 1990).

$$Q = w_0\theta_0 = \text{constant} \tag{3}$$

For a given numerical aperture of the focusing system (NA_f) and a rotationally symmetric beam with a total power of P, the average laser intensity in the focal plane (I_f) is given by:

$$I_f = \frac{PNA_f^2}{\pi}\frac{1}{Q^2} = \pi NA_f^2 B \qquad \text{where} \qquad B = \frac{P}{\pi^2 Q^2} \quad [\text{Wcm}^{-2}\text{steradian}^{-1}] \tag{4}$$

B denotes the brightness of the beam and is introduced into Equation 4 as an abbreviation. The physical meaning will be discussed later in this section. Equation 4 is widely used

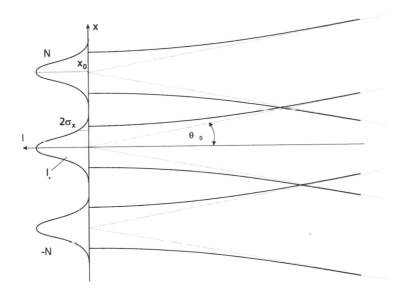

Figure 20. *Example for a one-dimensional array of individual, similar emitters, propagating from the beam waist at the left to the right.*

in many applications of conventional lasers with a single coherent and circular symmetric beam.

In contrast to conventional lasers, diode laser systems with incoherent beam combination are built up from one- or two-dimensional arrays of individual beams. For the calculation of the beam-parameter product of such beams it is necessary to determine the far-field divergence angle (θ_{tot}) as well as the radius of the beam in the waist (w_{tot}) for the total intensity distribution of the emitter array. In Figure 20 a typical geometry is shown for the simplified one-dimensional case, where $2N + 1$ similar intensity distributions $I_0(x)$ are equally spaced with a distance of x_0, each distribution having a radius of w_0.

Both x_0 and w_0 are defined, according to the ISO-standard (ISO 1997), in terms of the first and second order moments of the intensity distribution $I_0(x)$ by

$$
\begin{aligned}
x_0 &= \frac{1}{P_0} \int x I_0(x)\, dx & P_0 &= \int I_0(x)\, dx \\
\sigma_x^2 &= \frac{1}{P_0} \int x^2 I_0(x)\, dx & w_0 &= 2\sigma_x
\end{aligned}
\tag{5}
$$

where P_0 is the total power of a single beam. The intensity distribution (I_{tot}) and the power (P_{tot}) of the total beam are given by:

$$
I_{tot} = \sum_{n=-N}^{N} I_0(x + n x_0)
\tag{6}
$$

$$
P_{tot} = (2N + 1) \int I_0(x)\, dx = (2N + 1) P_0
$$

Examination of Figure 20 shows that the far-field divergence angle of the total beam (θ_{tot}) equals that of an individual beam:

$$\theta_{tot} = \theta_0 \tag{7}$$

The beam radius of the total beam (w_{tot}) can be calculated according to the ISO-standard from the second-order moment (σ_x) of the total intensity distribution ($I_{tot}(x)$) as:

$$w_{tot}^2 = 4\sigma_x^2 = \frac{4}{P_{tot}} \int x^2 I_{tot}(x) \, dx \tag{8}$$

Insertion of Equation 6 into Equation 8 and a coordinate transformation $x = x + nx_0$ gives the following relation for the radius of a total beam, consisting of M individual beams ($M = 2N + 1$):

$$w_{tot}^2 = 4\sigma_x^2 + \frac{1}{3}(M^2 - 1)x_0^2 \tag{9}$$

$$w_{tot} \cong 1.14 \times M w_0 F \qquad \text{(for } x_0 \gg \sigma_x, \; M > 2\text{)},$$

$$F = \frac{x_0}{2w_0} \qquad \text{(filling factor)}$$

As a rule of thumb, a beam consisting of M individual beams has a waist radius which is roughly M times larger than the waist radii of the individual beams, if the beams are densely packed (filling factor $F \approx 1$). The beam-parameter product of the total beam is then M times the beam-parameter-product of the individual beam:

$$Q_{tot} = w_{tot}\theta_{tot} \cong M w_0 \theta_0 = M Q_0 \tag{10}$$

Real beams are always two-dimensional. In many practical devices (Figure 17), arrays as shown in Figure 21 are used. With Equation 10 the beam quality of the total beam can be calculated separately for each dimension.

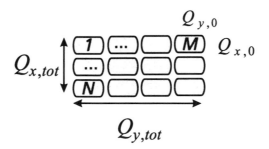

Figure 21. *Two-dimensional array of emitters as used in the devices shown in Figure 17.*

The brightness B of the total beam [in $Wcm^{-2}steradian^{-1}$] is the laser power per emitter area and solid angle. For the case of the rectangular symmetry in Figure 21 this is given by:

$$B_{tot} = \frac{P_{tot}}{16 Q_{x,tot} Q_{y,tot}} = \frac{M N P_0}{16 M Q_{x,0} N Q_{y,0}} = \frac{P_0}{16 Q_{x,0} Q_{y,0}} = B_0 \tag{11}$$

This is a form of the well-known brightness theorem (Born and Wolf 1980), which states that in the best case, that is, without any aberrations, the brightness of the total beam is the same as that of the individual beam. Real systems however suffer from several degradation mechanisms, that reduce brightness to

$$B_{\text{tot}} = \eta_{\text{FAC}} \cdot \eta_{\text{PL}} \cdot \eta_{\text{STQ}} \cdot \eta_{\text{FF}} \cdot B_0 \qquad (12)$$

where the η factors, all less than unity, are

$\eta_{\text{FAC}} = Q_x/Q_{x,0}$: optical quality of the *fast-axis collimation*, measured as the quotient of the beam-parameter products (BPP) behind (Q_x) and before $(Q_{x,0})$ the micro-optical lens.

$\eta_{\text{PL}} = P_{\text{tot}}/P_0$: the *power losses* at the micro-optical lens.

$\eta_{\text{STQ}} = \theta_{x,\text{tot}}/\theta_x$: the *stacking quality* defined as the ratio of the fast-axis divergence of the total beam and an individual beam.

$\eta_{\text{FF}} = F_x F_y$: the *filling factor* quality given as the product of the filling factors in x- and y-directions.

Typical values for the different factors and measures to minimise losses in beam quality and laser power will be discussed in the next sections.

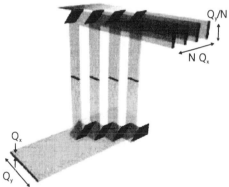

matching size and quality of the beam by means of step-mirrors

$$M \cdot Q_x \approx Q_y / M$$

principle   technical set-up

Figure 22. *Transformation of beam qualities: Schematic (left), technical set-up with two step-shaped copper mirrors (right). With this technique the coupling of 30W diode laser power into a fibre of 200μm core diameter (NA = 0.2) has been demonstrated.*

3.4 Examples of devices

Figure 22 shows a practical beam-transformation set-up for a low-power fibre coupling scheme of the type indicated in Figure 17 (upper right). Beam transformation is needed in many diode laser systems, as the beams of individual diode lasers and of stacked devices are usually asymmetric in terms of size and beam quality. A collimated beam from a diode laser bar, for instance, has a typical dimension of 1mm×10mm and a beam quality

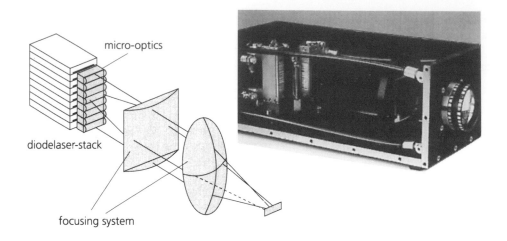

Figure 23. *Wavelength multiplexed diode laser stacks: The total output power is approx. 1,200W, the beam can be focussed down to a spot of 2mm x 4mm (NA approx. 0.2 in both directions).*

$Q_{x,0} \times Q_{y,0}$ of 1mm·mrad×2000mm·mrad. This will often require shaping, for example, in the case of coupling into a fibre.

The basic purpose of transformation techniques is to change beam size and divergence angle in the two lateral dimensions, while maintaining the brightness of the system, that is, the product $Q_{x,0} \times Q_{y,0}$. This objective can be achieved with several optical arrangements (e.g. Albers *et al.* 1992, Clarkson and Hanna 1996, Endriz 1992). In Figure 22 a simple and efficient solution with a step-shaped mirror is illustrated (Ehlers *et al.* 1997). In all systems the line-shaped beam from a diode laser bar after fast-axis collimation, is cut or divided into N parts, which are spatially rearranged. After rearrangement the beam quality in x- and y-direction has changed into $Q_{x,1} \times Q_{y,1}$. With Equation 10 the value for these modified beam qualities can be easily calculated as well as the number of beam parts, M, needed to produce a symmetric beam according to:

$$Q_{x,1} = MQ_{x,0} \qquad Q_{y,1} = Q_{y,0}/M \qquad (13)$$

$$M = \sqrt{Q_{y,0}/Q_{x,0}} \qquad \text{(for } Q_{x,1} = Q_{y,1})$$

In order to extend output powers into the kW-range, stacking techniques are commonly used (*e.g.* Product Information 1998, Krause 1998, Heineman and Leiniger 1998, De Odorico and Hewing 1997). The photograph in Figure 23 shows two stacks, each incorporating approximately 25 diode laser bars, each bar being individually collimated by a micro-lens. One of the stacks is operating at 808nm, the other at 980nm. The beams of both are combined by the edge filter in front of the stacks.

Conventional lasers for industrial applications such as CO_2- and lamp-pumped solid-state lasers are usually characterised by their output power and their beam-parameter product, which is taken for the rotationally symmetric case. Whether a laser is fit for the different fields of application which are discussed in Section 3.6, depends mainly on these

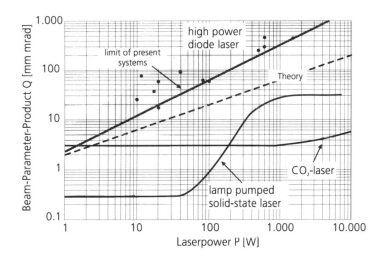

Figure 24. *Rated power and beam-parameter product of CO₂-, lamp-pumped Nd:YAG- and diode laser systems. The parameters of commercially available systems are all located above the indicated lines.*

parameters. In order to compare diode laser systems with conventional lasers, the beam-parameter products (Q_x, Q_y) of commercial diode laser systems have been converted to one single number Q_{ry}, characterising the equivalent rectangular-symmetric beam. This conversion is performed on the basis of Equation 13 according to the following relation:

$$Q_{xy} = Q_{x,0}M = Q_{y,0}/M - \sqrt{Q_{x,0}Q_{y,0}} \qquad (14)$$

These calculated data have been put together with data of commercial CO₂- and lamp-pumped solid state lasers into one diagram (Figure 24).

The lower limit for the beam-parameter product (or the highest beam quality) is always given by the diffraction limited Gaussian beam of wavelength λ (GB = Gaussian Beam):

$$Q_{GB} = \frac{\lambda}{\pi} = \begin{cases} 0.30 \text{ mm·mrad} & \text{for a Nd:YAG laser with } \lambda = 1.06\mu m \\ 3.00 \text{ mm·mrad} & \text{for a CO}_2 \text{ laser with } \quad \lambda = 10.6\mu m \\ 0.24-0.28 \text{ mm·mrad} & \text{for a diode laser with } \quad \lambda = 0.8-1.0\mu m \end{cases} \qquad (15)$$

In principle Q_{GB} does not depend on the output power, from which it follows that the brightness of a laser is unlimited in theory. In practical lasers, however, the maximum beam quality (or minimum beam-parameter product) can only be maintained up to a certain output power level. At present, diffraction-limited CO₂-lasers are available up to approximately 2kW, diffraction limited lamp-pumped Nd:YAG-lasers up to some 10W. Above these power levels beam quality decreases due to several physical and technical shortcomings such as wave aberrations in the active medium.

The maximum beam quality of diode lasers given with the Gaussian beam is also indicated in Equation 15 and can be produced with diode laser systems at low output powers. The method of power scaling by spatial multiplexing discussed in the previous

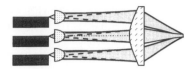 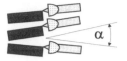

Figure 25. *Reasons for beam degradation in diode laser stacks: Aperture underfilling (left) and angular tolerances of the individual diode lasers (right).*

sections manifests itself in a shape of the curve in Figure 24 which is different from those of classical lasers. According to Equation 11 the one-dimensional beam-parameter product scales with the square-root of the laser power. The starting point for the line describing this scaling behaviour is given by the power and beam quality of one elementary emitter of the two-dimensional array. For the dotted curve, labeled "theory" in Figure 24 , current data for a typical broad-area emitter, which emits 1W at 2mm·mrad have been taken. At low output powers, real diode laser systems are available with a beam quality which closely matches this theoretical limit. At higher powers, where many individual emitters have to be combined in order to achieve the required power, actual systems have not yet reached the limiting curve, labelled "theory" in Figure 24.

The main reasons for this discrepancy were summarised in Equation 12. First, beam degradation and power losses can occur at the micro-optical collimation lens if non-optimised components are used or if alignment tolerances are not met. These aspects will be discussed in detail in Section 3.5.

A second source for beam degradation can be aperture underfilling in either direction of the two-dimensional array of diode laser emitters. Figure 25 illustrates this effect for the dimension in which the diode lasers are stacked. If the stacking geometry is not designed properly, the beams have considerable gaps and the filling factor F is well below one, thus decreasing beam quality according to Equation 12. Methods to overcome this limitation will be discussed in Section 3.6 as well as techniques to solve the problem of angular stacking tolerances, shown in the right-hand part of Figure 25.

In conventional stacks, where a number of diode lasers are put on top of each other, the unavoidable mechanical tolerances due to water sealings, spacers, electrical contacts etc. may sum up to quite large angular variations (α) of the individual diode lasers. If α is of the order of the divergence angle of the collimated beam, which is typically in the mrad-range, the divergence angle of the total beam will exceed that of the individual beam. In this case Equation 7 is violated and the beam quality is decreased. This problem can be overcome by properly modified stacking techniques, as will be discussed in Section 3.6.

3.5   Beam collimation

A typical property of diode lasers is the highly divergent emission in the direction perpendicular to the pn-junction, the "fast-axis". Numerical apertures of up to 0.8, that is, full divergence angles of approximately 100 degrees are not unusual. Therefore, in most cases where diode lasers are directly used, a collimation of this fast axis is necessary. A simple and low-cost solution for beam collimation is a cylindrical lens as shown in Figure 26. An

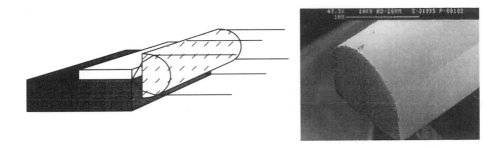

Figure 26. *Principle of fast-axis collimation with a cylindrical lens (left) and SEM-picture of a sample (right), which has been manufactured from high-index glass with an ultra-precision grinding technique.*

important requirement for this cylindrical lens is that as much beam quality and laser power as possible must be conserved, that is, the beam-parameter product in the fast axis behind the lens should be as close to the diffraction limit as it is at the emission facet of the diode laser.

In order to compare different lens types and to assess the lens performance in relation to the physical limit, ray-tracing calculations have been carried out as shown in Figure 27 (Sturm *et al.* 1997). For the calculations, a diode laser with a diffraction-limited beam, (i.e. with Gaussian intensity profile in the x-direction) has been used as a source. The

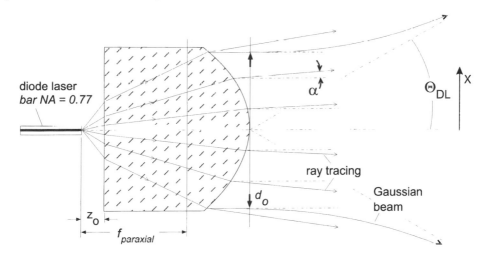

Figure 27. *Ray-tracing calculation for the collimation of the diode laser emission by a cylindrical lens.*

numerical aperture (NA=0.77 in all examples discussed below) is the divergence angle where the intensity is reduced to $1/e^2$ of the centre value. For the one-dimensional case approximately 95% of the total power falls within this numerical aperture. In addition to the fast-axis divergence, a numerical aperture in the y-direction ("slow axis") of 0.084

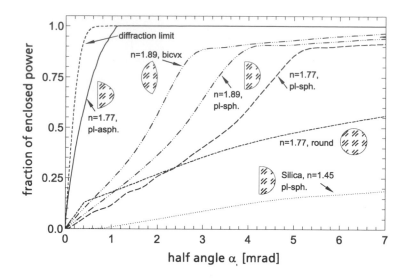

Figure 28. *Comparison of the quality of diode laser beams, after collimation by different types of cylindrical lenses (diode laser-$NA_{\text{fast}} = 0.77$, $NA_{\text{slow}} = 0.084$).*

(approx. 5 degrees) has been taken into consideration in all the calculations given below. This is a typical divergence angle for many diode laser emitters.

From the source point, a fan of equally spaced rays was traced through the lens, each ray carrying the local laser intensity. Behind the lens the intensities of the rays were summed up and plotted against the angle α (Figure 28). These curves are compared with the case of a diffraction-limited Gaussian beam with the same diameter d_0 as the fan of rays which leaves the lens (dotted line in Figure 28). If the curves calculated by ray-tracing are in the range of this dotted line, the lens is assumed to be sufficient for a high-quality collimation of the beam.

Figure 28 clearly shows that beams with high numerical aperture as discussed here can not be collimated adequately with standard plano-spherical or biconvex lenses. Even if a lens material with high refractive index is used, the numerical aperture of the collimated beam (defined as the angle with 95% power enclosure) is more than 10 times bigger than for the ideal case of a diffraction-limited beam (θ_{DL}). This is the reason why easy-to-manufacture plastic lenses can not be applied here, because plastics typically have low refractive indices. The lens with the highest performance in the diagram is a single lens made from high refractive index glass which has an aspherical surface, optimised for the emission characteristics of the diode laser bar. However, even with such an optimised lens the diffraction limit can not be achieved under all conditions. The reason for this is the coupling of the fast-axis and the slow-axis divergence due to Snell's law (Loosen *et al.* 1995).

At the time of writing, several manufacturing techniques are available to produce aspheric glass microlenses of the type shown in Figure 26, with the necessary accuracy in the sub micrometre range: ultra-precision grinding (Biesenbach *et al.* 1994) or ultra-sonic

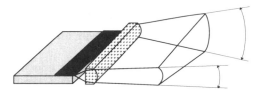

Figure 29. *Beam degradation due to axial misalignment between microlens and diode laser bar.*

shaping with a hard-metal mould (Product Information 1998). With such commercially available components a power throughput in excess of 90% and a beam quality behind the lens of approximately two times the diffraction limit can be achieved in the best cases (Sturm *et al.* 1997).

The requirements for positional accuracy of the microlenses are as high as those for the shape of the lens. The output beam is readily degraded or misaligned by axial defocussing and lateral beam twist. Axial defocusing, shown schematically in Figure 29, is a result of the cylinder lens and emitter facet of the bar not being perfectly arranged in parallel. While the emitters in the front part of the picture are assumed to be perfectly aligned in the back focal position of the lens (z_0), thus creating a beam with minimal divergence, the emitters in the rear part of the picture are axially out of focus, resulting in a beam with an enlarged divergence angle. If several such collimated diode lasers with statistically varying lens misalignment angles are combined, Equation 7 no longer holds. The divergence angle of the total beam is larger than that of an individual beam and the beam quality of the total beam is reduced.

With the ray tracing modeling discussed above, this process has been studied quantitatively, the result being summarised in Figure 29. The uppermost emitters in the picture (labelled with '1') are located in the back focal position of the lens ($z = 0$), the lowest emitters (labeled with '2') are shifted in the z-direction due to lens tilt as indicated by β. A local tilt of the lens without any defocus has negligible impact on the divergence angle of the emitters. This can be seen for the emitter '1', where the divergence angle remains constant over the whole displayed range of tilt angles. In contrast, the divergence angle of the emitter '2' strongly increases approximately linearly, due to the tilt-induced axial shift into a defocus position. If a variation of the divergence angle along the bar of some 10% is accepted at maximum, the axial position of the lens has to be maintained with tolerances well below $1\mu m$ in the case studied in Figure 30. This imposes high demands on alignment accuracy and on the long term mechanical and thermal stability of the fixtures of the lenses.

Apart from axial deviations, the back focal point of the lens may of course also deviate laterally (in the x-direction) from its nominal position, giving rise to a strong twist of the collimated beam in the x-direction. This process, however, can be compensated by an appropriate tilt of the diode laser as a whole, including the lens. Another difficult-to-handle phenomenon, for which compensation cannot be made, is the so-called "smile", shown in Figure 31. If either the diode laser bar or the micro cylinder lens is not perfectly straight, a part of the beam is shifted upwards or downwards with respect to its nominal position. The example in Figure 31 shows that the beam behind the lens then looks like

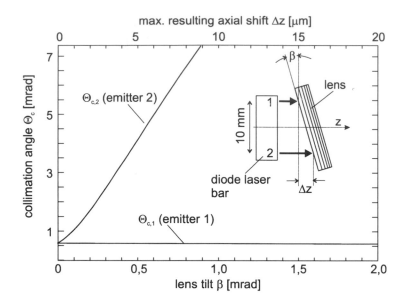

Figure 30. *Increase of the collimation angle due to a misalignment angle β between the microlens and the emitting facet of the diode laser bar ($NA_{fast}=0.77$, $NA_{slow}=0.084$, paraxial focal length of the lens f=0.88mm).*

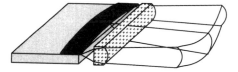

Figure 31. *Beam degradation where the diode laser bar is not perfectly straight. The exit beam looks like a "smile".*

a "smile".

The upper part of Figure 32 illustrates a "smile" in a typical experimental case. The picture has been taken by using an imaging system which strongly enlarges the x-dimension. If a couple of such beams with statistically varying smile are combined, the outgoing beam angle (or far-field image) of the total beam is enlarged as shown in the lower part of Figure 32.

In order to obtain quantitative data on the sensitivity of lateral bar position and geometry, ray-tracing calculations for a beam slightly shifted in the x-direction have been performed with the model and data discussed in the foregoing section. Figure 33 shows the small effect of this shift on the divergence angle Θ_c, which increases approximately 10%, and the much larger effect on the total tilt of the beam δ as it leaves the lens. This large tilt angle is caused by the short focal length of the lens as is indicated in the figure. For a discussion of the smile effect, the tilt angle in Figure 33 is considered to apply to the centre part of the bar, while the wings are assumed not to be tilted. In order to keep

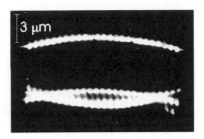

Figure 32. *Enlarged picture of a diode laser bar with a "smile" of approximately $2\mu m$ (upper picture); if a couple of those bars with statistically varying "smile" are superimposed, the far-field picture is enlarged (lower picture).*

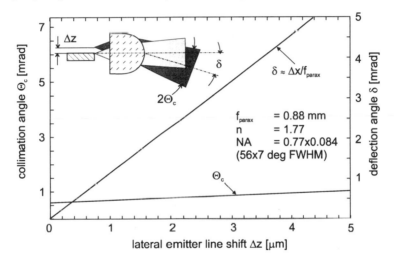

Figure 33. *Dependence of the collimation angle and the beam tilt on the lateral shift of the diode laser emitter in the x-direction.*

the tilt of the beam δ well below the divergence angle Θ_c, a maximum smile of the bar below $0.5\mu m$ has to be ensured.

3.6 Aperture filling and stacking accuracy

If the emission of a diode laser bar is used directly without any further optical processing of the slow axis, the beam quality in this direction is given by the product of the bar width (divided by 2) and the divergence angle of the individual emitter or the bar respectively. Depending on the filling factor of the bar, this beam quality may be considerably lower than theoretically possible due to aperture underfilling as illustrated in Figure 34. If the emitters shown were not spaced by the pitch of x_0, but were located more densely side by side with $x_0 = 2w_0$, the total beam width and thus the beam-parameter product could be reduced by a factor of $F = 2w_0/x_0$.

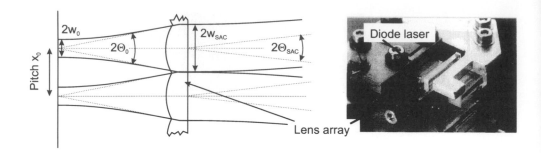

Figure 34. *Principle and experimental set-up for beam quality improvement by slow-axis collimation.*

An alternative way to achieve this improvement is shown in Figure 34 , where by means of a lens array, each individual emitter is individually collimated and the divergence angle thus reduced. The lenses are put at an axial position shortly before the beams start to overlap. The focal length of the lenses is chosen so as to transform the local phase front curvature of the individual beam into a plane phase front. The diameter of the individual beam is increased by a factor of $1/F$ and according to Equation 3 the divergence angle reduced by F, leading to an decrease of Q by a factor of F at best.

Table 2 summarises the results which have been achieved with the experimental device of Figure 34. The beam quality could be improved by a factor of approximately 3, corresponding to the figure expected for the filling factor of 0.3. However, the table shows also that considerable power was lost due to reflection losses at the non-coated lens surfaces and due to the fact that in the specific case shown here, the lens array geometry was not perfectly matched to the structure of the emitter array. Under optimised conditions a power throughput of more than 90% and thus a brightness increase slightly below a factor of 3 is achievable.

	before lens array	behind lens array
Divergence angle (slow axis)	4.5° (79mrad)	1.5° (26mrad)
Beam-parameter product (slow axis)	395mm·mrad	130mm·mrad
Laser power	P_0	0.7P_0
Brightness	B_0	2.2B_0

Table 2. *Improvement of the beam quality by slow-axis collimation illustrated with experimental results obtained with the set up of Figure 31 (bar width 10mm, emitter width 150μm, emitter spacing 500μm)*

Beam quality reduction due to aperture underfilling has to be taken into consideration not only in the slow direction but also in the case of stacked diode laser bars in the fast direction as well. In many cases a dense package of the diode lasers in the fast direction, such as shown in Figure 23 , is hard to achieve in practice or has serious technical drawbacks. In order to obtain high output power from compact systems, thin heat sinks in the range of 1–2mm are often inserted between emitting bars. However with such thin and mechanically sensitive heat sinks, the beam may be seriously degraded due to mechanical deformations which, in turn, lead to smile (Figure 31) or angular misalignment

Figure 35. *Beam staggering for increasing the filling factor in the fast direction of a diode laser stack: principle (left) and experimental set up (right).*

(Figure 25). Thick heat sinks on the other hand are mechanically stable but the stack height and thus the output power is limited by the size and handling capability of the optical systems needed to focus the beam on the workpiece.

This restriction can be overcome by beam staggering techniques as illustrated in Figure 35. In the specific case shown here, three stacks are used. The beams from the individual diode laser bars are each collimated in the fast axis and have a spacing which is nearly three times the height of the individual beams. By means of comb-shaped mirrors the emissions of the three stacks are staggered together into one outgoing beam, which is now densely packed. Filling factors in the fast direction up to about 90% have been proven to be achievable applying such technologies.

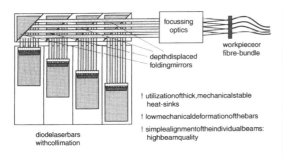

Figure 36. *"Optical stacking": Schematics (left) and experimental set up (right).*

Even with structures described above, the problem remains, that small mechanical tolerances from water sealings and electrical contacts, located between each pair of diode lasers, may sum up to considerable numbers, which are beyond acceptable tolerances for high beam quality systems. The stacking technique shown in Figure 36 ("optical" stacking) yields a solution for this problem as well as for the problem of aperture underfilling while using thick heat sinks. The individual diode lasers are located side by side instead of on top of each other, on a staircase-shaped mechanical holder. Each beam from a diode laser is folded by an angle of 90^{circ} by means of individually aligned folding mirrors. This mechanical arrangement ensures high filling factors, even if thick, mechanically stable heat sinks are utilised. Due to the adjustable folding mirrors, mechanical tolerances of each diode laser can be compensated for individually, thus ensuring highly parallel emission of the beams.

In Table 3 the results achieved with such an optical stack comprising 8 diode lasers and a total power of 240W, are compared with a stack, built from 8 diode lasers as well but according to conventional design. Due to the more precise alignment and the dense package in the fast direction, the brightness of the optical stack was increased by a factor of 3 over conventional technology.

	"conventional" stack (Figure 23)	"optical" stack (Figure 36)
Beam aperture (fast axis)	14.5mm	10.4mm
Divergence angle (fast axis)	4.1mrad	1.8mrad
Beam-parameter product (fast axis)	58mm×mrad	18.7mm×mrad
Laser power	240W	240W
Brightness	B_0	$3.1B_0$

Table 3. *Improvement of the beam quality by "optical" stacking as shown in Figure 36 ;*

4   Applications

In Section 2.2 it was stated that the present high-volume applications of marking, cutting and deep penetration welding of sheet metal require relatively high beam quality and, for cutting and welding, output powers above approximately 1kW. Typical process intensities are in the range above approximately $10^5 W/cm^2$. According to the foregoing sections, such intensities are hard to achieve with present high-power diode lasers. Current diode lasers are mostly used in the "low intensity applications" shown in Figure 11, where the beam merely has to be concentrated to moderate spot sizes and lasers with lower beam quality are adequate. Two examples of the welding of plastics and soldering are shown in Figure 37.

Plastics welding with lasers (Hanch *et al.* 1998) in most cases requires that the two parts to be welded together are chosen from differently pigmented plastics in order to create a geometry where the laser penetrates through the transparent top layer and is fully absorbed at the surface of the layer underneath. In the example shown in Figure 37 (left), a plastic cover has to be welded to the plastic casing. The cover is coloured with a pigment which appears black in the visible spectral range, but is transparent for the IR-radiation of the diode laser. The requirements on laser power are quite moderate. Because the process power is applied very effectively, it only heats the joining zone and the melting energy for plastic is quite low. At typical widths of the joining zone, and thus spot size about 1mm, a welding speed of 1m/min is typically achieved with a laser power of a few tens of Watts.

Figure 37. *Examples of low-power and low-intensity laser applications, using diode laser systems: Welding of the plastic casing of an electronic car-key (left) and soldering of an electrical connecting braid (right).*

Advantages of welding by laser instead of conventional methods such as ultrasonic welding are for example, flexibility, high quality of the weld (important in the case of visible seams), controllability of the power input and the possibility of integration into automated quality control.

Most of these advantages also hold for the second example shown in Figure 37, the soldering of electronic or electrical components. Geometries being soldered range from millimetres for electric parts to tenths of a millimetre for electronic components. Required spot sizes and laser powers respectively, vary from 0.1mm and tens of Watts up to hundreds of Watts for the soldering of the connecting braid in Figure 37. For soldering as well as for plastics welding, fibre coupled diode lasers are usually utilised, for instance according to the principle discussed in Figure 22.

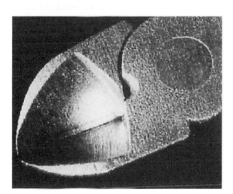
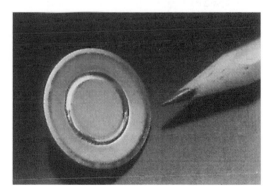

Figure 38. *Transformation hardening of the edges of a side-cutter (left) and heat conduction welding of a stainless steel coin (right).*

Considerably higher output powers are required in two further examples that are illustrated in Figure 38. The righthand side picture shows a side-cutter, the cutting edges of which have been hardened with diode lasers. Due to the relatively high temperatures required (approx. 750–1000°C) and the macroscopic geometry of typical parts the required laser power exceeds 1kW. Diode laser stacks, which have recently been employed for this application, are distinguished from classical lasers by already having a rectangular beam, which usually is the most desirable geometry. In the example of Figure 38 the beam of the diode laser stack was shaped so that the full length of the edge could be hardened in one step without any movement of beam or workpiece.

The other example of Figure 38 refers to welding of metals with diode lasers. A stainless-steel coin, 1.5mm thick, has been welded with a speed of 1.8m/min, using a diode laser stack of 0.7kW power. The diode laser beam was focused to a spot size around $1.5 \times 3.8 \text{mm}^2$, resulting in a processing intensity of approximately 10^4W/cm^2. At these intensities heat conduction welding instead of deep penetration welding takes place. Although welding speed is quite low with this process, an advantage of heat conduction welding with diode lasers is that the weld seams are very smooth and have a very high surface quality.

4.1   Discussion and perspectives

The discussion in the foregoing section showed that present diode lasers are excellent tools for a wide range of laser applications which require only moderate beam qualities or focussabilities. Even larger application fields would be opened up however, if both output powers and beam quality could be simultaneously raised to be comparable with current lamp-pumped solid state lasers. This would then enable cutting and welding of metal parts.

The sum of all the laser power and beam quality degradation processes which were summarised in Section 3.3, is the reason, why the beam-parameter product of present systems is above the theoretically achievable indicated in Figure 24. The more individual diode lasers are combined, the bigger is the difference between theory and real devices due to the increasing effect of degradation processes. Introduction and combination of all the technical means discussed in Section 3.6 can considerably enhance the beam quality. In addition to these means, several wavelengths and the two states of polarisation may be combined in future high-power systems.

If one also takes into account that output power and beam quality of diode lasers have been rising steadily over the past several years, it seems feasible to build diode laser systems in the future with an output power in the kW-range with a beam quality comparable to present lamp-pumped solid state lasers.

References

Albers P, Heimbeck H J and Langenbach E, 1992, *IEEE J. Quantum Electron.* **28**, 1088.

Beach R et. al., 1992, *IEEE J. Quantum Electron.*

Biesenbach J, Loosen P, Treusch H G, Krause V, Kösters K, Zamel S and Hilgers W, 1994, *SPIE Proc.* **2263** 152.

Born M and Wolf E, 1980, *Principles of Optics* p189, (Pergamon Press).

Clarkson W A and Hanna D C, 1996, *Opt. Lett.* **21**, 375.

Dausinger F, 1995, *Strahlwerkzeug Laser* (Teubner Verlag, Stuttgart, in german)

Dawes C, 1992, "Laser Welding", **17** (Abington Publishing, Cambridge).

De Odorico B and Hewing C, 1997, *Feinwerktechnik & Messtechnik* **105**–**6** 451 (in german).

Ehlers B, Du K, Baumann M, Treusch H-G, Loosen P and Poprawe R, 1997, *SPIE Proc.* **3097** (Lasers in Material Processing, Laser '97), 639.

Endriz J, 1992, *United States Patent 5,168,401*

Hänsch D, Pütz H, Treusch H-G, Gillner A and Poprawe R, 1998, "Welding of Plastics with Diode Lasers", Conf. ICALEO '98, Orlano/FL 1998, Proc. to be published by: Laser Institute of America, Orlando/FL.

Heineman S and Leiniger L, 1998, *SPIE Proc.* **3267** 116.

ISO/DIS-Standard 11 146, 1997, Optics and Optical Instruments ,lasers and laser related equipment, test methods for laser beam parameters, International Organization for Standardization,P O Box 65, CG1211, Geneve, Switzerland.

Krause V, 1998, Conf. ICALEO Orlano/FL 1998, Proc. to be published by: Laser Institute of America, Orlando/FL

Loosen P, Treusch G, Haas C R and Gardenier U, 1995, *SPIE Proc.* **2382** (Photonics West '95),78.

Loosen P, Ebert T, Jandeleit J and Treusch H-G, 1998, *Proc. LEOS '98* (Orlando/Fl), IEEE Catalog No 98CH36243, 434.

Mayer A, 1998, Optech Consulting, Bisingen, Germany (Priv.Comm.)

Petring D, 1994, PhD-Thesis, RWTH Aachen, Aachen (in german)

Product information, Limo Corp., 1998, Dortmund Germany, (http://www.limo.de)

Product information, Rofin-Sinar,1998, Hamburg/Dilas,Mainz (http://www.rofin-sinar.com.

Siegman A E, 1990, *SPIE Proc.* **1224** (Optical Resonators), Conf. OE-Lase, Los Angeles/CA.

Sturm V, Treusch H-G and Loosen P,1997, *SPIE Proc.* **3097** (Lasers in Material Processing, Laser '97), 717.

Treusch G, 1985, PhD-Thesis, Technische Hochschule Darmstadt/Germany, Darmstadt 1985 (in german)

Trumpf GmbH, 1996, *Faszination Blech* (Dr. Josef Raab Verlag, Stuttgart, in german)

Applications of ultrashort laser pulses

Wilson Sibbett[1] and Wayne H Knox[2]

[1]University of St Andrews, Scotland
[2]Bell Laboratories, Lucent Technologies, Holmdel, USA

1   Introduction

An impressive range of ultrashort-pulse lasers has been developed, based on the general techniques of active and passive modelocking, which date back to the mid-1960s (Siegman 1986, Keller this book). More recent noteworthy advances have resulted from the development in the 1980s of high quality titanium-doped sapphire as a practical broad-bandwidth laser crystal (Moulton 1982, 1986) and the subsequent demonstration of self- (or Kerr-lens) modelocking (Spence *et al.* 1990, 1991) which has afforded picosecond/femtosecond-pulse generation with unrestricted tunability for such a gain medium. When complemented by chirped-pulse amplification (Strickland and Mourou 1985, Keller this book), the modern "user-kit" now provides clear access to robust and versatile ultrashort-pulse sources that provide peak optical powers in the megawatt to petawatt regimes. With appropriate frequency up/down conversion procedures, spectral access can extend from X-rays to T-rays (terahertz radiation) with relative ease. Additionally, the established characterisation techniques of autocorrelation and optical/microwave spectral analyses have been enhanced recently by schemes such as frequency-resolved optical gating (Delong *et al.* 1994a,b) so that the phase-related features of these ultrashort pulses can be monitored in a quantitative manner. Thus, a range of well-characterised ultrashort-pulse sources is currently available to several broadly-based user communities.

In the following sections we will review a selection of applications that arise in current exploitations of ultrafast science and technology . We acknowledge at the outset, however, that it is not possible to include all of the reported applications and these examples are intended to be representative of the types of implementations that can be expected to become more commonplace in the future.

2 Imaging with optical and quasi-optical pulses

In the last century flash-photography using microsecond-pulse xenon lamps provided new insights into the dynamics of many physical processes. Now in the late twentieth century, the availability of versatile picosecond/femtosecond laser-based sources is serving to provide amazingly clear new insights in fundamental processes that occur in photophysics, photochemistry, photobiology and photomedicine. Progress in these areas of science is serving to open up new opportunities in ultrafast optoelectronic and photonic technologies.

The concept of ultrafast imaging is one that continues to be highly relevant to modern science. One aspect of this is the fact that objects which are optically imaged through scattering media have image-related information whose quality has a time dependence. Specifically, as illustrated in Figure 1, the earliest-arrival, least-scattered, optical signal will provide the best image fidelity. The recording of this "quasi-ballistic component" in preference to the lower quality more scattered signals, requires a suitably high-speed time-shuttering technique. One reported scheme used a passively modelocked dye laser as the illuminating source and involved holographic imaging. In this case the phase retention in the image beam favoured the least-scattered components (Chen *et al.* 1991).

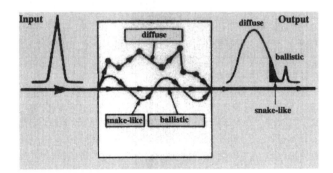

Figure 1. *Pulse-propagation through a scattering medium.*

The impressive example (shown in Figure 2) of the "prompt-imaging" of metal needles that are viewed through a 6mm-thickness of raw chicken breast serves to illustrate the advantages of ultrashort-pulse illumination combined with ultrafast gating (Chen *et al.* 1991). Excellent more recent results of 3-D imaging through turbid media using photorefractive holography have also been reported with a millimetre depth resolution and transverse resolution of $\sim 30\mu m$ (Hyde *et al.* 1995).

Optical coherence tomography (OCT) is a clever imaging technique that was originally designed to exploit low coherence illumination sources such as superluminescent light-emitting-diodes. The principle is based on the "white-light-fringe" configuration for a Michelson-type interferometer that is set up to have equal optical pathlengths in its two arms (Figure 3).

If this condition is maintained while a reference mirror that terminates one arm is translated in the y, z axes, for example, then spatial detail can be resolved within the coherence length of the source. Because a femtosecond laser provides a well controlled

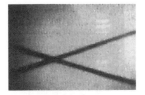

*Photograph of two crossed
metal needles
(0.5 mm diameter)*

*The needles viewed through a
6mm slab of raw chicken
breast in ordinary illumination*

*'Snapshot' image of the
needles using femtosecond
illumination and gating pulses*

Figure 2. *Example of 'prompt-imaging'. (Chen et al. 1991)*

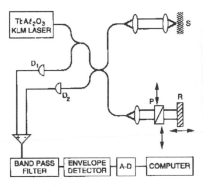

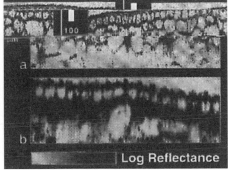

Figure 3. *OCT technique and OCT image of onion sample (Bouma et al. 1995).*

and relatively broad spectrum, such a source has a coherence length of less than 2μm. Thus, with a modified imaging scheme that incorporates a femtosecond Ti:sapphire laser, Bouma and co-workers (Bouma *et al.* 1995) have demonstrated the imaging of an onion sample with a 3.7μm resolution and 93dB dynamic range. This is illustrated by the image reproduction given in Figure 3 where the superiority of the femtosecond laser (Figure 3a) over a superluminescent diode source (Figure 3b) is clearly evident.

Another asset of the high intensity modelocked laser pulses is that the highly localised nature of multiphoton excitation enables fluorescence microscopy to be conveniently implemented. The basis of this technique, as illustrated in Figure 4, was devised by Webb and colleagues at Cornell University (Piston *et al.* 1992). The multiphoton excitation concentrated at the focal region affords excellent spatial discrimination and resolution. The relatively modest light dosage delivered as a periodic sequence of picosecond/femtosecond pulses at typically 10ns intervals means that live cells can be examined. Also, as the lasers are operating in the near-infrared, the scatter and sample damage is reduced and fluorescence imaging of structures $\sim100\mu$m below the sample surface is possible.

By way of an example, Figure 5 shows a comparison between image quality for two-photon picosecond-pulse excitation (at 1047nm; Nd:YLF) and single-photon continuous excitation (at 488nm; Ar-ion) with confocal imaging. (The sample is a naturally occurring enzyme which has been labelled with a safranine stain for two-photon excitation at 1047nm.) It follows, therefore, that multiphoton imaging enables fluorescence microscopy

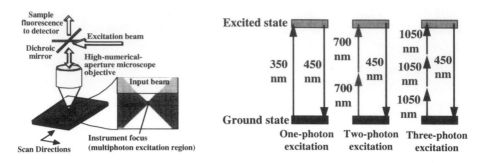

Figure 4. *Multi-photon fluorescence imaging. (Piston et al. 1992)*

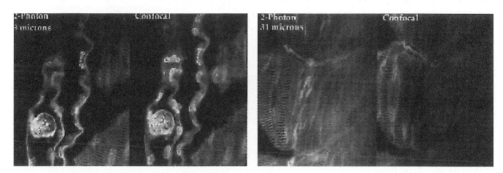

Figure 5. *Two-photon and single-photon (confocal) fluorescence images (J White, Univ. of Wisconsin).*

to be carried out without the expense or complexity of a confocal microscope!

Imaging at terahertz frequencies has the advantage of providing transparency for objects that are opaque to visible/near-infrared light. By building upon the initial work of Grischkowsky and colleagues (van Exter *et al.* 1989), which demonstrated how THz-pulses could be produced conveniently using semiconductor optoelectronic switches illuminated with femtosecond laser pulses, Hu and Nuss (1995) demonstrated that high quality "T-ray" images could be obtained. A fine example from this Lucent Technologies group is included as Figure 6. This illustrates that the plastic packaging of an electronic chip is

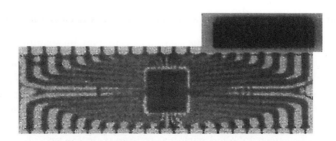

Figure 6. *T-ray image of packaged electronic chip (Hu and Nuss 1995).*

transparent to the THz waves while metal leads and the silicon chip absorb or reflect radiation in this frequency band. Strong absorption of water molecules in the 1THz spectral region also enables such sources to be relevant to the THz-spectroscopy of water-containing media.

At the high-frequency end of the electromagnetic spectrum, there are interesting possibilities for imaging in the soft X-ray "water-window" transmission region of biological materials. Such wavelengths are created by exploiting high harmonic generation of femtosecond pulses in the ∼3–5nm spectral region. Indeed, recent high harmonic generation of coherent soft X-rays in helium (2.7nm) and neon (5.2nm) gas jets has been reported using amplified (26fs, 20mJ) Ti:sapphire laser pulses (Chang *et al.* 1997). Interestingly, theoretical modelling of high harmonic generation using intense excitation in this femtosecond-pulse regime predicts the tantalising possibility of generating attosecond (10^{-18}s) X-ray pulses (Corkum *et al.* 1994) and so a new era of linear and nonlinear ultrafast X-ray science may be in prospect.

3 Two-photon photoemission spectroscopy in the femtosecond regime

As mentioned in Section 1, modern ultrashort-pulse lasers can be rendered more compatible with some applications if the output from the femtosecond laser oscillator is subjected to procedures that include regenerative amplification , self-phase modulation and frequency up/down conversion . This is especially the case when a frequency-tunable self-modelocked Ti:sapphire laser is employed as the oscillator stage of such a system. By exploiting an optical-source tool-kit of this type, Professor Harris and co-workers at the University of California, Berkeley Laboratory have demonstrated stunningly impressive results in two-photon photoemission spectroscopy (Ge *et al.* 1998).

Two-photon photoemission spectroscopy that is time-resolved in the femtosecond regime provides a means of investigating electron dynamics at metal-nonmetal interfaces. Given that such carrier dynamics are of primary importance in determining the performance of transistors within microprocessor chips, the scientific and technological relevance of these fundamental time-resolved studies cannot be overstated.

The specific example that is featured here relates to the dynamics of two-dimensional small-polaron formation at ultrathin alkane layers on a silver surface. These dynamics have been studied with femtosecond time and angle-resolved two-photon photoemission spectroscopy. Optical excitation creates interfacial electrons in quasi-free states for motion parallel to the interface. These initially delocalised electrons self-trap as small polarons in a localised state within a few hundred femtoseconds. A detailed study of carrier dynamics for a silver-alkane interface can be undertaken where the key processes are as follows: (i) UV femtosecond pulses pump electrons out of the silver surface into the overlayer of alkane molecules, (ii) visible pulses from an optical parametric amplifier probe excite electrons out of the interface into vacuum and (iii) electron kinetic energies are measured at various angles during the relaxation process by a time-of-flight analysis. By this means it is possible to deduce the distinctive wavefunctions for the electron transition from a delocalised state to a localised state in transferring from the metal to the dielectric interface. This is illustrated in Figure 7 where the timescale is such that

Figure 7. *Illustration of electron dynamics at a silver-alkane interface (Ge et al. 1998).*

within a few hundred femtoseconds the electron has induced a lattice distortion in the alkane overlayer and evolved into a localised state. In a further period of about 1ps the self-trapped electron is able to escape the trap by quantum mechanically tunnelling its way back to the metal. The optical excitation/probe-source versatility comes from regeneratively-amplified, self-modelocked Ti:sapphire laser pulses and a subsequent combination of self-phase modulated white light spectrum generation, second harmonic and optical parametric generation and temporal-compression of pulses.

4   Micromachining with amplified fs laser pulses

By applying the chirped-pulse amplification technique referred to earlier, it is now quite straightforward to enhance the energy and peak power from self-modelocked Ti:sapphire lasers. For example, 10nJ/100fs pulses produced at a pulse repetition frequency of about 80MHz can be regeneratively amplified with a temporal stretcher and compressor arrangement to provide $\sim$300μJ/200fs pulses at repetition rates ranging from 50Hz to 1kHz. When required, even higher peak powers, up to 100GW, can be obtained at this repetition rate for amplified 5fs (0.5mJ) Ti:sapphire laser pulses (Schnurer *et al.* 1998).

For micromachining applications, it is usual to employ self-modelocked Ti:sapphire lasers that are regeneratively amplified to the GW regime. For instance, for producing $\sim$50μm holes in 100μm-thick steel sheet, it is readily demonstrated that femtosecond pulses are much more effective than those with nanosecond durations due to the effects of thermal diffusion. A striking example is shown in Figure 8 where laser ablation results are given for a Ti:sapphire laser operating at 780nm. It can be seen clearly that with the 120μJ, 200fs pulses at a fluence of 0.5J/cm^2 the machining process avoids the melting and heat-affected zones that are associated with 3.3ns pulses at the higher fluence of 4.2J/cm^2 (Momma *et al.* 1996).

Figure 8. *Hole drilling in steel sheet. Left: 3.3ns at 4.2J/cm^2. Right: 200fs at 0.5J/cm^2. (Momma et al. 1996).*

It is noteworthy, also, that with higher energy (1.3mJ), 150fs pulses, these researchers demonstrated the usefulness of beam delivery via diffractive optical elements . For example, Nolte and co-researchers have shown that the energy throughput of diffractive optical elements is superior to conventional imaging systems and that this type of delivery minimises the nonlinear beam distortions prior to the incidence of the laser radiation at the workpiece. It has thus been possible to micromachine regular arrays of ~250μm burr-free holes in steel for fuel-injection nozzles as illustrated in Figure 9.

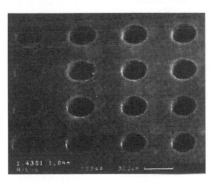

Figure 9. *Array of burr-free holes in steel. (Nolte et al. Laser Zentrum, Hannover)*

Another intriguing application of regeneratively-amplified, femtosecond Ti:sapphire laser pulses relates to the processing of energetic materials. As a demonstration of the feasibility of this approach Roeske and colleagues at the Lawrence Livermore National Laboratory have used 5mJ, 100fs, 820nm, pulses at a repetition rate of 1kHz to cut safely through pressed pellets of common explosives. Figure 10 shows that the charring which occurs for the 500ps pulses is avoided by using femtosecond pulses. This can be explained

326 Wilson Sibbett and Wayne H Knox

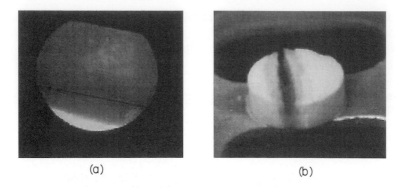

(a) (b)

Figure 10. *Cutting of High Explosive Pellets (a) Femtosecond laser cut in HMS (LX-15) (b) Cut in PETN (LX-16) with 500ps laser pulse.(Roeslie et al., Lawrence Livermore)*

on the basis that the energy from each femtosecond pulse is absorbed in a time that is shorter than the lattice-vibration period. Because heat is transferred by lattice vibrations, essentially no heat is absorbed. The hot plasma created expands and cools rapidly such that it transfers very little heat to the material and any short waves created by the laser pulses are too brief to induce any significant reaction. These initial feasibility studies are important and timely because they indicate that femtosecond-pulse laser systems may have a key role in some weapon-decommissioning operations in the future.

Another example, included as Figure 11, relates to the micromachining of tantalum and titanium stents (Nolte *et al.* 1998). These are cardiovascular implants that are used in the treatment of patients having arteriosclerosis. A primary requirement is to have absolutely burr-free micromachining of high strength, physiologically inert metals. The sub-millijoule, femtosecond pulses from a regeneratively-amplified, self-modelocked Ti:sapphire laser are exceptionally well suited to this type of application and the photographs confirm the quality of the laser-produced structuring of such stent materials

Micromachining of hard and soft biological tissue is also readily undertaken using regeneratively amplified, femtosecond pulses. The example shown in Figure 12 is of a nonthermal in-vitro ablation of bovine neural tissue. In this instance, 300μJ, ~200fs pulses

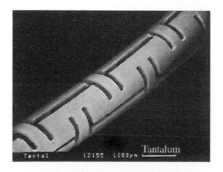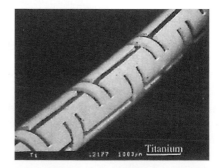

Figure 11. *Micromachining of Stents (Nolte et al. 1998).*

Figure 12. *Ablation of bovine neural tissue (Loesel et al. 1998).*

at 800nm with a repetition rate of 1kHz were used to induce a plasma-mediated ablation process (Loesel *et al.* 1998). The associated thermal damage (coagulation/carbonisation) and structural changes are minimal and limited to a zone smaller than 5μm around the ablation zone. This is about two orders of magnitude smaller than that observed with CO_2 or Nd:YAG lasers and implies an attractive applications-related potential of these femtosecond pulse-lasers in neurosurgery .

5   Ultrashort-pulse sources for tele- & data-comms

As optical communications systems and applications evolve in capacity, complexity, bandwidth and density, there are increasing opportunities for novel technologies to contribute. Ultrafast technology that is capable of generating pulses of 100fs or shorter in the standard telecommunications band around 1550nm can be used with different data encoding schemes, as illustrated in Figure 13. Typically when data are encoded at very high speeds

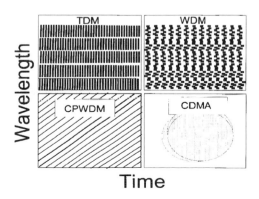

Figure 13. *Four different data formats for optical communications systems.*

such as 40, 100Gb/s or even higher, at a small number of wavelengths, people refer to it as TDM or time-division multiplexing . If the high speed data streams are optically time-delayed and recombined to form the data stream, if is called OTDM, or Optical Time-Division Multiplexed communications. In systems where many more wavelengths (perhaps hundreds) are encoded and the speeds are not as high, such as 622Mb/s, 2.5 or 10Gb/s, this is referred to as WDM, or Wavelength-Division Multiplexed communications. Recently, a new format has been introduced (Boivin *et al.* 1997) called CPWDM, or Chirped-Pulse WDM.

This approach allows for a flexible division of WDM and TDM, depending on the system requirements. A fourth format different in principle, that has been demonstrated is illustrated in Figure 13 as well: Code-Division Multiplexed Multiple Access (Chang 1997). In this format, all users can receive all wavelengths but the information is channelled to particular users by wavelength or temporal pattern coding.

A number of different systems demonstrations have been carried out with total transmission capacity of 1Tb/s or higher. If one wanted to transmit 1Tb/s capacity using only a single wavelength, it would require 1Tb/s OTDM, whereas transmitting 1Tb/s using data rates of only 1Gb/s would require 1000 wavelengths. As shown in Figure 14, experiments in each of the limits have been carried out recently, and interestingly, they both use ultrafast technology. NTT showed 640Gb/s OTDM with one wavelength channel and transmitted the data over 63km of fibre (Nakazawa *et al.* 1998). Because there is no electronics operating at 640Gb/s they used a nonlinear optical loop mirror for the demultiplexing. In the limit of many wavelength channels but lower data rates, a Bell Labs group demonstrated generation of 300 WDM channels and OTDM up to 2.5Gb/s using CPWDM (Boivin *et al.* 1997) In this case, standard WDM demultiplexers can be used to decode the data. Using ultrafast modelocked diode lasers, a group at Heinrich Hertz Institute used 80GB/s OTDM at 8 wavelengths to obtain a capacity of 640Gb/s (Ludwig *et al.* 1998). In the limit of very high speed OTDM, clearly ultrashort pulses are needed. But, in the case of WDM, ultrashort pulses have the interesting property of ultra-wide spectral bandwidth, so they can be used to cover many wavelengths as well.

But what of electronics? That, too is becoming ultrafast. We note that recently multiplexer/demultiplexer chips have been demonstrated at NTT (Otsuji *et al.* 1998) operating at 80Gb/s (Figure 15). Even at these levels, the authors report that they are interconnect-limited. We can expect further progress in ultrafast electronics in the future and so ultrafast optical technologies will always have to compete with electronic solutions.

It is clear from the above that ultrafast optics has contributed many interesting research results, but these techniques are not yet being employed in industry. More work is needed on applications, sources and markets. In particular, telecom-compatible femtosecond sources are possible, but they need to be engineered for particular applications.

6   Conclusion

The main objective in this chapter has been to illustrate that femtosecond, laser-based sources have opened up an exciting variety of applications in ultrafast science and technology. As the source options mature in terms of their capabilities, integrability and cost-effectiveness, it can be expected that the advantages of ultrashort-pulse laser-based

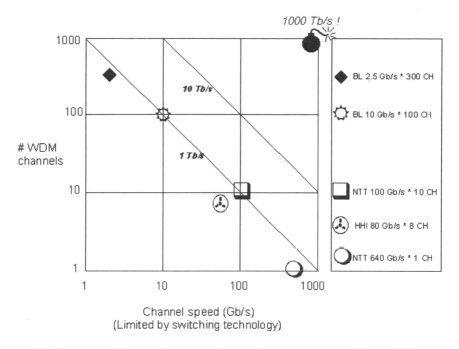

Figure 14. *Summary of recent demonstrations of optical systems with ∼ 1 Tb/s capacity.*

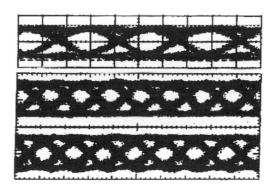

Figure 15. *Eye diagrams of ultrafast electronics operating at 80Gb/s (bottom trace). (Otsuji et al. 1998)*

sources will become increasingly compelling and wide-ranging. It is already clear that ultrafast lasers and related applications are attracting worldwide interest and relevance in many interdisciplinary sectors. As source configurations become more compatible with optoelectronic/photonic integrations their applications potential can be predicted to continue on an impressive growth curve.

References

Boivin L, Nuss M C, Stark J B, Knox W H and Cundiff S T, 1997, *OSA TOPS on Ultrafast Electronics and Optoelectronics* **13** 50.

Bouma B, Tearney G J, Boppart S A, Hee M R, Brezinski M E and Fujimoto J G, 1995, *Opt.Lett.* **20** 1486.

Chang C-C, Sardesai H P and Weiner A M, 1998, *IEEE PTL* **10** 171.

Chang Z, Runquist A, Murnane M M, and Kapteyn H C, 1997, *Phys.Rev.Lett.* **79** 2967.

Chen H, Chen Y, Dilworth D, Leith E, Lopez J, and Valdmanis J, 1991, *Opt.Lett.* **16**, 487.

Corkum P B, Burnett N H and Ivanov Y M, 1994, *Opt.Lett.* **19** 1870.

Delong K W, Trebino R and Kane D J, 1994a, *J.Opt.Soc.Am.B* **11** 1595.

Delong K W, Trebino R, Hunter J and White W E, 1991b, *J.Opt.Soc.Am.B* **11** 2206.

Ge N-H, Wong C M, Lingle Jr R L McNeill J D, Ganney K J and Harris C B, 1998, *Science* **279** 202.

Hu B B and Nuss M C, 1995, *Opt.Lett.* **20** 1716.

Hyde S C W, Barry N P, Jones R, Dainty J C, French P M W, Klein M B and Wechsler B A, 1995, *Opt.Lett.* **20** 1331.

Loesel F H, Fischer J P, Gotz M H, Horvath C Juhasz T, Noack F, Suhm N and Bille J F, 1998, *Appl.Phys.B* **66** 121.

Ludwig R, Diez S and Weber H G, 1998, *Proc. OFC 98*, Postdeadline paper PD-22.

Momma C, Chichkov B N, Nolte S, von Alvensleben F, Tunnerman A, Welling H and Wellegehausen B, 1996, *Opt.Commun.* **129** 134.

Moulton P F, 1982, *Optics News* **8**

Moulton P F, 1986, *J.Opt.Soc.Am.B* **33** 125.

Nakazawa M, Yoshida E, Yamamoto T, Yamada E and Sahara A, 1998, *Proc. OFC 1998* paper PD-14.

Nolte S, Momma C, Kamlage G, Chichkov B N, Tunnerman A, von Alvensleben F and Welling H, 1998, *Tech.Digest., CLEO '98 (OSA)*, Paper CFD3, p 510.

Otsuji T, Murata K, Enoki T and Umeda Y, 1998, *IEEE J.Solid State Circuits* **33** 1321.

Piston D W, Sandison D R and Webb W W, 1992, "Time-resolved Laser Spectroscopy in Biochemistry III", ed. J R Lakowicz, *Proc. SPIE* **1640** 379.

Schnurer M, Cheng Z, Santania S, Hentschel M, Tempea G, Brabec T and Krausz F, 1998, *Appl.Phys.B* **67** 263.

Siegman A E, 1986, *Lasers*. Chapters 27, 28, (Oxford University Press, Oxford)

Spence D E, Kean P N, and Sibbett W, 1990, *Tech.Digest of CLEO '90 (OSA)* Paper CPDP10, 619.

Spence D E, Kean P N, and Sibbett W, 1991, *Opt.Lett.* **16** 42.

Strickland D and Mourou G, 1985, *Opt.Commun.* **56** 219.

Van Exter M, Fattinger Ch, Grischkowsky D, 1989, *Opt.Lett.* **14** 1128.

Presentations by participants

The titles of papers presented by participants at the School, either orally or as posters, are listed below.

- **Influence of proton-exchange on ferroelectric inverted Lithium Niobate**
  Ms Irene Aboud – (Université de Nice Sophia-Antipolis)

- **Transversally pumped hexagonal core-doped rods**
  Mr Lucianetti Antonio – (Institut für Angewandte Physik, Bern)

- **Broadband CW lasers**
  Mr Dale Atkin – (DERA)

- **High average power femtosecond diode-pumped Nd:glass laser**
  Mr Juerg Aus der Au – (Swiss Federal Institute of Technology, Zürich)

- **Optical parametric oscillator of quasi-phasematched LiNbO$_3$ pumped by a high repetition rate single frequency passively Q-switched Nd:YAG laser**
  Mr Uwe Bäder – (Universität Kaiserslautern)

- **Development and properties of the diode-pumped Yb:S-FAP laser crystal**
  Mr James Baldwin – (University of Manchester)

- **Control of wavepacket dynamics of I$_2$ in Kr by chirped OPA-pulses**
  Mr Matias Bargheer – (AG Schwentner Experimentalphysik FU-Berlin)

- **Coherent zone-folded acoustic phonons in semiconductor superlattices**
  Mr Albrecht Bartels – (Institut fuer Halbleitertechnik II, RWTH Aachen)

- **Theoretical modelling and simulation of Er:LiNbO$_3$ waveguides for optical amplification**
  Dr Giuseppe Bellomonte – (Palermo)

- **Measurement of gain spectra in as-grown and intermixed GaAs/AlGaAs quantum well lasers**
  Miss Claire Blay – (University of Bath)

- **High power operation of a continuously-tunable doubly resonant optical parametric oscillator**
  Mr Markus Bode – (Laser Zentrum Hannover)

- **In plane diode-bar pumped, multiwatt, Nd:YAG planar waveguide lasers**
  Miss Catherine Bonner – (University of Southampton)

- **Quantum images in cavityless parametric down-conversion**
  Dr Enrico Brambilla – (Universita degli Studi di Milano)

- **Energy transfer process involved in the 1.5 microns and 2.3 microns laser transitions in the Yb-Tm LiYF4 and KY$_3$F$_{10}$ single crystals**
  Mr Alain Braud – (ISMRA Caen)

- **The laser system for the gravitational wave detector GE0600**
  Mr Oliver Sascha Brozek – (Max-Planck Institut fr Quantenoptik, Hannover)

- **Quantum state design using superpositions of a few coherent states: a study of the interference using Weyl**
  functions Mr Spiros Chountasis – (University of Liverpool)

- **Low-index coating for cladding-pumped optical fibre lasers**
  Ms Åsa Claesson – (Institute of Optical Research, Stockholm)

- **Holographic storage**
  Mr Thomas Clausen – (Riso National Laboratory, Roskilde)

- **Conversion of a prototype interferometric gravitational wave detector from operating with green light to infra red**
  Mr David Clubley – (Glasgow University)

- **Whole field fluorescence lifetime imaging with ps resolution for biomedicine**
  Miss Mary Cole – (Imperial College, London)

- **Q-switched microchip lasers**
  Dr Richard Conroy – (University of St Andrews)

- **Nd:groundstate laser with directly exitation in upper laser manifold**
  Mr Christoph Czeranowsky – (Universität Hamburg)

- **Medium power, passive Q-switched diode pumped Neodymium lasers**
  Dr Stefano Dell'Acqua – (Universita Degli Studi Di Pavia)

- **Light beams for atom guiding**
  Dr Kishan Dholakia – (University of St Andrews)

- **Development of a high power thulium-doped silica-based fibre laser operating at 2 microns**
  Mr Ben Dickinson – (University of Manchester)

- **Characterisation and modelling of a single-frequency Tm:YAG microchip laser with external feedback**
  Mr Bernd-Michael Dicks – (Universität Hamburg)

- **Characterisation and modulation of amplified femtosecond pulses**
  Mr Christopher Dorrer – (Laboratoire d'Optique Appliquée, Palaiseau)

- **Lasers in Medicine**
  Miss Eleni Drakaki – (National Technical University of Athens)

- **The sequential recording of holographic interferograms**
  Ms Dimitra Fadithou – (FORTH/IESL)

- **Bloch oscillations in InGaAs/InGaAs superlattices**
  Mr Michael Foerst – (Institute of Semiconductor Electronics II, RWTH Aachen)

- **The influence of energy-transfer upcoversion on thermal lensing in end-pumped Nd:YLF and Nd:YAG lasers**
  Mr Paul Hardman – (University of Southampton)

- **Passively Q-switched microchip lasers with semiconductor saturable absorber mirrors**
  Mr Reto Haring – (Swiss Federal Institute of Technology, Zürich)

- **Intracavity pumping as a route to a high-power 2.1 micron source for LIDAR**
  Mr Rob Hayward – (University of Southampton)

- **Time-resolved investigation of photoinduced birefringence in polymers**
  Mr Manual Hegelich – (Laser Laboratorium Göttingen)

- **Demonstration of a new ultrashort pulse amplification technique**
  Miss Cristina Hernandez-Gomez – (Rutherford Appleton Laboratory)

- **Towards a VUV coherent source for surface photochemistry**
  Mr Luis Hernandez-Pozos – (University of Birmingham)

- **Intensity noise control in solid-state lasers**
  Ms Elanor Huntington – (The Australian National University)

- **Guiding effects in microchip lasers operating well above threshold**
  Mr Alan Kemp – (University of St Andrews)

- **Diode pumped Yb:KYW Laser**
  Mr Alexander A Lagatsky – (International Laser Center, Minsk)

- **Coupled Tandem OPO: OPO within OPO**
  Dr Kin Seng Lai – (DSO National Laboratories, Singapore)

- **A femtosecond CPA system**
  Mr Stephen Lee – (Strathclyde University)

- **Excited-state absorption of Er^{3+}-doped LiNbO3**
  Dr Rozenn Loison – (Université Bernard Lyon 1)

- **Optical to microwave conversion using optical rectification in semiconductor waveguides.**
  Mr V. Loyo – (University of Glasgow)

- **Generation and application of high-order harmonics**
  Ms Claire Lyngå – (LTH, Lund)

- **Parametric generation excited by the Bessel beam**
  Mr Andrius Marcinkevicius – (Vilnius University)

- **A comparative study of parameters involved in laser marking process**
  Ms Mihaela-Anca Marian – (National Institute for Laser, Plasma and Radiation Physics, Bucharest)

- **Ti:sapphire pumped PPLN OPO at 1.5μm**
  Ms Dawn Marshall – (University of St Andrews)

- **Modular and Educational solid state laser (Nd:YAG)**
  Mr Miguel C Q S M Martins – (Lisboa Codex)

- **Nonlinear transformations of the optical vortices**
  Mr Aidas Matijosius – (Vilnius University)

- **On resonance effects on the temporal and spatial profile of laser produced plasmas**
  Mr Tom McCormack – (UC Dublin)

- **Cerium lasers pumped by frequency doubled copper vapour lasers**
  Mr Andrew McGonigle – (University of Oxford)

- **Optical systems for the early detection of cancer**
  Ms Tracy McKechnie – (University of St Andrews)

- **Ultrafast time-resolved two-photon-photoemission from metal surfaces**
  Mr Gunnar Moos – (Fritz-Haber-Institut der MPG, Berlin)

- **Laser deposition of indium-tin-oxide thin films and spectroscopical analysis of the plume**
  Dr Mauro Mosca – (Palermo)

- **Ultrafast nonlinear spectrometer for material characterisation**
  Ms Raluca Negres – (University of Central Florida)

- **Comparative study of Yb^{3+} Er^{3+} codoped phosphate glasses**
  Ms Anne-Francoise Obaton – (Université C Bernard Lyon)

- **Application of ultrafast lasers to ablation**
  Mr Duncan Parsons-Karavassilis – (Imperial College London)

- **Comparison of spectroscopic properties of erbium in yttria, scandia and lutetia**
  Mr Volker Peters – (Universität Hamburg)

- **Design and development of a femtosecond Cr^{4+}:forsterite laser source**
  Mr Enrico Piccinini – (Universita Degli Studi Di Pavia)

- **Spectroscopic studies of Er:YAG laser irradiated tissue**
  Mr Mark C Pierce – (University of Manchester)

- **Excited-state absorption and laser efficiency of Cr:forsterite**
  Mr Alexander V Podlipensky – (International Laser Center, Minsk)

- **Correlation of noise sources of monolithic diode pumped non-planar ringlasers**
  Mr Volker Quetschke – (Institut für Atom and Moleklphysik, Hannover)

- **Fine line 3D Device fabrication by laser patterning and related processing**
  Mr Luigi Raffaele – (Palermo)

- **The AMOLF terawatt laser system - a tool for XUV studies**
  Ms Florentina Rosca-Pruna – (AMOLF, Amsterdam)

- **Travelling wave optical parametric amplification of fs-pulses in $MgO:LiNbO_3$**
  Mr Fabian Rotermund – (Max-Born Institute, Berlin)

- **Upconversion-pumped high-power Pr,Yb:ZBLAN fibre amplifier at 635nm**
  Mr Hanno Scheife – (Universität Hamburg)

- **Comparison of Pr,Yb- doped host lattices**
  Ms Kathrin Sebald – (Universität Hamburg)

- **Erbium doping of $LiNbO_3$ by the Ion Exchange Process**
  Dr Francesco Segato – (University of Padova)

- **Intracavity intensity stabilisation via electro-optic feedback**
  Mr Daniel Shaddock – (Australian National University)

- **Passively Q-switched microchip lasers at 1μm using semiconductor saturable absorber mirrors. Theory, experiments, and design guidelines**
  Mr Gabriel Spühler – (Swiss Federal Institute of Technology Zürich)

- **Lasers in opthalmology**
  Mr Andreas Thoss – (Aesculap Meditic, Jena)

- **The dynamic of holographic recording in InOx films**
  Ms Eleftheria Tzamali – (FORTH/IESL)

- **Nanosecond Optical Parametric Oscillator (OPO) model**
  Mr Stephane Victori – (IOTA, Orsay Cedex)

- **Beam reformatting of 2-D waveguide array lasers**
  Dr Francisco Villarreal – (Heriot-Watt University)

- **Type I phase matching for UV generation in PP KTP**
  Ms Shunhua Wang – (Royal Institute of Technology, Stockholm)

Participants

●Ms Irene Aboud
Laboratoire de physique de
la Matière Condensée
CNRS UMR 6622
Université de Nice Sophia-Antipolis
Parc Valrose, 06108 Nice Cedex 2
FRANCE

●Mr Lucianetti Antonio
Institut für Angewandte Physik
Sidlerstrasse 5
3012 Bern
SWITZERLAND

●Mr Dale Atkin
DERA Portsdown West
Portsdown Hill Road
Fareham PO17 6AD
UK

●Dr J Aus der Au
Ultrafast Laser Physics
Institute of Quantum Electronics
Swiss Federal Institute of Technology
ETH Hönggerberg HPT D20
CH-8093 Zürich
SWITZERLAND

●Mr Uwe Bäder
Universität Kaiserslautern
Erwin-Schrödigner-Strasse 46
67663 Kaiserslautern
GERMANY

●Mr James Baldwin
Laser Photonics Group
Schuster Laboratory
Brunswick Street
Manchester M13 9PL
UK

●Dr John Barr
Pilkington Optronics
1 Linthouse Road
Glasgow
UK

●Mr Matias Bargheer
AG Schwentner
Experimentalphysik FU-Berlin
Arnimallee 15
14195 Berlin
GERMANY

●Mr Albrecht Bartels
Instituet für Halbleitertechnik II
RWTH Aachen
Sommerfeldstr. 24
D-52056 Aachen
GERMANY

●Dr Giuseppe Bellomonte
Dipartimento di Ingegneria Elettrica
Universita' degli studi di Palermo
Palermo
ITALY

●Miss Claire Blay
School of Physics
University of Bath
Bath, BA2 7AY
UK

●Mr Markus Bode
Laser Zentrum Hannover
Hollerithallee 8
D-30419 Hannover
GERMANY

●Miss Catherine Bonner
ORC, Big 46
University of Southampton
Southampton
UK

●Mr Frederic Borgeois
Université Claude Bernard Lyon 1
43 bd du 11 novembre 1918
F-69622 Villeurbanne Cedex
FRANCE

●Dr Enrico Brambilla
Universita' degli Studi di Milano
Dipartimento di Fisica
Via Celoria 16
20133 Milano
ITALY

●Mr Alain Braud
Lab de Spectroscopie Atomique,
UPRES A 6084 ISMRA
6 Boulevard Marechal Juin
14050 Caen cedex
FRANCE

●Mr Oliver Sascha Brozek
MPI für Quantenoptik,
Calinstr. 38
D-30167 Hannover
GERMANY

•Mr Spiros Chountasis
Dept of Electrical Eng/Electronics
The University of Liverpool
Liverpool L69 3BX
UK

•Ms Åsa Claesson
Institute of Optical Research,
AB IOF
SE-100 44 Stockholm
SWEDEN

•Mr Thomas Clausen
Riso National Laboratory
Optics and Fluid Dynamics Dept
PO Box 49 DK-4000 Roskilde
DENMARK

•Mr David Clubley
Dept of Physics and Astronomy
Glasgow University
Glasgow, G12 8QQ
UK

•Miss Mary Cole
14 Blackett Street
London SW15 1QG
UK

•Dr Richard Conroy
School of Physics and Astronomy
St Andrews, Fife KY16 9SS
UK

•Mr Johannes Courtial
School of Physics and Astronomy
University of St Andrews
St Andrews, Fife KY16 9SS
UK

•Mr Christoph Czeranowsky
Universitat Hamburg
Institut für Laserphysik
Jungiusstrasse 9a
20355 Hamburg
GERMANY

•Dr Stefano Dell'Acqua
Universita Degli Studi Di Pavia
Dipartimento Di Elettronica
via Ferrata 1
27100 Pavia
ITALY

•Dr Kishan Dholakia
School of Physics and Astronomy
University of St Andrews
St Andrews, KY16 9SS
UK

•Mr Ben Dickinson
Dept of Physics and Astronomy
Schuster Laboratory
University of Manchester,
Manchester M13 9PL
UK

•Mr Bernd-Michael Dicks
Institut für Laser-Physik
Jungiusstrasse 9a
20355 Hamburg
GERMANY

•Mr Christopher Dorrer
Laboratoire d'Optique Appliqué
Batterie de l'yvette
91761 Palaiseau
FRANCE

•Miss Elini Drakaki
Physics Dept
National Tech Univ. Athens
Zografou Campus
15780 Athens
GREECE

•Prof Malcolm Dunn
School of Physics and Astronomy
University of St Andrews
St Andrews, Fife KY16 9SS
UK

•Dr Majid Ebrahimzadeh
School of Physics and Astronomy
University of St Andrews
St Andrews Fife KY16 9SS
UK

•Ms Dimitra Fadithou
FORTH/IESL
Laser Applications Laboratory
PO Box 1527
GR-711 10 Heraklion
GREECE

•Dr Carsten Fallnich
Laser Zentrum Hannover
Hollenthallee 8
D-30419 Hannover
GERMANY

•Dr David Finlayson
School of Physics and Astronomy
University of St Andrews
St Andrews, Fife, KY16 9SS
UK

•Mr Michael Foerst
Institut für Halbleitertechnik II
RWTH Aachen
Sommefeld str. 24
D-52056 Aachen
GERMANY

•Prof Dave Hanna
Optoelectronic Research Centre
University of Southampton
Southampton SO17 1BJ
UK

•Mr Paul Hardman
Optoelectronics Research Centre
University of Southampton
Southampton SO17 1BJ
UK

•Mr Reto Haring
Swiss Federal Institute of Technology
Institute of Quantum Electronics
Ultrafast Laser Physics
ETH Hönggerberg HTP E9
CH-8093 Zurich
SWITZERLAND

•Mr Rob Hayward
Optoelectronics Research Centre
Building 46
University of Southampton
Southampton S017 1BJ
UK

•Mr Manuel Hegelich
MPI for Quantum Optics
Laser Plasma Group
Hans-Kopfermann-Str. 1
D-85748 Garching
GERMANY

•Miss Cristina Hernandez-Gomez
Central Laser Facility
Rutherford Appleton Laboratory
Chilton, Didcot OX11 0QX
UK

•Mr Luis Hernandez-Pozos
The University of Birmingham
Nanoscale Physics Research
Birmingham B15 2TT
UK

•Dr Jean Pierre Hirtz
Thomson-CSF Laser Diodes
Route departemantale 128, BP46
91401 Orsay Cedex
FRANCE

•Mr John-Mark Hopkins
School of Physics and Astronomy
University of St Andrews
St Andrews, Fife KY16 9SS
UK

•Prof Jim Hough
Dept of Physics and Astronomy
University of Glagow
Glasgow G12 8QQ
UK

•Prof Günter Huber
Institut für Laser-Physik,
Universität Hamburg
Jungiusstr. 99
20355 Hamburg
GERMANY

•Ms Elanor Huntington
Dept of Physics
Faculty of Science
The Australian National University
Canberra ACT 0200
AUSTRALIA

•Mr Andrew Irving
GEC Marconi Electro-Optics Ltd
NEOSD
Edinburgh EH4 4AD
UK

•Prof Ursula Keller
ETH Hönggerberg
HPT Institute of Quantum Electronics
CH-8093 Zurich
SWITZERLAND

•Mr Alan Kemp
School of Physics and Astronomy
University of St Andrews
St Andrews, Fife KY16 9SS
UK

•Dr Wayne Knox
Bell Labs
Lucent Technologies
101 Crawford Corner Road
Holmdel NJ 07733
USA

•Mr Lukas Krainer
Swiss Federal Institute of Technology
Institute of Quantum Electronics
Ultrafast Laser Physics
ETH Hönggerberg HPT E17
CH-8093 Zurich
SWITZERLAND

•Mr Laurent Krummenacher
Spectra Precision AB
Box 64
18211 Daneryd
SWEDEN

•Dr Bill Krupke
Lawrence Livermore National Labs
PO Box 808
Livermore
California CA 94550
USA

•Mr Alexander A Lagatsky
International Laser Center
Belarus State Polytech Academy
Scoryna ave 65
Minsk 22027
BELARUS

•Dr Kin Seng Lai
DSO National Laboratories
20 Science Park Drive
S118230
SINGAPORE

•Mr Stephen Lee
Institute of Photonics,
Strathclyde University
Wolfson Centre
106 Rottenrow, Glasgow
UK

•Dr Rozenn Loison
Laboratoire de Physico-Chimie
des Matériaux Luminescents
Université C Bernard Lyon 1
43 bd du 11 novembre 1918
F-69622 Villeurbanne Cedex
FRANCE

•Dr Peter Loosen
Fraunhofer Inst für Lasertechnik
Steinbachstr 15
D-52074 Aachen
GERMANY

•Mr Valentin Loyo
Dept Electronics/Elec Engineering
Rankine Building
Oakfield Avenue
Glasgow G12 8LT
UK

•Ms Claire Lyngå
Atomfysik, LTH, Box 118
22100 Lund
SWEDEN

•Mr Andrius Marcinkevicius
Vilnius University
Laser Research Centre
Sauletekio 10
Vilnius 2040
LITHUANIA

•Ms Mihaela-Anca Marian
National Institute for Laser,
Plasma and Radiation Physics
Quantum Electronics Laboratory
PO Box MG-36
RO-76900 Bucharest
ROMANIA

•Ms Dawn Marshall
School of Physics and Astronomy
University of St Andrews
St Andrews, Fife KY16 9SS
UK

•Mr Miguel C Q S M Martins
Est. do Paco do Lumiar
1699 Lisboa Codex
PORTUGAL

•Mr Aidas Matijosius
Vilnius University
Laser Research Centre
Sauletekio 9 Bldg 3
Vilnius 2040
LITHUANIA

•Mr Tom McCormack
Spectroscopy Lab
Physics Dept., UCD
Belfield
Dublin 4
IRELAND

•Mr Andrew McGonigle
Clarendon Laboratory
University of Oxford
Parks Road,
Oxford OX1 3PU
UK

•Ms Tracy McKechnie
School of Physics and Astronomy
University of St Andrews
St Andrews, Fife KY16 9SS
UK

•Prof Alan Miller
School of Physics and Astronomy
University of St Andrews
St Andrews, Fife KY16 9SS
UK

•Mr Gunnar Moos
Fritz-Haber-Institut der MPG
Dept of Phys. Chemistry
Faradayweg 4-6
14195 Berlin
GERMANY

•Dr Mauro Mosca
Via Scordia 5
I 90147 Palermo
ITALY

•Dr Peter Moulton
Q-Peak
135 South Road
Bedford, Massachusetts 01730
USA

•Dr Larry Myers
Lightwave Electronics
2400 Charleston Road
Mountain View
California CA 94306
USA

•Ms Raluca Negres
CREOL-UCF
4000 Centre Florida Blvd
PO Box 162700
Orlando FL32816-2700
USA

•Ms Anne-Francoise Obaton
Laboratoire de Physico-Chimie
des Matériaux Luminescents
Universit C Bernard Lyon 1
43 bd du 11 novembre 1918
F-69622 Villeurbanne Cedex
FRANCE

•Mr Duncan Parsons-Karavassilis
Flat 2
70 Shepherds Hill
Highgate
London N6 5RH
UK

•Mr Volker Peters
Institut für Laserphysik
Universität Hamburg
Jungiusstr. 9a
20355 Hamburg
GERMANY

•Mr Enrico Piccinini
Universita Degli Studi Di Pavia
Dipartimento Di Elettronica
via Ferrata 1
27100 Pavia
ITALY

•Mr Mark C Pierce
Laser Photonics Group
Dept of Physics and Astronomy
University of Manchester
Manchester M13 9PL
UK

•Prof Algis Piskarskas
Vilnius University
Quantum Electronic Dept
and Laser Research Centre
Sauletekio al. 9, build. 3
LT-2040 Vilnius
LITHUANIA

•Mr Alexander V Podlipensky
International Laser Center
Belarus State Polytech. Academy
Scoryna ave 65
Minsk 22027
BELARUS

•Mr Volker Quetschke
Inst für Atom and Molekülphysik
Abteilung Spektroskopie
Calinstr. 38
D-30167 Hannover
GERMANY

•Dr Luigi Raffaele
via Mercè, 4
I-98077 Santo Stefano di Camastra
Messina
ITALY

•Dr Derryck Reid
School of Physics and Astronomy
University of St Andrews
St Andrews, Fife KY16 9SS
UK

•Ms Florentina Rosca-Pruna
FOM Inst Atomic/Molecular Physics
(AMOLF)
Kruislaan 407
1098 SJ Amsterdam
THE NETHERLANDS

•Mr Fabian Rotermund
Max-Born Institute
Rudower Chaussee 6
12489 Berlin
GERMANY

•Dr Sheila Rowan
Applied Physics
Stanford University
California 94305
USA

•Mr Hanno Scheife
Institut für Laser-Physik
Universitaet Hamburg
Jungiusstrasse 9A
D-20355 Hamburg
GERMANY

•Mr Robert Scott
GEC Marconi Electro-Optics Ltd
NEOSD
Edinburgh EH4 4AD
UK

●Ms Kathrin Sebald
Institut für Laser Physik
Universität Hamburg
Jungiusstr 9a
D-20355 Hamburg
GERMANY

●Dr Francesco Segato
Dipartimento di Fisica
via Marzolo8
35131 Padova
ITALY

●Mr Daniel Shaddock
Dept of Physics
Faculty of Science
Australian National Laboratory
Canberra 0200
AUSTRALIA

●Prof Wilson Sibbett
School of Physics and Astronomy
University of St Andrews
St Andrews, Fife KY16 9SS
UK

●Dr Bruce Sinclair
School of Physics and Astronomy
University of St Andrews
St Andrews, Fife KY16 9SS
UK

●Ms Gabriel Spühler
Swiss Federal Inst Technology
Institute of Quantum Electronics
Ultrafast Laser Physics
ETH Hönggerberg HPT D20
CH-8093 Zurich
SWITZERLAND

●Mr Andreas Thoss
Fritz Ritter Strasse 18
D - 07747 Jena
GERMANY

●Dr Anne Tropper
Optoelectronics Research Centre
University of Southampton
Southampton SO17 1BJ
UK

●Ms Eleftheria Tzamali
FORTH/IESL
Laser Applications Laboratory
PO Box 1527
GR-711 10 Heraklion
GREECE

●Dr Rudolf Verdaasdonk
Medical Laser Centre
and Clinical Physics
University Hospital Utrecht
PO Box 85500
Utrecht
NETHERLANDS

●Mr Stephane Victori
IOTA Bt 503-BP 147
91403 Orsay
Cedex
FRANCE

●Dr Francisco Villarreal
Dept Physics,
Heriot-Watt University
Riccarton, EDINBURGH
UK

●Ms Shunhua Wang
Physics Dept
Royal Institute of Technology,
Lindstedstvagen 24
10044 Stockholm
SWEDEN

Index